HUMAN GENETICS AND SOCIETY

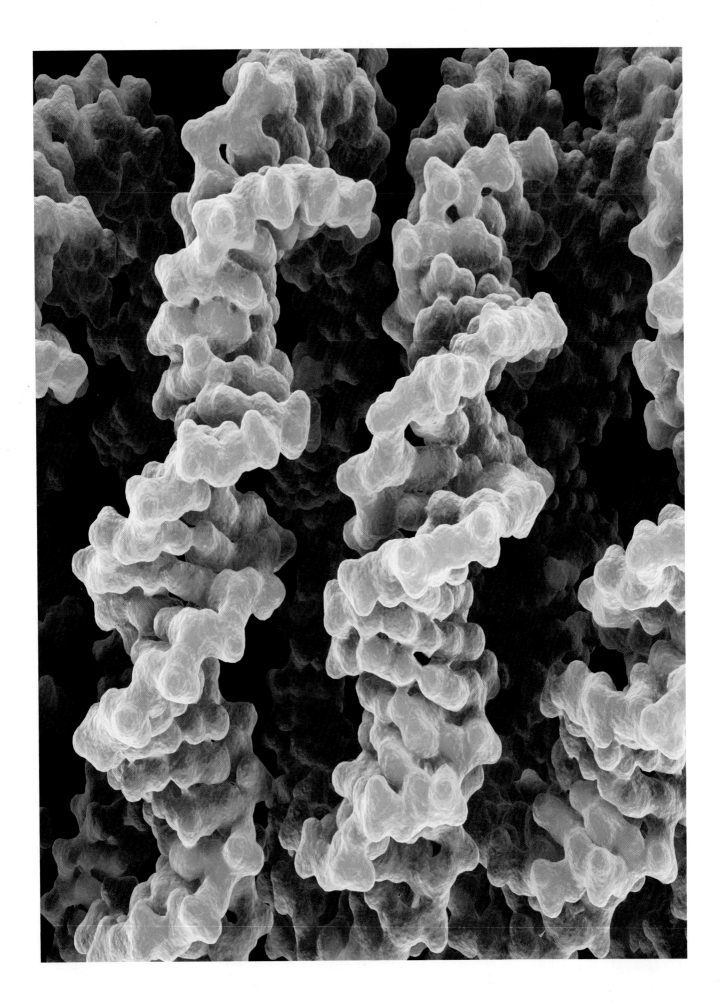

HUMAN GENETICS AND SOCIETY

RONNEE K. YASHON
*Experimental College of Tufts University
and Boston University School of Medicine*

MICHAEL R. CUMMINGS
Illinois Institute of Technology

BROOKS/COLE
CENGAGE Learning™

Australia • Brazil • Japan • Korea • Mexico • Singapore • Spain • United Kingdom • United States

BROOKS/COLE
CENGAGE Learning™

Human Genetics and Society
Ronnee K. Yashon and Michael R. Cummings

Publisher: Yolanda Cossio

Development Editor: Christopher Delgado

Assistant Editor: Lauren Oliveira

Editorial Assistant: Samantha Arvin

Media Editor: Kristina Razmara

Marketing Communications Manager: Linda Yip

Project Manager, Editorial Production: Andy Marinkovich

Creative Director: Rob Hugel

Art Director: John Walker

Print Buyer: Judy Inouye

Permissions Editor: Mardell Glinski-Schultz

Production Service: Lachina Publishing Services

Text Designer: Jeanne Calabrese

Photo Researcher: Susan Van Etten; Prepress PMG

Copy Editor: Amy Schneider, Lachina Publishing Services

Illustrator: Suzannah Alexander; Lachina Publishing Services

Cover Designer: John Walker

Cover Image: DNA Sequencing Gel, Closeup © Stephen Puetzer/Getty Images; Group Shot © Randy Faris/Corbis

Compositor: Lachina Publishing Services

Printer: Transcontinental Printing/Interglobe

For product information and technology assistance, contact us at **Cengage Learning Academic Resource Center, 1-800-423-0563**

For permission to use material from this text or product, submit all requests online at **cengage.com/permissions** Further permissions questions can be emailed to **permissionrequest@cengage.com**

Library of Congress Control Number: 2008933533

ISBN-13: 978-0-495-11425-3

ISBN-10: 0-495-11425-1

Brooks/Cole Cengage Learning
10 Davis Drive
Belmont, CA 94002-3098
USA

Cengage Learning is a leading provider of customized learning solutions with office locations around the globe, including Singapore, the United Kingdom, Australia, Mexico, Brazil, and Japan. Locate your local office at **international.cengage.com/region**.

Cengage Learning products are represented in Canada by Nelson Education, Ltd.

For your course and learning solutions, visit **academic.cengage.com**.

Purchase any of our products at your local college store or at our preferred online store **www.ichapters.com**.

Printed in Canada
1 2 3 4 5 6 7 12 11 10 09 08

We dedicate this book to our teachers and students—
past, present, and future.

About the Authors

Ronnee K. Yashon—Tufts University

Ronnee K. Yashon is a nationally known expert in teaching genetics, ethics, and the law on all levels—including high school, undergraduate, graduate, and law school. A recipient of the Presidential Award for Excellence in Science Teaching and the Outstanding Biology Teacher Award in Illinois, she has directed numerous workshops for science teachers and disseminated interdisciplinary lessons at local and national conventions, including NSTA and NABT. A genetics seminar at Ball State University in 1985 sparked her interest in law and inspired her to study the field. Her case study methodology for introducing bioethics and law in the curriculum uses simple, personalized, and current scenarios that involve the student in decision making. Yashon has presented this case study method all over the country. She has six case study books, including two mini-books that focus on genetics and environmental issues. The implementation of science-oriented law courses in current law school curriculum sparks her interest, and she has run workshops for jurists and attorneys on the subject of genetics. She now teaches at the Experimental College of Tufts University and Boston University School of Medicine.

Michael R. Cummings—Illinois Institute of Technology

Michael R. Cummings is the author or coauthor of several leading college textbooks, including *Human Heredity: Principles and Issues*, *Concepts of Genetics*, and *Essentials of Genetics*. He also authored an introductory biology text for non-majors, *Biology: Science and Life*, as well as articles on genetics for the *McGraw Encyclopedia of Science and Technology*. For several years, he wrote and published a newsletter on advances in human genetics for instructors and students. He was a faculty member at the University of Illinois at Chicago for over 25 years. While there, he was recognized as an outstanding teacher by his peers on the University faculty and mentored junior faculty in undergraduate teaching. He was twice named by graduating seniors as the best teacher in their in years at the University, and in several years, was selected as an outstanding instructor by the Biology Student Organization. His research interests center on the physical organization of repetitive DNA sequences in the heterochromatic short arm/centromere region of human chromosome 21 and the role of these sequences in generating chromosomal aberrations. He now teaches general biology, cell biology, and genetics at the Illinois Institute of Technology.

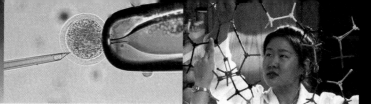

Brief Contents

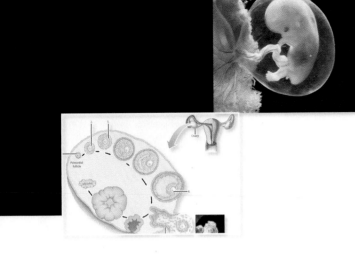

Contents

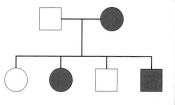

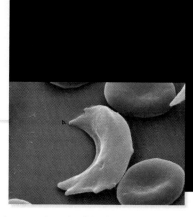

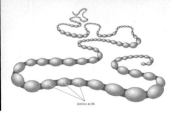

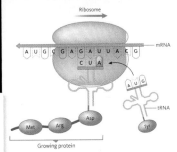

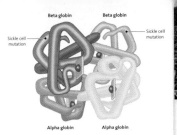

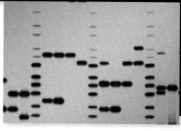

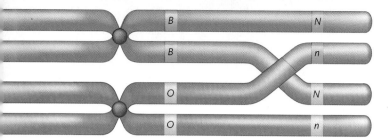

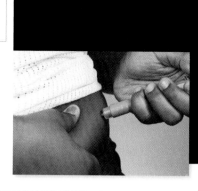

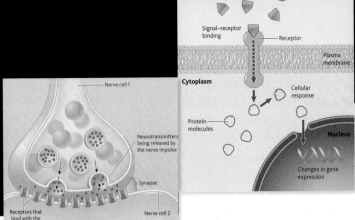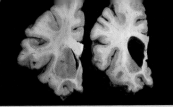

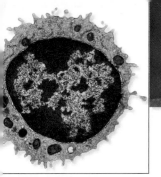
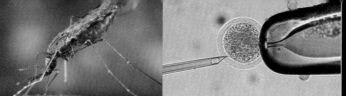

Preface

To put it simply, we wrote this book to directly place genetics and its emerging role as a social and cultural entity into the everyday life of students. The book is written for use in undergraduate courses filled with students from a wide range of disciplines with little or no background in science.

In teaching, as in so many of life's other activities, putting your subject into context is important. In our subject area, this includes relating the rapid advances in genetics to parts of everyday life. With the growing interest of the Internet and print media in genetics, the distance between findings in the research lab and new products, medicines, and foods in the marketplace gets shorter and shorter every day. As this distance shortens, it is clear that students are not exposed to genetics just in the classroom. Today, genetics and other aspects of science permeate many facets of our lives. As individuals and as members of society, we need to reach a consensus about the social, political, cultural, ethical, and legal impact of genetics and its related technologies and how to use these advances for the benefit of everyone. As part of this discussion, an understanding of the new choices we face and the social consequences of using these technologies must be discussed on an equal footing with the relevant genetic and scientific concepts.

Current textbooks do not ignore the social, legal, or personal implications of advances in genetics. However, these important topics are often separated from the science by being placed at the end of chapters where they are represented only by a series of questions or Web-based activities. Occasionally, boxes dealing with narrowly drawn single issues are scattered through chapters.

Separating these issues does a disservice to the students and understates the growing importance of the non-scientific aspects of human genetics.

Consumers are faced daily with a bewildering number of choices that are based on genetics and recombinant DNA technology. These include paternity tests available at drug stores that use DNA from a cheek swab, Internet sites offering to trace your ancestry back to people who lived thousands of years ago, DNA identification cards carrying someone's DNA profile, and genome scans that offer to test your genome for disease risk factors.

In addition, an ever-increasing variety of reproductive options offers ways to treat infertility, select the sex of a child before pregnancy, screen for genetic disorders, and even to choose embryos genetically compatible with older siblings so that the younger child can serve as a tissue or organ donor for an older sibling with a genetic disorder.

As information from the Human Genome Project is used to develop more diagnostic tests and more treatments, the options in medical care are beginning to expand in a nonlinear fashion. It is not hard to imagine that in the next few years, medical records will include a personal genome sequence coupled with information about diseases an individual has, may get, or is at risk for. These data will be used to develop plans for personalized medicine, with treatments and drug selection and dose based on allele profiles, metabolic rates, and predicted outcomes derived from genetic status.

Rationale

For students to navigate through the changes and choice that society will face in reproduction, medical care, the workplace, and other aspects of our personal and public lives, science education must make connections between the science and its broader implications in our lives. Changes generated by genetics and biotechnology will force the generation now in school to rethink many present-day social and cultural conventions. There will be new laws drafted to protect information from misuse and to protect individuals affected with genetic disorders. This book was written specifically because:

- Today's students will be the consumers of these new technologies and will be actively involved in making decisions about its use in ways that affect themselves and others.
- In order to cultivate informed decisions made by students, especially those who are not science majors, must be based on some understanding of human genetics.
- Students must have some understanding of the social impact and consequences associated with the use of genetic technology.

This Book's Goals

1. **Readability.** Students are faced with enough new terms and jargon in science courses to make the vocabulary equivalent to learning a foreign language. This is often a major stumbling block in learning. As much as possible, we use everyday language to explain genetic concepts and give understandable examples of these concepts. This not only demystifies the science and makes it easier for the student to understand, it promotes retention by omitting the need to remember dozens of acronyms and their meaning along with the concepts. Students will find this to be a text they will want to read and keep as a reference.

2. **Relevance.** Each chapter begins with a case based in real life, describing a situation that many people might face everyday. This is followed by an explanation of the science relevant to the situation to provide background for further discussion of the issues raised by the opening scenario. With the scientific knowledge in hand, students are equipped to explore the relationship between the genetic concepts and the questions asked about the case. By applying the science concepts to the case, they can easily see how genetics is relevant to them.

3. **Context.** Each chapter contains a section dealing with legal and ethical issues related to the genetic concepts presented in the chapter, and most contain details of a related court case. These cases have shaped society's response to the challenge of defining how, when, and if new genetic discoveries and technologies will be used. A review of these issues and decisions provides the opportunity to further discuss the science as it applies to the lives of individuals and to society at large.

4. **Interest.** By reading the opening case in a chapter, students will become engaged with the people in the case and their decision-making process. Students will ask themselves how they might react. This curiosity will motivate them to read on and learn about the science involved and help them integrate this information into their own lives.

Features of the Book

To help prepare students to make informed decisions, we have chosen a straightforward approach to teaching human genetics. We feel it is important to bring the social, legal, ethical, and personal aspects of genetics out of the ancillary material and into the text to give them equal standing with the science. This alignment achieves several important goals:

- It shows how specific advances in genetics have impacts that go far beyond the scientific laboratory and influence our everyday decisions, such as whether or not to buy genetically modified food, what medical tests to have, what technology to use in having a baby, and even which workplace environment may be best for us.

- It anchors the discussions about the social, ethical, and political issues generated by the discoveries in human genetics, we have chosen to use the law as a point of reference. Science and the law are increasingly in collision. Juries and judges are being bombarded with scientific questions in law suits, murder trials, and even divorce and custody cases. These court decisions provide a mirror with which to examine how we as a society perceive and cope with genetics technologies. These decisions may help us decide how to translate the use of genetics and related technology into our everyday lives. In addition, the more decisions the public must make about these issues, the more difficult they become. These decisions have a cumulative impact that spreads through many levels of our society. Because many of these decisions interlock with one another, they become more complex. As a result, we are increasingly turning to the courts for guidance and as a forum for making these decisions.

- Using the law as a guide, we can discuss relevant social and ethical topics related to genetics without letting opinion, politics, or religious beliefs influence or direct the discussion. Not everyone may agree with the enacted laws or the decisions, but even that disagreement can serve as a point of discussion, always using established facts of science as the foundation.

- By looking through the lens of laws already enacted and cases already decided, we can make human genetics more meaningful and enable students see how genetics will change our lives in the future. In addition, discussing unresolved questions and asking questions that may be raised in the future will make students better consumers and more educated voters.

- Our opinions are important, and we encourage users of the text to read, discuss, and form their opinions about the impact of genetics on individuals and society. To help students do this, we encourage them to start and maintain a blog. To help with blogging, we have placed questions throughout the chapters that interconnect the cases and the scientific content. Questions at the end of chapters are often open-ended and encourage students to form educated opinions and put those opinions to use in making decisions.

- Once students have learned the science, they can examine the cases, legal decisions, and end-of-chapter questions in the light of their own politics, family background, or religious beliefs. In discussions with classmates, or through their blogs, they will learn about the opinions of others, and may re-evaluate their own opinions, but in any case, they will learn to use fact-based critical thinking to make science-based decisions that will affect their future.

The Art Program

When opening the book, you will notice the extensive art program. This is no accident. Given the competition from other media that students use, we have placed emphasis on illustrations and photographs that support and enhance learning of the concepts and bring the cases and examples to life. To directly integrate the art into the text, the art program is presented in a more informal way, with many of the figures unnumbered and without captions.

The art program has been designed to provide the non-science student with dynamic photos and illustrations that make the concepts accessible and easy to understand. This interaction between art and text reinforces the concepts and helps the student with retention and learning. We have made an effort to select new and more recent photographs to give the text a fresh perspective on teaching genetics.

The layout of the art has been carefully crafted to enhance reading and provide a visual way to learn the material. We use featured illustrations to emphasize the basic genetics students will encounter, or may have already encountered, in their lives.

Pedagogical Features and Chapter Organization

Following a short preview, we begin the book with chapters on sex, development, and reproductive technologies before delving into chapters that place genetics more in the foreground. We did this because the topics in these initial chapters are of import and interest to almost everyone and represent areas in which genetic technology has had a significant influence. In addition, the social impact and the ethical and legal issues raised by the development and use of this technology are easily defined. This provides a way of allowing students to adjust to how the book works and to see the interrelated nature of the science and its applications.

The book is organized into three sections: Chapters 1–4 cover reproduction, development, meiosis, gamete formation, and transmission genetics. Chapters 5–9 emphasize the molecular aspects of genetics and include coverage of gene expression, mutation, recombinant DNA technology, biotechnology, genetic testing, DNA forensics, and the human genome project. Chapters 10–14 discuss the relationship between genes and the environment. The topics include multifactorial inheritance, cancer, behavior, the immune system, and populations. Chapter 15 considers the past, present, and future of human genetics, showing how genetic information has been used and misused, and offers a glimpse into what may be the future of genetic technology.

Each group of chapters is preceded by a short section entitled Biology Basics. Chapters 1–4 are preceded by Biology Basics: Cells and Cell Structure. Chapters 5–9 are preceded by Biology Basics: DNA. Chapters 10–15 are preceded by Biology Basics: Genes, Population and the Environment. These sections will serve as a review for those who have had previous biology courses and give those with no previous knowledge an opportunity to learn the basics.

We believe this organization is an important pedagogical tool for both the student and the instructor. In each Biology Basics section, we equip the student with the underlying biological concepts needed to read and understand the material in the following chapters. This organization has two advantages:

- It organizes the basic biological concepts into one section. This gives the student a single source to learn and review the biology behind the genetics discussed in the following chapters. It also allows the instructor to assign Biology Basics as a reading or use as lecture material to ensure that all students are familiar with the biology behind the material in the next group of chapters.
- It allows each chapter to focus on the essential genetic concepts instead of interspersing material on the fundamental biology with genetics. This avoids clutter within the chapters and gives the student a clear understanding of the genetic-related topics discussed in the chapter.

Within each chapter, features have been carefully selected and organized to help the non-science student clearly identify and understand the relevant genetic concepts and to consider the related social, legal, and ethical issues. These features include:

Central Points, Section Essentials, and The Essential Ten

Each chapter begins with a short list of the topics covered in the chapter. Students can use this list of the Central Points to preview the material. It also provides a handy way for the student to organize a review of the chapter material in preparation for examinations. Each Central Point corresponds to a section in the text and ties into the Essentials listed at the end of the section. At the end of each chapter is the Essential Ten. These are the ten essential points that a student should come away with after reading the chapter.

Cases

The chapters open with a story written in everyday language about an individual or an entire family faced with learning something about genetics: a disorder, a diagnostic test, a therapy, or a decision to be made. The opening story has been written to provide the foundation for a presentation of the relevant genetics concepts that follow.

After students read the opening case, they can work through the science and relate it to the people in the case and their problem.

In addition, near the end of the chapter, a second case is presented. This might be a continuation of the first case or a new one designed to make students think through what they have learned in the light of ethics and the law.

Case Questions

Following the opening case and a review of the genetic concepts, a series of questions probe the relationship between the situation presented in the opening case and the related genetics. Often these questions ask the student to make decisions about the social and genetic issues raised. This drives home to the student the need to understand the science in order to make sound decisions in one's personal or professional life.

By the Numbers

This list draws on data about the frequency of genetic disorders, risk factors for disease, or other information to illustrate how many people are affected by specific problems or situations related to genetics. These charts draw one's eye to how numbers play a part in genetics. In many of today's media, statistics are presented in this way to bring attention to important information.

Legal and Ethical Issues

Following the science and the case questions, we consider the legal and ethical questions that are relevant to the genetics discussed in the chapter. This outlines the legal and ethical issues that arise from the use or failure to use genetic technology. A chart addresses the basic questions that often examine the issue from the patient's, physician's, and attorney's point of view, providing the student with a balanced approach to the topics. The chart shows how some of these questions have been addressed by courts and legislatures on both the state and federal levels.

The Spotlights

A feature called *Spotlight* is presented at the end of each chapter. These features are of three types: Spotlight on Law, Spotlight on Society, and Spotlight on Ethics. Each includes a true situation or legal case related to genetics. The legal case or societal situation in this section highlights the genetic concepts in the chapter and the issues generated from the application of this knowledge. The Spotlights analyze landmark cases and situations from real life and delve into how they have affected genetics or how genetics has affected them. They draw together the threads of the chapter to illustrate how the use and effects of genetic information and technology are interpreted by the courts and the public. The Spotlights help the student learn what the courts have found to be acceptable and unacceptable about genetics and then apply this information throughout subsequent chapters. Some of the topics covered in the Spotlights include the legal status and custody of frozen embryos, ownership of cell lines, how the study of some populations has helped genetics, and how public opinion and politics can change a society's attitude.

End-of-Chapter Features

These features include innovative elements that highlight the chapter topics and reinforce the ideas presented. Some of these are:

What Would You Do If. . . .?

The main ideas in the chapter are reinforced by posing hypothetical questions that ask the student to make decisions using all the information from the chapter, including the genetic, ethical, legal, and social concepts and issues. The questions place the student in various situations, including the people in the opening case, legislators, physicians, judges, and others who often make personal and professional decisions that affect themselves and others.

Review Questions

Each chapter ends with two sets of questions. Review questions ensure that the student has a grasp of the basic genetics concepts presented in the chapter. This gives students an opportunity to clarify their understanding about the concepts as a way of preparing for examinations and answering the second set of questions.

Application Questions

Application questions relate specific concepts in the chapter to situations where individuals are faced with decisions based on the results of genetic diagnosis and/or risks of genetic disorders. These questions direct students to calculate, research, and analyze the latest material available on the Internet and in journals by asking the student to investigate a topic as a way of evaluating the pros and cons as well as the limitations of genetic technology.

Making a Blog

In various chapters a blog component entitled *Learn by Writing* is inserted in the *Application Questions*. We suggest that students set up a blog either on their institutional server or a public blogging site and in the questions at the end of several chapters we suggest possible subjects for blog discussions and remind students that they may want to invite others to participate in this online activity.

In the Media

This feature uses recent news items from newspapers, magazines, television, and the Internet to personalize the stories of people and their actions, reactions, and understanding of genetics and its applications. These features can be used as the basis of class discussions or assignments for students to find how the issues raised in the article interact across different religions, political systems, or cultures.

Key Terms

An alphabetized list of all the bolded terms includes page numbers where the term first appears and is defined in the text. These terms are defined in the glossary, and the list provides students with a way of organizing and accessing the important genetic terms.

Ancillary Materials

For Instructors

PowerLecture DVD with JoinIn

This one-stop digital library and presentation tool includes preassembled Microsoft® PowerPoint® lecture slides. In addition to a full Instructor's Manual and Test Bank, Power-Lecture also includes all of the media resources organized by chapter: an image library with art and photos from the book, animations and video clips, and more.

The JoinIn Student Response System allows you to pose book-specific questions and display students' answers within the Microsoft® PowerPoint® lecture slides, in conjunction with the "clicker" hardware of your choice.

CengageNOW

CengageNOW™ is an online teaching and learning resource that gives you more control in less time and delivers better outcomes. CengageNOW™ offers you teaching and learning resources in one intuitive program organized around the essential activities you perform for class—lecturing, creating assignments, grading, quizzing, and tracking student progress and performance.

Online Instructor's Manual

Each chapter has suggestions for videos and news clips that can be incorporated into your class, in addition to suggested topics for discussions, papers, and activities. Also available on PowerLecture.

- Summary and Lecture outlines organize the material and have the key terms integrated and highlighted.
- Teaching tips give ideas on additional ways to engage the students in the material.
- Suggested answers are given for all the questions in the chapter.

Test Bank

The test bank contains a variety of questions, such as Multiple Choice, True/False, Matching, and Short Answer to help you create quizzes and exams easily.

- More than 750 new questions written specifically for this text.

ExamView

This full-featured program helps you create and deliver customized tests (both print and online) in minutes. Includes all the questions from the Test Bank.

Transparencies

This set features all the important figures and photos from the text making it easy to incorporate the art into your lecture.

ABC News: Genetics in the Headlines

The informative and short video clips cover current news stories in genetics. Covering topics such as the use of PGD to pick the sex of your child or living with the cancer gene, these clips can spark discussions in class and provide opportunities for students to analyze the news.

For Students

CengageNOW

CengageNOW™ is an easy-to-use online resource that helps students study in less time to get the grade they want. Featuring CengageNOW™ Personalized Study (a diagnostic study tool containing valuable text-specific resources) students focus on just what they don't know to learn more in less time and get a better grade.

The CengageNOW printed access card can include:

- Interactive eBook. This complete online version of the text is integrated with multimedia resources and special study features, providing the motivation that so many students need to study and the interactivity they need to learn.
- *InfoTrac® College Edition.* This fully searchable online library gives users access to complete genetics articles from several hundred current periodicals and others dating back over 20 years.
- Personal Tutor. This one-on-one online tutoring service with SMARTHINKING gives students live access to a biology expert. Whiteboarding and text messaging features give students help with homework and review for exams.

Study Guide

This study guide has been designed to complement the textbook by including additional case studies for every chapter to allow students more opportunities to think about how the science of genetics can apply to their own lives. A fill-in-the-blank summary of the chapter and various types of questions further challenge the students' grasp of the material.

- Brand new case studies that are not included in the textbook enhance learning.
- A fill-in-the-blank summary review helps students review the chapter, while allowing them to check their understanding of the content.

Student Companion Web site

The book-specific Web site that can be located at *www.cengage.com/biology/yashon/hgs1* offers students a variety of study tools and useful resources such as quizzes, flash cards, glossary, Internet activities, InfoTrac and weblinks, and more.

Acknowledgments

This book began with a conversation between one of us (MRC) and Peter Adams, who at the time was the executive editor at Brooks/Cole. Peter became an early and enthusiastic supporter of the idea, and was the original editor on this project. He was also responsible for getting us together as authors. Peter's input, gleaned from many years of experience and conversations with instructors and students, helped shape the book. We are very grateful for his efforts and in particular, his role as author matchmaker. Yolanda Cossio, the publisher for Biology at Brooks/Cole stepped in at an early stage of this project. Her fresh perspective and attention to quality and pedagogy helped guide the project through its crucial stages of development and helped us meet our goals. It was our good fortune to have Christopher Delgado as our development editor. Working with Christopher has been one of the most enjoyable aspects of writing this book. His upbeat, can-do attitude and skills at nudging two often picky authors back on track are remarkable. Especially enjoyable was his subtle and clever sense of humor that made the time spent on the project speed by. He shared our vision, and was tireless in his pursuit of quality at every level in the book. Much of the way the book looks and reads is a result of his efforts. Our assistant editor, Lauren Oliveira, who handled print ancillaries, Kristina Razmara, the media editor, and Amy Schneider, the copy editor, rounded out the editorial team. They adapted to the ever-changing schedule deadlines and moved the book toward production in a seemingly effortless fashion.

Special thanks are owed to Suzannah Alexander, who supervised art development and the illustration program. She created many of the outstanding figures that grace the book and brought artistic flair to the presentation figures in each chapter. She seemed to know what we wanted even before we could visualize it. In addition, the staff at Lachina Publishing Services patiently worked through several last-minute changes in figures.

The many tasks involved in photo research were handled by Susan Van Etten and Pre-Press PMG.

We owe a debt of gratitude to Gerry McNeil for his painstaking efforts in reading the entire manuscript for accuracy. He undertook this task with patience, and his nuanced approach and suggestions about terminology have made the book more student-friendly. Karen Kurvink and Mary King Kananen also made valuable suggestions about accuracy, all of which sharpened the book's focus and precision. Thanks also to Gwendolyn Kinebrew, who suggested the name of the "Biology Basics" sections in the book. In spite of their efforts, we as authors remain responsible for any errors the book might contain.

Once the book made the transition to production, Andy Marinkovich, the production manager, Katherine Wilson, the project manager, and Judy Inouye, the print buyer, took over, and with energy and careful attention to detail, moved the book through all stages of the production process.

The material in the appendix is excerpted from *Intervention and Reflection: Basic Issues in Medical Ethics*, 8th edition, by Ronald Munson.

We extend thanks to all others who played a role in creating this book, and as we move into future editions, we hope that instructors and students will help improve the book as a resource for those who want to teach and learn about the interactions between human genetics and society.

Reviewers

The following reviewers were instrumental in the development of this text through their feedback and suggestions: William P. Baker, Midwestern University; Bruce Bowerman, University of Oregon; Cherif Boudaba, Tulane University; Jay L. Brewster, Pepperdine University; Mary Bryk, Texas A&M University; Barry Chess, Pasadena City College; T.B. Cole, Kent State University; Patricia L. Conklin, SUNY-Cortland; Drew Cressman, Sarah Lawrence College; William Cushwa, Clark College; Thomas R. Danford, West Virginia Northern Community College; Kathleen Duncan, Foothill College; Cheryld L. Emmons, Alfred University; Michael L. Foster, Eastern Kentucky University; Gail E. Gasparich, Towson University; Urbi Ghosh, Triton College; Nabarun Ghosh, West Texas A&M University; Meredith Hamilton, Oklahoma State University; Deborah Han, Palomar College; Jennifer Herzog, Utica College; Barbara Hetrick, University of Northern Iowa; Robert Hinrichsen, Indiana University of Pennsylvania; Mary King Kananen, Penn State University-Altoona; Gwendolyn M. Kinebrew, John Carroll University; Jennifer Knight, University of Colorado; Karen Kurvink, Moravian College; Clint Magill, Texas A&M University-College Station; Gerard P. McNeil, York College CUNY; Tyre J. Proffer, Kent State University-Salem; Michael D. Quillen, Maysville Community and Technical College; Laura Rhoads, SUNY-Potsdam; Michael P. Robinson, University of Miami; Jeanine Sequin Santelli, Keuka College; Monica M. Skinner, Oregon State University; Sue Trammell, John A. Logan College; Jose Vazquez, New York University; Mary Ann Walkinshaw, Pima Community College; Dan Wells, University of Houston; Robert Wiggers, Stephen F. Austin State University; Denise Woodward, Pennsylvania State University-University Park; and Calvin Young, Fullterton College.

Contacting the Authors

We want to hear from instructors and students about the book, about new findings they would like to see in future editions, and things they don't like, or they think are not working, with suggestions on how to fix the problem. Please contact Ronnee Yashon at yashon.boston@gmail.com and/or Michael R. Cummings at cummings.chicago@gmail.com and let us know how we can improve the text or its pedagogy.

Preview: Genetics in Your Life

PHOTO GALLERY
Twins have identical genes.
Digital Vision/Getty Images

Have you ever wondered why you look like your parents? Have you gone into a coffee shop and seen families talking and noticed how members of the family resemble each other? Have you seen ads on television selling tests for the breast cancer gene? Do you know any identical twins? As you walk through the aisles of the drug store, have you seen the DNA testing kits to determine paternity? If the answer to any of these questions is yes, this is the book for you.

Some people say they don't like science. They have no interest in it. It doesn't have anything to do with their lives. But human genetics is more than a science. It has become an important part of our society that will touch your own life and the lives of your family and friends in many ways. Some of the ways that this will happen will be discussed throughout this book. You will be able to take what you learn from this course and apply it to real life.

PHOTO GALLERY
Lab technicians work with DNA.
Phanie/Photo Researchers

We will examine genetics through the eyes of case studies based on the lives of real people—possibly even people you know. Hopefully, you will talk to your friends and families about the cases we will be reading. They are interesting and thought-provoking situations, as well as conversation starters.

These cases and the questions they generate will allow you to analyze your own knowledge, thoughts, feelings, and opinions on many issues related to genetics and to share them with others. This will include information about your own heredity. Some of your questions will be answered; others will not. Science is like that; it is an ongoing process in which answers are the starting points for more questions.

Can you put yourself in the place of these people?

- a woman about to give birth to her first child
- a man having a test for infertility
- a scientist with a great discovery he wants to share with the world
- a doctor giving sad news to a patient

PHOTO GALLERY
Doctors often have important genetic decisions to make with their patients.
Thinkstock Images/Jupiter Images

If you can, you'll be able to gain new insight through the cases in each chapter. If you can't, the cases may bring questions to mind that will help you understand the problems faced by these people, and start you wondering what you would do in their situation.

In addition, as the course proceeds, you will be analyzing your own opinions and the opinions of others as well as talking to others about issues related to genetics that cut across medicine, politics, laws, and social norms. Hopefully, as you acquire more information and examine and discuss these issues, your opinions will be refined, and you might even change your mind about some things. You'll find that many people have questions about genetics and the uses of the new technologies.

After using this book, you'll be able to answer your own questions and those of others, and even help members of your family understand their genetic background. You will see references in social networking sites, on blogs, on Web sites, and in magazines that will intrigue you and remind you of many things from this book. That is the goal: to put human genetics into your everyday life, and to equip you with the knowledge to make sound decisions about the use of genetics in your personal life and in our society. That's why this book is called *Human Genetics and Society*.

To share your opinions and your knowledge, think about setting up your own blog and invite others in your class, your family, and your friends to participate to see what they are thinking, and to work through questions and issues together. Watch for some blogging ideas in Chapter 1 and in other chapters throughout the book.

As you can see from the chart below, the topics that we will cover throughout the book will include questions to get you thinking about human genetics and what you might do when you come face to face with certain questions.

Genetics is everywhere! You see it on television, in

PHOTO GALLERY
Questions about human genetics are found on the Internet, in blogs, and on Web sites.
Digital Vision/Getty Images

advertising, on talk shows, and even in comedy routines. The problem is that a lot of the information you have been seeing and hearing is an oversimplification of human genetics. In this book, we have made every effort to provide essential information that is understandable and accessible. We have done this by starting each chapter with a case study presenting real problems faced by real people, followed by a discussion of the genetic concepts related to the problems. These are followed by questions, discussions about legal, ethical, and social issues that arise out of the use and application of genetics and its related technologies.

SOME GENERAL QUESTIONS YOU MAY ENCOUNTER . . .

	One of Our Topics	A Personal Question
 Cheryl Casey/iStockphoto.com	Your own genetics	Do I carry a disease gene?
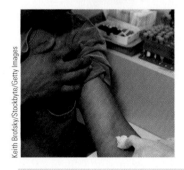 Keith Brofsky/Stockbyte/Getty Images	Genetic testing now and in the future	Will I be asked to take a test to find out information about my genetics?
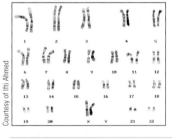 Courtesy of Ifti Ahmed	Analyzing all the human genes and finding out what they do	When will we know all about human genes and how they work?
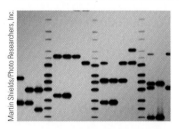 Martin Shields/Photo Researchers, Inc.	Applying genetics to legal actions such as criminal proceedings and civil ones	Should a man in prison be released if DNA evidence proves he is innocent?

In the very near future, analysis of our genomes may affect our insurance rates, our medical treatment, our lifestyles if we wish to lower the risk of certain disorders, our political views, and even more. This book will allow you to be ahead of others who are not interested in how genetics is changing our lives. It's a book you'll find yourself using long after the course is over. Welcome to the world of *Human Genetics and Society*.

Will you want to be a science major when you are done with this book? Maybe

PHOTO GALLERY
Families will be talking about cases from *Human Genetics and Society*.
Creatas Images/Jupiter Images

not, but you'll acquire useful scientific skills and information. You may be able to interest your family and friends in the material, and you'll be able to provide enough background to help them through their own decision making.

We are sure that what you learn from this book will stay with you throughout your life. So read on . . .

PHOTO GALLERY
Advertisments for genetic testing are on television, and products are available to all who want them.
Courtesy/Myriad Genetics, Inc.

	One of Our Topics	A Personal Question
 Elyse Lewin/Getty Images	Applying genetics to laws enacted by a state	Should my state pass a a law that all married couples be tested for genetic conditions?
 Alice Edward/Getty Images	Applying genetics to infertility and the treatments for it	When a couple finds they are infertile, what can be done to help them?
 Billy Hustace/Getty Images	Applying genetics to society as a whole	Are genes more common in certain groups?
WHAT SHOULD WE DO?	Looking at the ethical questions that surround human genetics	What would you do if you had to answer the ethical questions raised by human genetics?

BIOLOGY BASICS:
CELLS AND CELL STRUCTURE

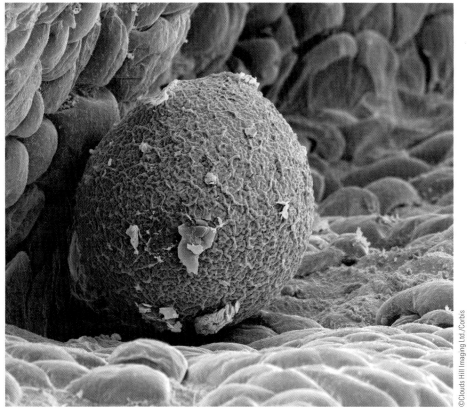

©Clouds Hill Imaging Ltd./Corbis

Fertilized human egg traveling through the fallopian tube.

BB1.1 Why are cells important in human genetics?

Humans begin life as a single cell: a fertilized egg, or **zygote.** The instructions for making all the cells, tissues, and organs of an adult are encoded in the DNA of that single cell. About 36–39 weeks after this cell is created, the newborn entering the world contains approximately 40 billion cells, all of which originated from the zygote and were created by cell division. The infant's cells are organized into many organ systems, each associated with highly specialized functions, and all controlled by the genetic information inherited from each parent.

In spite of the differences in size and shape of the more than 200 different cell types in the human body, all cells carry the same set of genetic information and share a basic architectural plan. Each cell is surrounded by a membrane and, inside, each possesses a structure known as the **nucleus.** In addition, cells contain a number of internal structures that act as tiny organs, known as **organelles.**

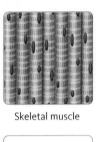
Skeletal muscle

Cardiac muscle

Smooth muscle

Red blood cells

Kidney cells

Lung cells

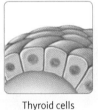
Thyroid cells

Pancreatic cells

Some of the more than 200 cell types in the human body.

The study of cells is a key part of human genetics; our genetic information is carried in the chromosomes of our cells. Errors in this genetic information (called mutations) have an impact on the structure and function of cells and can result in genetic disorders. For example, sickle cell anemia is a genetic disease that changes the shape of red blood cells. Most of the resulting physical symptoms, such as heart failure, stroke, pneumonia, brain damage, and pain, are a direct effect of the changes in cell shape. This disorder afflicts many of those who live in or have ancestors from West Africa, regions around the Mediterranean Sea, or parts of the Middle East and India.

New cells are constantly needed in our body for growth and/or replacement of dead or worn-out cells. These new cells are formed by cell division, a process in which one cell divides to form two cells. As they prepare to divide, the genetic information they carry must be copied exactly so it can be distributed to the daughter cells formed by division. This is accomplished by structures in the nucleus known as **chromosomes.** Chromosomes carry the cell's genetic information in the form of DNA. The process of cell division that produces these exact copies of the chromosomes is known as **mitosis** (discussed in Chapter 11)**.** Although

most cells divide by mitosis, the sperm and egg that fuse to form the zygote are produced by a special form of cell division known as **meiosis,** which will be discussed in detail in Chapter 3. Mistakes in either form of cell division can have serious genetic consequences, many of which can cause life-threatening or fatal disorders.

BB1.2 How is the cell organized?

Cells are the building blocks of the body. A description of the basic features of a human cell will provide a background for later discussions about what happens to the structure and function of cells or cell organelles in many genetic disorders. A photo of a human cell is shown below.

1. Plasma membrane   3. Cytoplasm

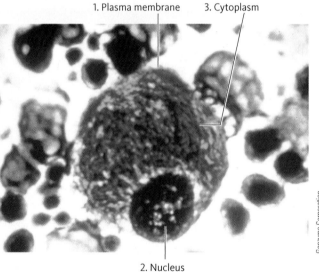

Genzyme Corporation, Cambridge MA

2. Nucleus

Although human cells differ widely in their size, shape, functions, and life cycles, they have three components in common: the **plasma membrane** (1), that separates a cell from its external environment, the **nucleus** (2) and the **cytoplasm** (3).

BB1.3 What does the plasma membrane do?

Follow along with the cell drawing as we discuss the major parts of a cell.

The double-layered plasma membrane has many important functions. It controls the movement of materials such as water, oxygen, and nutrients into and out of the cell. Chemical markers in and on the membrane give the cell an identity. These markers are determined by our genes and are responsible for many important properties of cells. In organ transplants, the HLA (human leukocyte antigen) marker identifies organs that are foreign to the body. A mismatch in HLA markers when an organ is being transplanted can cause the recipient's body to reject it.

The plasma membrane also contains molecular sensors or receptors that receive and process chemical signals from the cell's environment. These signals are important in regulating many critical cell functions. Mutations that affect these receptors play an important role in some diseases such as cancer. The plasma membrane encloses and protects the cytoplasm, a watery mixture that contains many types of molecules and structural components.

BB1.4 What is in the cytoplasm?

The cytoplasm contains membrane-bound compartments called organelles (small organs). Cell function is directly related to the number and type of organelles it contains.

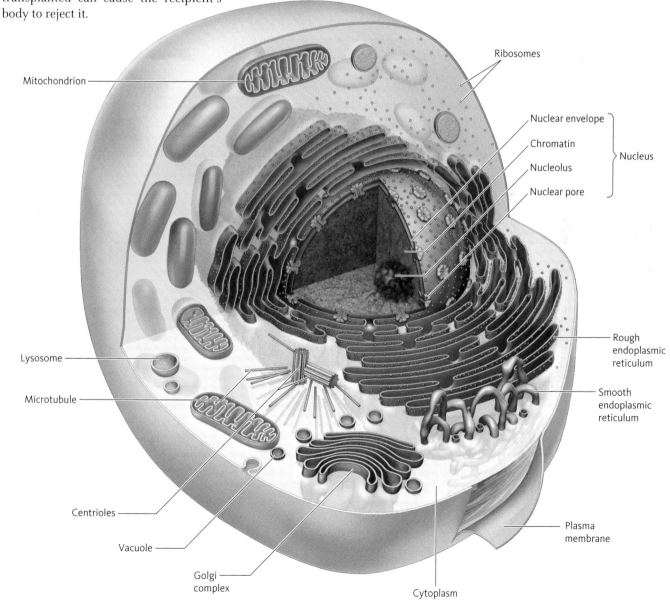

Mitochondrion

Ribosomes

Nuclear envelope

Chromatin

Nucleolus

Nuclear pore

Nucleus

Lysosome

Microtubule

Rough endoplasmic reticulum

Smooth endoplasmic reticulum

Centrioles

Vacuole

Golgi complex

Cytoplasm

Plasma membrane

Each cell can be viewed as a protein factory. These proteins are created in the cell and the organelles play a large part in their production. Many organelles create, modify, and transport the proteins. The figures below summarize the major organelles and their functions.

Chapter 1 examines the events that follow the fusion of the sperm and egg and considers how new technology allows parents to select the sex of their children. Chapter 2 explores how reproductive technology is used to help couples with fertility problems. Chapter 3 follows chromosomes through

PARTS OF A CELL

The **endoplasmic reticulum (ER)** is a network of membranes that form channels in the cytoplasm. Parts of the outer ER surface are covered with **ribosomes,** another organelle. They appear as small dots on the ER. Proteins are made by the ribosomes and enter the ER to be folded. (Protein synthesis is discussed in Chapter 5.)

Inside the ER, proteins are not only folded, but modified and made ready to move to other locations inside or outside the cell.

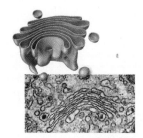

After leaving the ER, proteins move to the **Golgi apparatus (GA).** There they are sorted and distributed to their destinations inside and outside the cell. Abnormalities of the Golgi apparatus are responsible for certain genetic disorders, including one called Menkes disease.

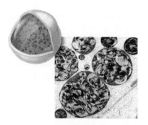

Lysosomes are the processing centers of the cells. They are membrane-enclosed vesicles that contain digestive enzymes. Materials brought into the cell that are marked for destruction end up in the lysosomes, where they are broken down. More than 40 genetic disorders are associated with defects in lysosomes.

Mitochondria make the energy for the cell. Cells such as liver cells that require a lot of energy can contain more than 1000 mitochondria. Mitochondria carry their own genetic information in the form of circular DNA molecules.

The largest organelle in a cell is the **nucleus.** It is enclosed by a double membrane called the nuclear envelope. The envelope contains pores that allow molecules to move in and out. Within the nucleus, dense regions known as **nucleoli** make ribosomes.

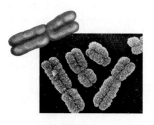

Chromosomes are found in the nucleus and exist in pairs. Each member of a chromosome pair is called a **homologous chromosome.** Most human cells carry 46 chromosomes (23 pairs). This is called the **diploid number ($2n$).** Certain cells, such as sperm and eggs, carry only one copy of each chromosome (23 chromosomes). This is called the **haploid number (n).** At fertilization, the fusion of a sperm with an egg's nucleus (each with the haploid number) forms a diploid zygote.

meiosis and reviews the problems that arise when errors in meiosis produce offspring with an abnormal number of chromosomes. Chapter 4 explores how genes are transmitted from parents to offspring and how this information is used in human genetics.

Glossary

chromosomes The DNA-containing threadlike structures in the nucleus that carry genetic information.

cytoplasm The viscous material contained between the inner surface of the plasma membrane and the outer nuclear membrane. It contains organelles, each with specialized functions.

diploid number (2n) The condition in which chromosomes are present as pairs. In humans, the diploid number is 46.

endoplasmic reticulum (ER) A series of cytoplasmic membranes arranged as sheets and channels that function in the synthesis and transport of gene products.

Golgi apparatus (GA) A membranous organelle composed of a series of flattened sacs. It sorts, modifies, and packages proteins produced in the ER.

haploid number (n) The condition in which each chromosome is present once, unpaired. In humans, the haploid number is 23.

homologous chromosomes Members of a chromosome pair.

lysosomes Membrane-enclosed organelles that contain digestive enzymes.

meiosis A form of cell division in which haploid cells are produced.

mitochondria (singular: mitochondrion) Membrane-enclosed organelles that are the site of energy production.

mitosis A form of cell division that produces two daughter cells that are genetically and chromosomally identical to the parent cell.

nucleoli (singular: nucleolus) A region in the nucleus that functions in the synthesis of ribosomes.

nucleus The membrane-enclosed organelle in cells that contains the chromosomes.

organelles Membrane-enclosed structures with specialized functions found in the cytoplasm of cells.

plasma membrane The outer border of cells that serves as an interface between the cell and its environment.

ribosome Cytoplasmic organelles that are the site of protein synthesis.

zygote The fertilized egg that develops into a new individual.

1 Sex and Development

✪ CENTRAL POINTS

- A fetus' chromosomal sex is determined at fertilization.

- The development of a fetus has many stages.

- During development, a fetus becomes visibly male or female.

Sandra Warick/Photonica/Getty Images

CASE A Choice of Baby's Sex Now Available

Susanna Carter had been thinking for the last few months about having another baby. She was happy with her family of three boys but had always wanted a girl. Already in her late 30s, she would have to decide soon. She wanted to discuss the possibility with her husband at dinner. She had seen an article in the paper about a new medical procedure called **sex selection.** Using this method, Susanna and her husband, Bob, could pick the sex of their baby before it was even growing in her uterus.

After talking about it, the Carters made an appointment with their physician, Dr. George Leon, to discuss sex selection. Dr. Leon knew that some methods of sex selection are more successful than others. He had a number of patients who wanted this type of procedure for various reasons. In his mind he went over some of them:

The first patient is a color-blind male. His wife's family has no history of color blindness;

therefore, all their daughters will carry the color blindness gene they receive from their father, but they will also get a normal copy of the gene from their mother and will have normal vision. However, they will pass the color blindness gene on to their children, and their sons will have a 50% chance of being colorblind. None of the sons of Dr. Leon's patient will get the colorblindness gene from their father, and all will have normal vision. The man and his wife do not want to pass this gene along to future generations of the family, and so would prefer to have only boys.

The second patient is a woman who carries the gene for Hunter syndrome, while her husband does not. Children with this disorder have severe facial, heart, and brain deformities. Because of the way this gene is inherited, daughters will not have Hunter syndrome, but may get the gene from their mother and pass it on to their children. Sons have a 50% chance

of getting this deadly disorder. This patient wants to know that an embryo is healthy before implanting and carrying it to birth.

The husband of the third patient believes that men are physically and mentally superior to women. He wants only sons. His wife does not agree with him, but she worries about how he would treat a girl.

The fourth patient was left several million dollars by an eccentric uncle. He will receive the money only if he and his wife have a son and name him after the uncle. The man already has a daughter and wants only two children. Obviously, he wants his next child to be a boy.

Some questions related to biology, law, and ethics come to mind when reading about these cases. Before we can address those questions, let's look at the biology behind sex selection.

1.1 How Is the Sex of a Child Determined?

In humans, as in many other animals, we can see obvious physical differences between males and females. These differences appear early in a fetus's development. Whether the fetus develops male or female sex organs depends on a complex interaction between the genes and environment of the fetus. Some differences—such as body size, muscle mass, patterns of fat distribution, and body hair—are not directly related to reproduction. These are called **secondary sex characteristics.**

Do chromosomes help determine sex?

In "Biology Basics: Cells and Cell Structure," we learned that the cells of humans carry a set of 46 chromosomes (■ Figure 1.1a, b). Females have two **X chromosomes** (XX) and males have an X chromosome and a **Y chromosome** (XY). These chromosomes

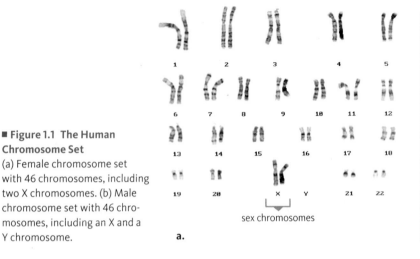
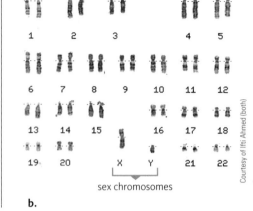

■ Figure 1.1 The Human Chromosome Set (a) Female chromosome set with 46 chromosomes, including two X chromosomes. (b) Male chromosome set with 46 chromosomes, including an X and a Y chromosome.

a.

sex chromosomes

b.

sex chromosomes

Courtesy of Ifti Ahmed (both)

(X and Y) are called **sex chromosomes** because they help determine the sex of an individual. The other 22 pairs of chromosomes are called **autosomes.** In humans, sex determination begins when a sperm, carrying an X or a Y chromosome, fuses with an egg (which carries an X chromosome). This is called **fertilization** and produces a cell called a **zygote** that carries either an XX or XY set of sex chromosomes in addition to the 22 pairs of autosomes.

Although saying that females are XX and that males are XY seems straightforward, there is more to it than that. Is a male a male because he has a Y chromosome or because he does *not* have two X chromosomes? Can someone be XY and develop as a female? Can someone be XX and develop as a male? These questions are still not completely resolved, but about 40 years ago individuals with only one X chromosome (45,X)* were discovered. Those with only one X chromosome are female. At about the same time, males who carry two X chromosomes and a Y chromosome were discovered (47,XXY).*

Anyone who carries a Y chromosome is almost always male, no matter how many X chromosomes he may have. However, the X and Y chromosomes are not the only determinants of one's sex. As we will see later in this chapter, the outcome depends on interactions between genes as well as internal and external factors.

How does the human sex ratio change over time?

The **sex ratio** is the proportion of males to females in a population. All eggs produced by human females carry a single X chromosome; roughly half the sperm produced by males carry an X chromosome and half carry a Y (formation of gametes will be explored in Chapter 3). An egg fertilized by an **X-bearing sperm** results in an XX zygote, which will develop as a female. Fertilization by a **Y-bearing sperm** produces an XY zygote, which will be male (■ Figure 1.2).

*The number indicates the total number of chromosomes present; the letters indicate which of the sex chromosomes are present.

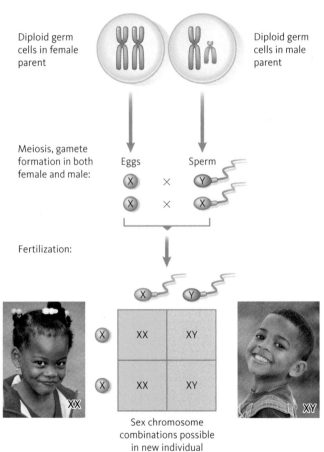

Diploid germ cells in female parent

Diploid germ cells in male parent

Meiosis, gamete formation in both female and male:

Eggs   Sperm

X × Y

X × X

Fertilization:

	XX	XY
X	XX	XY
X	XX	XY

Sex chromosome combinations possible in new individual

■ Figure 1.2 X and Y Chromosomes in Sex Determination
The random combination of X- or Y-bearing sperm with an X-bearing egg produces, on average, a 1:1 ratio of males to females.

Obviously, Susanna Carter wants to guarantee fertilization of one of her eggs by an X-bearing sperm to have the daughter she has always wanted. The couple who wish to satisfy the terms of their inheritance from the eccentric uncle need to achieve fertilization by a Y-bearing sperm, or a way to select XY embryos in order to have a son and inherit several million dollars.

Because males produce approximately equal numbers of X- and Y-bearing sperm, it seems that equal numbers of males and females should be conceived. If equal numbers of male and female zygotes are produced, the sex ratio should be 1:1. Although we cannot be completely certain about the numbers or the reasons, estimates indicate that more males are conceived than females. The sex ratio at birth is about 1:1.05 (100 females for every 105 males). For several reasons, more males than females die in childhood, and between ages 20 and 25 the ratio is close to 1:1. After that, the ratio of females to males keeps increasing because men have a shorter life span than women.

PHOTO GALLERY
Human sperm
© Eye of Science/Photo Researchers, Inc.

What procedures are used to select the sex of a baby?

For centuries, people have wanted to control the sex of their children. Many rituals and traditions have been used to ensure that a child will be a boy or girl, including having sex during certain phases of the moon, putting a knife or an egg under the bed before sex, or eating certain foods during pregnancy. Although many of these folklore-inspired methods are interesting and even ingenious, only two reliable, scientific procedures for sex selection exist: **sperm sorting** and **preimplantation genetic diagnosis (PGD).**

Sperm sorting, shown below, is possible because sperm carrying an X chromosome have about 2% more DNA than sperm carrying a Y chromosome. They are heavier because the X chromosome is larger than the Y (■ Figure 1.3). After collection, the sperm are isolated using a centrifuge (1) and treated with a non-toxic fluorescent dye that binds to the DNA in the chromosomes (2). The sperm are then sorted using a laser beam (3). As sperm flow past the laser, the dye glows.

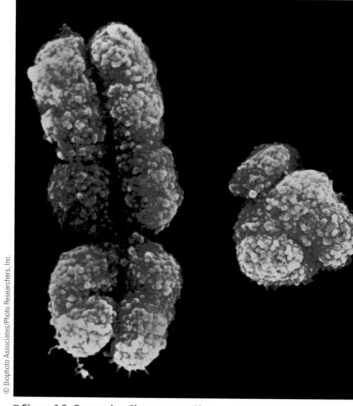

© Biophoto Associates/Photo Researchers, Inc.

■ **Figure 1.3 Comparing Chromosome Size**
The human X chromosome (left) is much larger than the Y chromosome (right).

Sperm with X chromosomes contain slightly more DNA and glow a little brighter than those with Y chromosomes. Based on how brightly they fluoresce, the sperm are sorted into two test tubes (4). One tube collects sperm with an X chromosome; the other collects sperm with a Y chromosome.

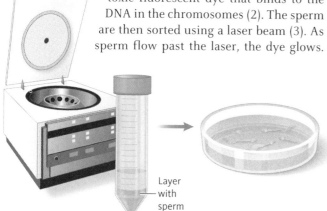

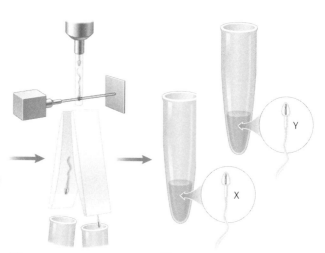

Layer with sperm

1 Sperm are separated from surrounding fluids, using a centrifuge.

2 Sperm are then placed in a saline solution containing a flourescent dye, which binds to the DNA in the sperm.

3 As the sperm pass one by one through a detector, a laser bounces light off the dyed DNA.

4 Because X-bearing sperm have more DNA and reflect more light than Y-bearing sperm, each can be separated into different test tubes.

Sperm from either tube can be placed into the uterus to fertilize the egg. This procedure costs $4,000–$6,000. If eggs are removed from the body and fertilized in a dish with sperm, the cost increases greatly.

Because some sperm can be misidentified during sorting, this procedure is not 100% accurate. Sperm sorting has a success rate of about 90% for female births and about 73% for male births. This might be an acceptable method for Susanna Carter, but probably not for the couple seeking to avoid having a son with Hunter syndrome.

The second method of sex selection, preimplantation genetic diagnosis (PGD), is shown below. In this method, a woman receives hormone treatments to stimulate her ovaries to produce many eggs at the same time. Next, a number of eggs are surgically removed from her ovary and placed in a dish with sperm until the eggs are fertilized (a process called *in vitro* fertilization (IVF), discussed in Chapter 2).

When the newly formed embryos have developed to the eight-cell stage (1), a single cell is removed and its chromosomes are analyzed to determine the sex of the embryo (2). Once male and female embryos have been identified (3), only embryos of the desired sex are placed into the mother's uterus (4). Unused embryos are discarded or frozen for later use or donation to others. PGD is a more invasive procedure than sperm sorting and is riskier for the mother. PGD costs $12,000–$15,000. Although it is more costly, the success rate for sex selection using PGD approaches 100%. Because of its high success rate, PGD would likely be the method of choice for Susanna Carter and the other patients mentioned earlier.

The use of sex selection and the fate of unused embryos created in PGD raise many social and ethical issues, as you can see from the Carters' situation and those of Dr. Leon's other patients. Some physicians refuse to use PGD for sex selection except when one spouse carries a genetic disease that affects one sex, whereas others embrace the technology.

1.1 Essentials

- Two methods of sex selection are sperm sorting and preimplantation genetic diagnosis (PGD).
- Females have two X chromosomes; males have an X chromosome and a Y chromosome.
- Chromosomal sex is determined at fertilization.

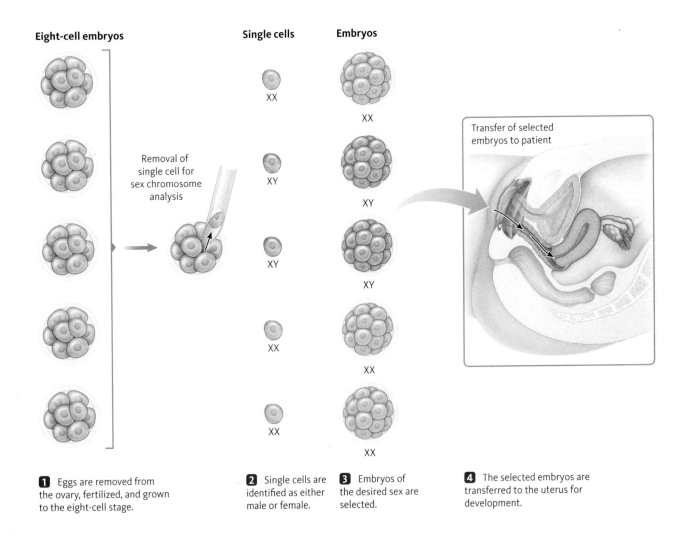

Eight-cell embryos **Single cells** **Embryos**

Removal of single cell for sex chromosome analysis

XX
XY
XY
XY
XX
XX

XX
XY
XY
XX
XX

Transfer of selected embryos to patient

1 Eggs are removed from the ovary, fertilized, and grown to the eight-cell stage.

2 Single cells are identified as either male or female.

3 Embryos of the desired sex are selected.

4 The selected embryos are transferred to the uterus for development.

1.2 How Does a Baby Develop from Fertilization to Birth?

To understand other factors that determine maleness and femaleness, let's look at events that occur during development of the fetus.

Fertilization, the fusion of sperm and egg, usually occurs in the upper third of the fallopian tube (■ Figure 1.4). Sperm deposited in the vagina move through the cervix, up the uterus, and into the oviduct; this

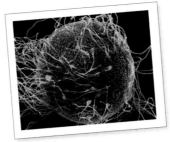

PHOTO GALLERY
These sperm are clustering around the egg, but only one will enter it.
©David M. Phillips/Photo Researchers, Inc.

trip takes about 30 minutes. Sperm travel this distance of about seven inches (approximately 17 cm) using whip-like contractions of their tails, assisted by muscular contractions of the uterus.

Usually only one sperm enters the egg, but other sperm (■ Figure 1.4[1–2]) help by triggering chemical changes in the egg that prevent more than one sperm from entering the egg. Once in the egg, the sperm's nucleus fuses with the egg's nucleus (■ Figure 1.4[3–4]), forming a zygote with 46 chromosomes (22 pairs of autosomes and 1 pair of sex chromosomes).

■ **Figure 1.4 Stages in Human Fertilization**
(1), (2) In fertilization, many sperm surround the egg and secrete enzymes that dissolve the outer barriers surrounding it. Only one sperm enters the egg. (3) The sperm tail degenerates, and its nucleus enlarges and fuses with the nucleus of the egg. (4) After fertilization, a zygote has formed.

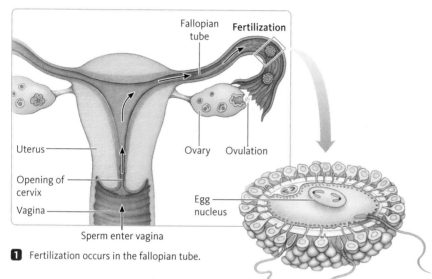

1 Fertilization occurs in the fallopian tube.

2 Several sperm surround the egg (above), but only one enters it.

3 The sperm nucleus fuses with the egg nucleus.

4 A diploid zygote forms.

The zygote begins the cell divisions that create an embryo and travels to the uterus over the next three to four days (■ Figure 1.5). When the embryo reaches the uterus, it remains unattached for several days. Cell division continues during this time, and the embryo enters a new stage of development; it is now a large hollow ball of cells called a **blastocyst** (■ Figure 1.5, Day 5). A blastocyst has several distinct parts: the **inner cell mass** (the source of embryonic stem cells), an internal cavity, and an outer layer of cells which form the membranes that surrounds the embryo. While the embryo is growing, the cells lining the uterus enlarge, preparing for the embryo to attach, a process called **implantation** (■ Figure 1.5, Days 6–7).

By about 12 days after fertilization, the embryo is firmly embedded in the wall of the uterus and has formed a protective membrane called the **chorion.** The chorion makes and releases a hormone called **human chorionic gonadotropin (hCG),** which prevents the lining of the uterus from breaking down and expelling the embryo. Excess hCG is eliminated in the urine. Home pregnancy tests work by detecting elevated hCG levels in urine.

A series of fingerlike projections called **villi** extend into the spaces in the uterine wall that are filled with maternal blood. The villi and maternal tissues form the placenta, a disc-shaped structure that nourishes the embryo throughout pregnancy. Membranes connecting the embryo to the placenta form the umbilical cord.

The major stages of embryonic and fetal development are shown in the photos on the following page.

Rapid growth takes place in the third trimester

As shown on page 13, the fetus grows rapidly in the third trimester (weeks 25–36), and the circulatory system and respiratory system mature to prepare for breathing air.

The fetus doubles in size during the last 8 weeks, and chances for survival outside the uterus increase rapidly during this time. At the end of the third trimester, the fetus is about 19 inches (48 cm) long, weighs from 5.5 to 10.5 lb (2.5–4.7 kg), and is ready to be born.

1.2 Essentials

- Many developmental changes occur in the fetus from fertilization to birth.
- Sex is only one of the thousands of traits determined during the gestation period in humans.

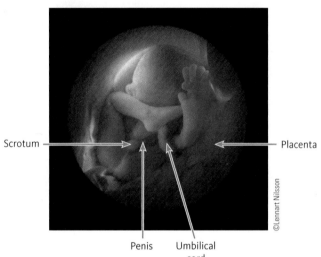

Scrotum   Placenta

Penis   Umbilical cord

©Lennart Nilsson

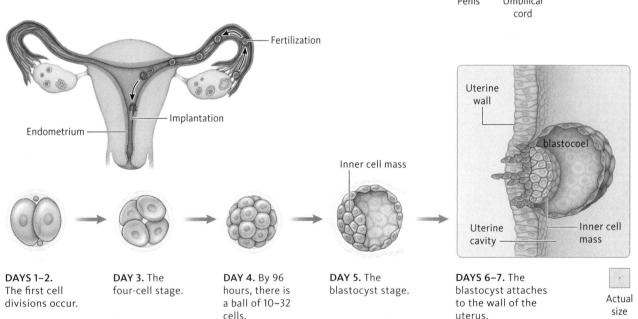

Fertilization

Endometrium   Implantation

Inner cell mass

Uterine wall

blastocoel

Uterine cavity   Inner cell mass

Actual size

DAYS 1–2. The first cell divisions occur.

DAY 3. The four-cell stage.

DAY 4. By 96 hours, there is a ball of 10–32 cells.

DAY 5. The blastocyst stage.

DAYS 6–7. The blastocyst attaches to the wall of the uterus.

■ **Figure 1.5 Human Development from Fertilization through Implantation**
As a blastocyst forms, its inner cell mass gives rise to a disk-shaped early embryo. As the blastocyst implants into the uterus, cords of chorionic cells start to develop. When implantation is complete, the blastocyst is buried in the endometrium.

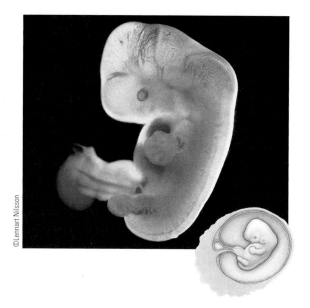

©Lennart Nilsson

In weeks 5 and 6, the embryo grows dramatically to a length of about 11 inches (28 cm). Most of the major organ systems, including the heart, are formed. Limb buds develop into arms and legs, complete with fingers and toes. The head is very large relative to the rest of the body because of the rapid development of the brain.

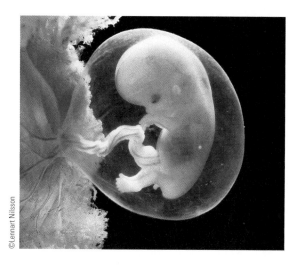

©Lennart Nilsson

By about 8 weeks, the embryo is large enough to be called a fetus. Although chromosomal sex (XX in females and XY in males) is determined at the time of fertilization, the fetus appears to be neither male nor female at the beginning of the third month. The sex organs cannot be seen in ultrasound scans until the 12th to 15th week. All the major organs have formed and are functional.

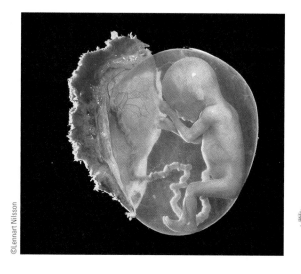

©Lennart Nilsson

By 16 weeks, major changes include an increase in size and further development of organ systems. Bony parts of the skeleton begin to form, and the heartbeat can be heard with a stethoscope. Fetal movements begin in the third month, and by the fourth month the mother can feel movements of the fetus's arms and legs. It has a well-formed face, its eyes can open, and it has fingernails and toenails.

CASE A QUESTIONS

Now that we understand how the XX and XY makeup of a fetus is determined and how it can be selected, let's look at some of the issues raised in the case of the Carters and Dr. Leon's other patients and apply the science we have discussed. A discussion of how courts of law interpret difficult scientific questions will follow.

1. Knowing how sex selection is accomplished, what do you think the Carters should decide to do? Why?

2. Which one of the Carters should make this decision? Susanna? Bob? Both? Why?

3. If you were Dr. Leon, would you perform sex selection for the Carters? Why or why not?

4. If you were Dr. Leon, would you perform sex selection for the other four patients described in the case? Give an answer and a reason for each patient.

5. Should sex selection be available to anyone, no matter their reason? Why or why not?

6. Should ability to pay for the procedure limit access to sex selection? Should insurance pay for it? Under what circumstances?

7. In India and China, there are problems with the use of sex selection. Historically, families have favored boys over girls, and with the use of sex selection, the sex ratio is changing. In China's 2000 census, it was 100 girls to 120 boys. That census also showed 19 million more males than females in the 0–15 age group in China. What problems will this create? If everyone were allowed to have access to sex selection, would this trend continue and affect the world population? How?

1.3 What Are the Stages of Sex Development?

As discussed previously, having an XX or XY chromosome pair is only part of becoming a male or a female. The formation of male or female reproductive organs depends on several factors including:

- How genes function
- Hormones
- The maternal environment

Interaction among these factors is important to the development of an individual's sex. Between fertilization and birth, sex can be defined at several levels: the sex chromosomes, the internal organs, and external appearance. In most cases, all of these are the same, but in other situations, as in the case of Maria Patino, they are not. Understanding what can go wrong requires a look at what normally happens in development.

The sex of an individual is formed in stages. Follow along with the chart on the facing page. The first stage, called chromosomal sex, begins at fertilization when the zygote has either an XX or an XY chromosome pair.

Although chromosomal sex is established at fertilization, the internal and external sex organs do not begin to

BY THE NUMBERS

3

Number of trimesters in a human pregnancy.

© Brand X Pictures/Jupiter Images

12

Number of weeks in one trimester.

36

Number of weeks of fetal development.

44

Number of rounds of mitosis in a human pregnancy.

Trillions

Number of cells in a human newborn.

develop until the seventh or eighth week of pregnancy. Up until this time, the external genitals of the embryo are neither male nor female. Internally, two nonspecific gonads and two sets of duct systems (male and female) are present (see the figure on the facing page).

In the second stage, called gonadal sex, expression of a gene on the Y chromosome causes the gonads to become testes. The cells in the newly formed testes secrete **testosterone** which controls the development of the external and internal male reproductive organs.

If two X chromosomes are present and no Y chromosome is present, the gonad develops as an ovary. In the absence of a Y chromosome, the default developmental pathway is female. Thus, without testosterone secretion, the gonads form ovaries; the female duct systems develop to form the fallopian tubes, uterus, and parts of the vagina; and the male duct systems degenerate. Then, in the third stage, phenotypic sex develops, and the fetus is identifiable as either male or female by its external organs.

Review the stages of sexual differentiation in the illustration on page 16. Note the three stages: chromosomal, gonadal sex, and phenotypic sex. Before reading about Maria Patino's condition, study the diagram.

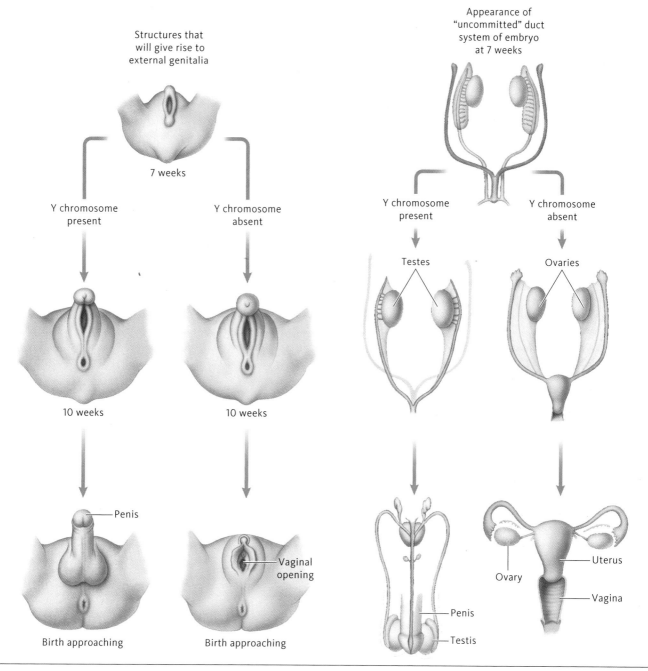

Structures that
will give rise to
external genitalia

7 weeks

Appearance of
"uncommitted" duct
system of embryo
at 7 weeks

Y chromosome
present

Y chromosome
absent

Y chromosome
present

Y chromosome
absent

Testes

Ovaries

10 weeks

10 weeks

Penis

Vaginal
opening

Birth approaching

Birth approaching

Ovary

Uterus

Vagina

Penis

Testis

As mentioned, if a Y chromosome is present, the gonads become testes and the resulting testosterone promotes the development of the male reproductive organs and duct systems. Secretion of a second hormone inhibits development of the female duct system.

In embryos with two X chromosomes, the absence of a Y chromosome, and the secretion of estrogen, cause the gonad to develop as an ovary. In the absence of testosterone, the default developmental pathway is female. Thus, without testosterone, the gonads form ovaries, the female duct systems develop to form the fallopian tubes, uterus, and parts of the vagina, and the male duct systems degenerate.

After the internal sex organs begin forming, the third phase, development of the external genitals, begins. In males, male hormones direct formation of the penis and the scrotum. In females, the clitoris, the labia minora, and the labia majora are formed from the same tissues.

The development of internal and external sex organs result from different genetic pathways. In males, this pathway involves the action of a gene (SRY) on the Y chromosome, the presence of at least one X chromosome, and expression of other genes, carried on the other 22 chromosomes. In females, this pathway involves the presence of two X chromosomes, the absence of Y chromosome genes, and expression of a female-specific set of genes on the other 22 chromosomes.

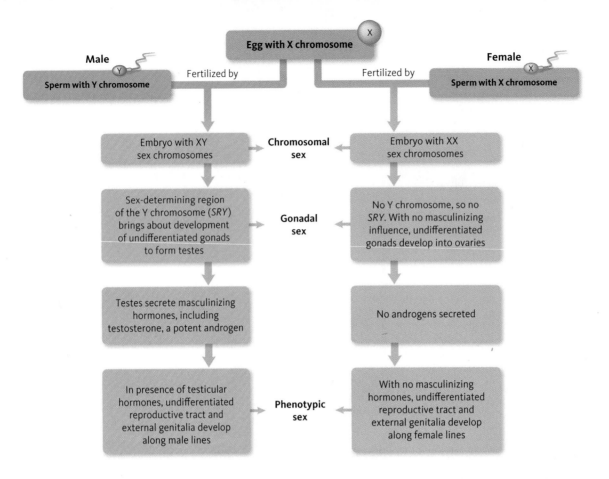

Embryo with XY sex chromosomes	**Chromosomal sex**	Embryo with XX sex chromosomes
Sex-determining region of the Y chromosome (*SRY*) brings about development of undifferentiated gonads to form testes	**Gonadal sex**	No Y chromosome, so no *SRY*. With no masculinizing influence, undifferentiated gonads develop into ovaries
Testes secrete masculinizing hormones, including testosterone, a potent androgen		No androgens secreted
In presence of testicular hormones, undifferentiated reproductive tract and external genitalia develop along male lines	**Phenotypic sex**	With no masculinizing hormones, undifferentiated reproductive tract and external genitalia develop along female lines

Courtesy, Professor Maria José Patino

CASE B Athlete Fails Sex Test

Sex testing was used in the Olympics and other world-class competitions following charges that some countries were having men compete as women in events reserved for women.

Maria Patino was a first-rate athlete, a hurdler who represented Spain. At the World University Games in Japan in 1985, she was instructed to report to "Sex Control," where cells were scraped from the inside of her cheek to examine her sex chromosomes. That didn't bother her; she had already passed such a "sex check" at the Helsinki World Championships in 1983. But she had forgotten to bring her "certificate of femininity" (■ Figure 1.6) to Japan and had to be retested.

A few hours after the test, an official told her the results were abnormal. "What does that mean?" she wondered. She was asked to go to the hospital in the morning, before her race, to be retested.

Maria was in and out of the hospital in a few minutes. The doctors spoke only Japanese, and no one said a word to her. When she was being driven to the stadium, just before the start of her race, Maria was told to fake an injury and withdraw from the games. Stunned, she did. She had failed the sex test.

■ **Figure 1.6 Certificate of Femininity**
A certificate of femininity issued to Maria Patino.

After returning home, Maria went to an endocrinologist, who diagnosed her condition as **complete androgen insensitivity (CAI)**, also called androgen insensitivity syndrome (AIS). This meant that although she was raised as a female and looked like a woman, her sex chromosomes were XY, not XX. Chromosomally, she was a man!

Some scientific, legal, and ethical questions come to mind when reading this case. Before addressing those questions, let's look at the biology behind the determination of sex in humans.

What was Maria Patino's condition?

Complete androgen insensitivity (CAI) is caused by a mutation in a gene on the X chromosome. This gene is called the **androgen receptor (*AR*) gene**. People with this condition lack molecular sensors (called receptors) for testosterone or hormones derived from testosterone on the surface of all their cells, including those that form the internal and external sex organs. The presence of an XY chromosome set normally causes the gonads to develop as testes and produce testosterone, but in CAI, cells in the gonad cannot respond to the presence of the hormone. Therefore, development proceeds as if no testosterone were present, which means that the female duct system and external genitals develop, although the undescended testes remain in the abdomen.

Individuals with this condition are chromosomal males (XY), but at birth, they appear to be females and are raised as girls. As adults, they do not menstruate, are infertile (because they have testes in their abdomens, not ovaries), and have well-developed breasts and very little pubic hair (■ Figure 1.7). This condition takes several forms with some variations in female development. Years later, after finding out about her condition, Maria Patino was quoted as saying, "It was obviously devastating to me. I had devoted my life to sport. But it was never a question for me of my femininity. In the eyes of God and medicine I am a woman."

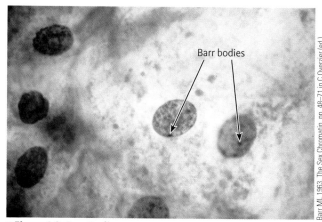

Barr ML 1963. The Sex Chromatin, pp. 48–71 in C Overzier (ed.) "Intersexuality," Academic Press, New York, NY; SM Carr after Barr

■ **Figure 1.8 Barr Bodies**
Inactive X chromosomes (Barr bodies) in a cell from a human female. XY males do not form Barr bodies.

How did Maria Patino pass one sex test, but not another?

In human females, one X chromosome in all cells (except developing eggs) becomes tightly coiled and inactive early in development. This coiled chromosome is referred to as a **Barr body** (■ Figure 1.8). The inactivated X chromosome can come from the mother or the father; once formed, it is a permanent condition. Males have only one X chromosome and do not form Barr bodies. Because females have two X chromosomes in each cell, and one or the other becomes inactivated, some cells express X chromosome genes from the mother's X chromosome and some cells express genes from the father's X chromosome.

The sex test required of Maria Patino and all other female athletes uses cells scraped from inside the cheek. The scrapings are placed on a microscope slide and stained with a dye that makes the Barr body visible. A certain number of cells are examined with a microscope to see if Barr bodies are present. If they are, the athlete is certified as eligible to participate in competition as a female. Males who are XY do not form Barr bodies, so this test was thought to be a reliable way of preventing males from gaining an advantage by competing in events for female athletes.

Unfortunately, this sex test presents several problems. First, the test is not completely reliable. If Barr bodies do not stain or are not visualized for a variety of reasons, an XX female can fail the test. Second, XXY males form Barr bodies and can pass a chromosomal sex test designed to identify females. Third, some females have only one X chromosome (called an XO condition) and do not form Barr bodies. More importantly, for our case, Marina Patino was XY and her cells did not form Barr bodies. The fact that she passed one sex test is an indication of how unreliable the Barr body test is in detecting XX individuals and screening out XY individuals.

Courtesy Cynthia P. Stone

■ **Figure 1.7 Androgen Insensitivity Syndrome**
An individual with complete androgen insensitivity.

In the face of evidence that the Barr body test was unreliable, organizers of the Olympic Games and other competitions turned to a series of other tests in a continuing effort to screen out males who might attempt to compete as females, even though there was little evidence that this had ever occurred. Finally, after recognizing that testing caused emotional stress, stigmatization, and discrimination, sex testing was abolished before the 2000 Olympic Games and has also been discontinued in most athletic events.

1.3 Essentials

- The internal sex organs begin developing in the seventh week of pregnancy.
- The most important developmental events occur in the first trimester.
- One X chromosome is inactivated in XX female cells and forms a Barr body.
- Development of the sex organs of a fetus is controlled by the hormones testosterone and estrogen.
- Disorders of sexual development such as CAI (complete androgen insensitivity) are genetically controlled.

CASE B QUESTIONS

Now that we have learned about the stages of sex determination for a fetus and what can go wrong during development, let's look at the issues raised in Maria Patino's case.

1. Knowing what you do about sex determination, do you think Maria Patino is female? Why or why not?

2. In the first test, when cells from her cheek were examined, she passed. Should this have been enough to verify that she was a female? Why or why not?

3. Did Maria have a muscular advantage over other female runners because she was XY? If so, would this have made a significant difference?

4. When Olympic and world championship competition required sex testing, only females were tested. Is this fair? Why or why not?

5. Should all newborns have their sex chromosome status determined? Would this avoid any problems that might arise later? Why or why not?

6. What problems may this testing cause for those with abnormal sex chromosomes, such as those we discussed in this chapter?

7. At a hearing, you are Maria's attorney.
   a. What argument would you make using the science related to sex determination? Be sure to make your argument simple enough for a juror to understand.
   b. What argument would you make using the law?
   c. What would you ask for as compensation?

8. At a hearing, you are the Olympic Committee's attorney.
   a. What argument would you make using the science related to sex determination? How would you make it understandable?
   b. What argument would you make to defend the actions of the committee?
   c. What action would you recommend the committee take? Why?

9. Based on Maria's case, should any changes in sex testing be instituted?

10. Why didn't Maria find out about this condition earlier? Do you think the fact that some female athletes in strenuous sports do not menstruate may have complicated her diagnosis?

11. Could Maria sue her doctors for malpractice? Why or why not?

12. Who else might she sue?

1.4 What are the Legal and Ethical Issues Associated with Sex Determination?

As you can see from Maria Patino's situation, people with some genetic conditions are not always treated fairly. In fact, they are often discriminated against. When this occurs, it can be handled in a number of ways. One way is to file a lawsuit that forces employers and others to treat them equally. Another way is to pass laws that make it clear that society will not tolerate this type of behavior.

There have been laws passed in every state of the United States that protect people from being discriminated against because of their race, color, sex, origin, religion, or age. This type of discrimination usually occurs in critical areas such as housing, public accommodations, education, transportation, communication, recreation, health services, voting, and access to public services.

It seems as though Maria Patino was discriminated against because she was competing as a woman but could not pass the "sex test." This meant that her chromosomes showed that she was genetically a male, although she was physically a female.

People with complete androgen insensitivity (CAI) fall into a medical category called *intersexuality*. A medical definition of intersexuality is "a condition in which chromosomal sex of a person is not consistent with his or her phenotypic sex, or in which the phenotype is not classifiable as either male or female."

Some other conditions fall under this category. We will be discussing chromosomal abnormalities called Klinefelter syndrome (47,XXY), and Turner syndrome (45,X) in later chapters.

In addition, the Americans with Disabilities Act (ADA) addresses the problems disabled Americans have in employment, insurance, and other areas. This Federal act was signed into law in 1990 and protects the rights of people with disabilities. Many of these disabilities are caused by genetic conditions.

Under the ADA, a person with a disability is defined as an individual who:

a. has a physical or mental impairment that substantially limits one or more major life activities; or

b. has a record or history of such an impairment; or

c. is perceived or regarded as having such an impairment

Do people with intersexuality fall under the ADA? This has been difficult to determine but individuals with other genetic conditions are clearly protected by the ADA. One of these is achondroplasia (see Chapter 4), a form of dwarfism. Because of their height, people with this condition may be discriminated against in employment, education, or insurance. The ADA is designed to keep this type of discrimination from happening.

The chart below addresses some of the ethical and legal questions regarding discrimination because of sex determination:

Question	How are these questions decided?	Related case or legal issue
Can a person be denied a job because of their sex?	Laws require one to hire without discrimination. Lawsuits are filed and decided by individual judges.	It is difficult to determine why a person is not hired unless it is specifically stated.
Was Maria Patino a victim of discrimination?	The rules of the committee and the Olympics did not allow someone to compete as a woman if they could not pass the chromosomal test.	No lawsuit was filed by Maria Patino but others have forced the removal of the "sex test" rules in international athletic competitions.

If Maria Patino had sued in a court of law, the court would have had to decide whether she was a woman or a man. As you saw when you answered the case questions, this situation does not always have a simple answer, and it involves an understanding of scientific principles. The following case was extremely important in helping courts determine what science is relevant to trials and what is not. Because Frye v. U.S. was the first case to address what science should be accepted in court, any lawsuit filed by Maria in the United States would have used it. The questions formulated in this case were used for more than 70 years as the standard for accepting or rejecting scientific evidence in U.S. courts.

FACTS OF THE CASE

James Alphonso Frye was on trial for murder. He had maintained his innocence since his arrest. Early on, his attorney suggested that he take a test to determine whether he was telling the truth. The test, the systolic blood pressure deception test, was an early version of the lie detector test used today.

The court was asked to accept Frye's "truthful" results as evidence to support his plea of innocence. The technique was new, but his attorney had an expert witness. William Marston, an attorney and psychologist, had done a great deal of research on how changes in the body's physiology correlate with lying. He claimed that the test measured whether the defendant was telling the truth by monitoring his systolic blood pressure. The systolic blood pressure goes up and down depending on how hard the ventricles of the heart are contracting. Marston said that fear always produces a rise in systolic blood pressure, and lying, along with fear of detection, raises the blood pressure. The judge of the trial court in the District of Columbia excluded Marston's testimony, stating that the science of deception testing was too new to be admitted.

Frye's lawyer took the case to the Court of Appeals of the District of Columbia. The appellate court ruled that the systolic blood pressure deception test had "not gained standing and scientific recognition," and therefore it upheld the lower court's decision not to admit Marston's testimony.

More importantly, the court set guidelines for other courts on admitting new scientific evidence. The ruling stated that a scientific principle would have to have gained "general acceptance" in the "particular field to which it belongs." The so-called general acceptance rule asked three questions:

1. To which scientific field does the evidence or testimony belong?
2. Is the evidence or testimony generally accepted in this field?
3. What constitutes general acceptance?

To be admissible, the scientific testimony must be scrutinized by a relevant scientific community and be significantly tested.

QUESTIONS

1. How much testing should be done before science is admitted into evidence?

2. What three questions would you ask an expert witness to determine whether he or she is an expert?

3. Should expert witnesses be paid? Why or why not?

4. All science begins as "new" or experimental. How long should it take before it becomes "real" science?

5. Fingerprints are now considered good evidence after about 50 years of use in trials. But they are not as ironclad as we might think. Often partial fingerprints are left and are difficult to identify. But the public likes fingerprints. Should public opinion have anything to do with admission in court? Why?

6. Today, lie detectors are not allowed in evidence. Why? Should they be?

7. In his book *The Truth Machine*, James Halperin envisions a future with a machine that can determine lying with no error. What effect do you think this would have on courtrooms?

8. Recently, the Supreme Court has given judges more control in determining what science should be allowed to be presented in trials. Name two advantages and two disadvantages of this decision.

★ THE ESSENTIAL TEN

1. Two methods of sex selection are sperm sorting and preimplantation genetic diagnosis (PGD). [Section 1.1]
   ▌ Sperm sorting involves separation of X- and Y-bearing sperm.

2. Females have two X chromosomes; males have an X chromosome and a Y chromosome. [Section 1.1]
   ▌ The X comes from either the mother or father, but the Y must come from the father.

3. Chromosomal sex is determined at fertilization. [Section 1.1]
   ▌ The sperm that joins with the egg contains either an X or a Y.

4. Many developmental changes occur in the fetus from fertilization to birth. [Section 1.3]
   ▌ The complicated process of development is directed by chromosomes, the genes they carry, and both internal and external factors.

5. Sex is only one of the thousands of traits determined during the gestation period in humans. [Section 1.2]
   ▌ Many parts of sex development are hormonally controlled.

6. The internal sex organs begin developing in the seventh week of pregnancy. [Section 1.3]
   ▌ Sexual development begins with the gonads.

7. The hormones testosterone and estrogen are secreted by the testis and ovary. [Section 1.3]
   ▌ Testosterone, estrogen, and their derivatives control development of the gonads, the duct system, and the genitals.

8. The most important developmental events occur in the first trimester. [Section 1.3]
   ▌ The first trimester is three months.

9. One X chromosome is inactivated in XX female cells and forms a Barr body. [Section 1.3]
   ▌ The Barr body is formed when the X chromosome coils and attaches to the inside of the nuclear membrane.

10. Disorders of sexual development such as CAI (complete androgen insensitivity) are genetically controlled. [Section 1.3]
    ▌ Mutations in many genes can affect formation of the sex organs.

KEY TERMS

androgen receptor (AR) gene (p. 17)
autosomes (p. 8)
Barr body (p. 17)
blastocyst (p. 12)
chorion (p. 12)
complete androgen insensitivity (CAI) (p. 15)
fertilization (p. 8)
human chorionic gonado-tropin (hCG) (p. 12)
implantation (p. 12)
in vitro fertilization (p. 10)
inner cell mass (p. 12)
preimplantation genetic diagnosis (PGD) (p. 9)

secondary sex characteristics (p. 7)
sex chromosomes (p. 8)
sex ratio (p. 8)
sex selection (p. 7)
sperm sorting (p. 9)
testosterone (p. 14)
villi (p. 12)
X-bearing sperm (p. 8)
X chromosome (p. 7)
Y-bearing sperm (p. 8)
Y chromosome (p. 7)
zygote (p. 8)

REVIEW QUESTIONS

1. The physical traits that we identify as belonging to males or females usually appear when a child is a teenager. What are these attributes called?

2. If a child has a Y chromosome, what is his or her sex? What if the child has two X chromosomes and a Y chromosome?

3. What is the sex ratio?

4. Can the sex of a baby be chosen during pregnancy? Why or why not?

5. At what point during pregnancy does each of the following structures develop?
   a. heart
   b. skeleton
   c. penis

6. Does the sex ratio in a population change over time?

7. What is PGD and why is it successful in sex selection?

8. How much does a fetus grow in the third trimester?

9. When sperm are formed, what percentage of them carry X chromosomes?

10. What percentage of the sperm carry Y chromosomes?

What Would You Do If . . . ?

1. You were given the option to use sex selection to pick the sex of your baby?

2. You were asked to write a rule that would govern sex determination in the Olympics?

3. You were asked to vote on a law that would require sex chromosome testing of all newborns in your state?

4. You were married to a woman who could not have a baby and the doctor told you that her chromosomes were XY?

5. Your daughter did not menstruate by age 15 and the doctor told you that her chromosomes are XY?

APPLICATION QUESTIONS

1. Do you think the chromosomes of every baby born should be examined? Why or why not?

2. In history, men have often been furious with their wives for not giving them male children. Research this topic and find out when it became clear that women were not at fault.

3. When a baby is developing, doctors can operate on the fetus to correct certain errors. When the repair is finished, the fetus is put back into the uterus to finish developing. What do you think might be the biggest problem in this type of operation?

4. Suppose you are an attorney, and a woman comes to you with the following problem. She had an ultrasound test early in her pregnancy, and her doctor said she was having a boy. But the woman gave birth to a girl, and now she wants to sue. How would you use Frye v. U.S. to explain that you cannot represent her?

5. Which do you think is the most dangerous time in the development of the fetus? Why?

6. There is a 91% chance of producing a female using sperm sorting to select sperm carrying an X chromosome, and a 71% chance of producing a male using sperm sorting to select sperm carrying a Y chromosome. Sex selection using PGD is 100% effective, but produces extra embryos that must be stored for future use, donated, or destroyed. As a genetic counselor, you meet with a couple who wants to use sex selection to have a boy in order to balance a family with four girls. What questions would you ask them, and what options would you explore with them?

7. You are a physician. A couple with five girls comes to you for a second opinion. They want to have another child, but for personal reasons, do not want to use sex selection. They consulted a reproductive expert who explained to them that the chances of having a boy are 50%, but since they have already had five girls, the odds are overwhelmingly in favor of their next child being a boy. They would like to know more about the odds of having a boy. What would you tell them? Would your answer change if you were an attorney? Why or why not?

8. Females carry two X chromosomes (XX) and males carry an X and a Y chromosome (XY). This means that females have two copies of all X chromosome genes and males have only one (in general, genes on the X are not found on the Y chromosome). If each gene produces one unit of product, does this mean that females make two units of product for X chromosome genes and males make one unit of all X chromosome gene products? Why or why not?

Learn by Writing

If you are interested in sharing your ideas about the questions raised in Chapter 1, we suggest you start a blog. Throughout the book we will give you some suggestions of discussion questions that you may want to include in your blog to share with members of your class or others.

Ask your instructor if your school has a blogging site or go to one of the free sites available on the internet (blogger.com, googleblog.com, or others) and invite members of your class or others to contribute. This way you can write your opinions, share them with others, and get comments back.

Watch for other "Learn by Writing" suggestions in the Application Questions in various chapters, but don't wait to see it. Address interesting cases, questions and ideas in every chapter. . . see what develops.

Here are some ideas to address in your blog from Chapter 1:

- The sex of a baby is a very personal subject among families. It is very important in some societies to have a child of a certain sex. Does that mean we should give everyone the opportunity to pick the sex of their child?

- Intersex patients often say they are discriminated against if they tell others about their condition. Should this information remain private or made public?

- A number of years ago, a child was undergoing a circumcision and a mistake was made. His penis was cut off. The doctors told the parents they would do surgery to make the child an external female and they should take him home and raise him as a girl. Do you think this would work? Why did it work with Maria Patino?

- Should the law actually be involved in any of these personal decisions?

- Address any other question or comment that came to mind as you read this chapter.

ONLINE RESOURCES

Preparing for an exam? Log on at academic.cengage.com/login for a pre-test, a personalized learning plan, and a post-test in CengageNOW's Study Tools to help you assess your understanding.

If assigned by your instructor, the Case A and Spotlight on Law activities for this chapter, "Sex Selection" and "Lie Detector Testing," will be available in CengageNOW's Assignments.

Dr. Phil (CBS), January 16, 2007

AIS Child on Dr. Phil Show

Kayleigh, whose parents brought her to the *Dr. Phil* show, has been diagnosed with androgen insensitivity syndrome (AIS) and is now four years old. Having the same syndrome as Maria Patino, Kayleigh looks female but has XY sex chromosomes.

Her mother said, "I was having to daily reinforce to Kayleigh that she was a girl, and if I talked to her about being a girl and not a boy, she would throw a temper tantrum. She cries and says she is a boy." Kayleigh also wanted to potty-train standing up.

To see highlights of this episode online, go to academic.cengage.com/biology/yashon/hgs1.

QUESTION:

Do you think it is possible for a child this young to understand whether she is male or female?

PHOTO GALLERY
Santhi Soundarajan, an Indian runner who failed the sex test because of androgen insensitivity syndrome (AIS).
© AP Photo/M. Lakshman

USA Today, December 19, 2006

Female Runner Disqualified after Failing Gender Test

An Indian runner, Santhi Soundarajan, was stripped of her medal in the 800-meter run in the 2006 Asian games when her sex test came back abnormal. It was reported that she "appeared to have abnormal chromosomes." The official also said the test revealed more Y chromosomes than allowed. Athletes are not required to take a gender test but may be asked to do so.

To access this article online, go to academic.cengage.com/biology/yashon/hgs1.

QUESTION:

If this is the same condition (AIS) that Maria Patino has, should Santhi be allowed to continue as a female runner?

New York Times, May 13, 2007

Genetic Testing + Abortion = ????

Amy Harmon

This article discusses the use of sex selection to choose the sex of a child and its effect on the percentage of females in a population, as well as whether genetic testing, in general, can increase the number of abortions. The director of the pro-choice Reproductive Health Technologies Project said that when her staff members recently discussed whether to recommend prenatal diagnosis for sex selection, they found it morally repugnant but would not ban it.

To access this article online, go to academic.cengage.com/biology/yashon/hgs1.

QUESTION:
Do you agree with the project's staff?

CBC News, May 18, 2007

Canadians Buying 6-week Gender Determination Test

A new home test that can determine the sex of a fetus as early as six weeks has become quite popular among Canadians. This test, which requires a sample of the mother's blood, can determine whether male DNA is present.

The testing company, DNA Worldwide, said the results can come back in as soon as four to six days. The company has studied the possibility of refusing to send the test to countries where couples practice sex selection and abortion.

To access this article online, go to academic.cengage.com/biology/yashon/hgs1.

QUESTION:

If this test became available worldwide, do you think many people would use it to determine the sex of their child?

2 Assisted Reproductive Technology

✴ CENTRAL POINTS

- Sperm and egg cells are formed in specialized organs.

- Males and females have different reproductive organs.

- Infertility is a serious problem in many parts of the world.

- Medical techniques are available to help people who are infertile.

- Legal cases are deciding how frozen embryos should be treated.

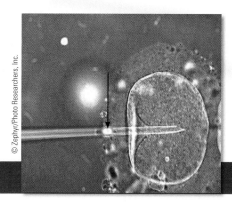
© Zephyr/Photo Researchers, Inc.

CASE A   A Couple Considers Fertility Treatment

Brian and Laura have been married for 10 years. During the first five years of their marriage, they didn't really think about having children, mainly because both of them were very busy; Laura was an attorney and Brian was finishing his master's degree.

But then Laura's sister became pregnant, and that started them thinking. They assumed that getting pregnant would be easy, but that was not the case. After a year of trying to conceive, they decided to talk to Laura's doctor. Shortly before the appointment with the gynecologist, Dr. Franco, the couple watched a television show that talked about new techniques to help infertile couples. Laura and Brian were amazed to learn that 5 million American couples are infertile, and clinics all over the country specialize in helping them. The procedures were expensive, though, and they wondered what they would do if they could not get pregnant the "normal" way.

Brian thought they should adopt and not put themselves through the tests and procedures. Laura disagreed; she wanted a genetically-related child.

Some questions come to mind when reading about Brian and Laura's situation. Before we address those questions, let's look at the biology behind fertility.

2.1 How Are Sperm and Eggs Made?

As discussed in "Biology Basics: Cells and Cell Structure," all humans begin life as a single cell, the zygote, produced by the fusion of a sperm and an egg (also called an **oocyte**). These two gametes are produced in the gonads of the respective sexes.

The **testes** of males produce **sperm** and male sex hormones called **androgens**. The **ovaries** of females produce **eggs** and female sex hormones called **estrogens**. Sperm and egg cells are produced by a process called **meiosis**, which will be discussed in Chapter 3.

2.1 Essentials

- Sperm are produced by meiosis in the testes of the male.
- Eggs are produced by meiosis in the ovaries of the female.

2.2 What Organs Make Up the Male Reproductive System?

Follow along with the figure on the next page. Testes (1) form in the abdominal cavity during male embryonic development; before birth, they descend into the scrotum (2), a pouch of skin located outside the body. The male reproductive system also includes the following:

- a duct system that transports sperm out of the body
- three sets of glands that secrete fluids to maintain sperm viability and motility
- the penis (3)

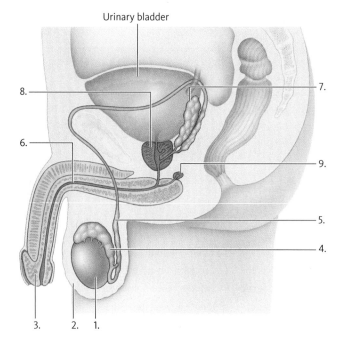

Urinary bladder

8.
6.
3. 2. 1.
7.
9.
5.
4.

Table 2.1	The Male Reproductive System
Component	**Function**
Testes	Produce sperm and male sex steroids
Epididymis	Stores sperm
Vas deferens	Conducts sperm to the urethra
Prostate gland	Produces seminal fluid that nourishes sperm
Urethra	Conducts sperm to the outside
Penis	Organ of copulation
Scrotum	Holds testes away from the body

The testes contain tightly coiled **seminiferous tubules,** where sperm are produced in a process called **spermatogenesis.** After they are formed, sperm move through the male reproductive system in stages. In the first stage, they move to the **epididymis** (4) for storage.

In the second stage, when a male is sexually aroused, sperm move from the epididymis into a connecting duct called the **vas deferens** (5). The walls of the vas deferens are lined with muscles that contract rhythmically to move sperm forward. In the final stage, sperm are propelled by muscular contractions through the **urethra** (6) and expelled from the body.

As sperm are transported through the duct system, secretions are added from three sets of glands. The **seminal vesicles** (7) contribute **fructose,** a sugar that serves as an energy source for the sperm, and **prostaglandins,** locally acting chemical messengers that stimulate contractions of the female reproductive system to assist in sperm movement. The **prostate gland** (8) secretes a milky, alkaline fluid that neutralizes acidic vaginal secretions and enhances sperm viability. The **bulbourethral glands** (9) secrete a mucus-like substance that provides lubrication for intercourse. Together, the sperm and these various glandular secretions make up semen, a mixture that is about 95% secretions and about 5% sperm (Table 2.1). After a **vasectomy,** in which the vas deferens are cut and closed off, the secretions of these three glands make up the ejaculate, and the sperm are absent.

As shown in the drawing at right, spermatogenesis occurs in the seminiferous tubules. Cells called spermatogonia divide by mitosis to form

spermatocytes (with 46 chromosomes). These cells undergo meiosis to produce intermediate cells called **spermatids.** Spermatids have 23 chromosomes, but they are not yet sperm and do not have tails. Spermatids undergo structural changes and form mature sperm with tails.

Spermatogenesis begins at puberty and continues throughout life.

2.2 Essentials

- **The major parts of the male reproductive system are the testes, urethra, and penis.**

2.3 What Organs Make Up the Female Reproductive System?

Follow along with the figure on the next page. The female gonads are a pair of oval-shaped ovaries (1) about 1.5 inches (3 cm) long, located in the abdominal cavity. Each ovary produces eggs in a process called **oogenesis.** On the outer side of each ovary are the **fallopian tubes** (2). They have fingerlike projections that partially surround the ovary. After the egg is released from the ovary, fertilization occurs in the fallopian tubes, and the fertilized egg (now called a zygote), moves into the uterus (3).

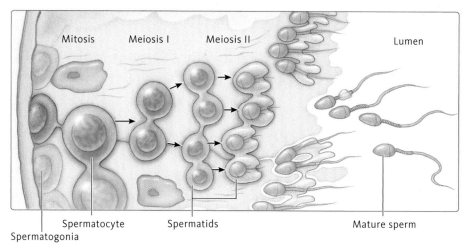

Mitosis   Meiosis I   Meiosis II   Lumen

Spermatogonia   Spermatocyte   Spermatids   Mature sperm

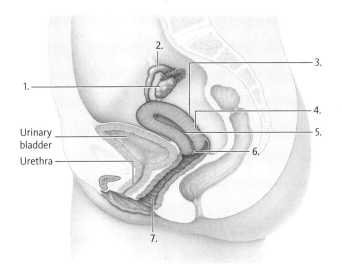

Urinary
bladder

Urethra

Table 2.2	The Female Reproductive System
Component	Function
Ovaries	Produce eggs and female sex steroids
Fallopian tubes	Transport sperm to ova and ova to uterus
Uterus	Holds embryo and fetus
Vagina	Receptacle for sperm; birth canal

The fallopian tubes open into the **uterus,** a hollow, pear-shaped muscular organ about the size of your fist (about 3 in. [7.5 cm] long and 2 in. [5 cm] wide). The uterus consists of a thick, muscular outer layer called the **myometrium** (4) and an inner tissue layer called the **endometrium** (5).

The innermost layers of the endometrium are shed at menstruation if fertilization has not occurred and are rebuilt during the next menstrual cycle. The lower neck of the uterus, the **cervix** (6), opens into the **vagina** (7). The vagina receives the penis during intercourse and also serves as the birth canal. It opens to the outside of the body behind the urethra (Table 2.2).

Egg production (oogenesis) occurs in the ovary. As shown in the figure below, an ovary contains a few hundred **follicles** (1), each consisting of a developing egg surrounded by an outer layer of follicle cells (2). Cells called oogonia (with 46 chromosomes) begin meiosis in the third month of a female's embryonic development, and then stop. In this state, they are called oocytes (3) and remain arrested in meiosis until just before they are released from the ovary. Meiosis is not completed until after they are fertilized. This means that at birth, a girl has a lifetime supply of developing eggs in her ovary.

At puberty, a developing egg, called a secondary oocyte (4), is released from a follicle (5), during a process called

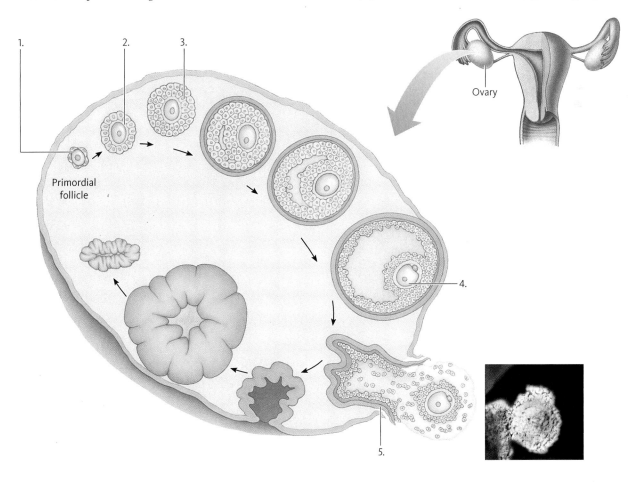

Primordial
follicle

Ovary

ovulation. Over a female's reproductive lifetime, about 400–500 oocytes will be released from the ovary and move into the oviduct.

If an egg is fertilized, the action of the sperm entering the egg triggers the completion of meiosis. When the sperm and egg nuclei fuse, a diploid cell carrying 46 chromosomes and called a zygote, is formed.

2.3 Essentials

- The major parts of the female reproductive system are the ovaries, fallopian tubes, and uterus.

2.4 Is Infertility a Common Problem?

Essentially, two things are needed for a successful conception: a healthy sperm and a healthy egg. After that, a place for fertilization to occur (a fallopian tube) and a place for the fetus to grow (a uterus) must be present.

PHOTO GALLERY
A normal uterus.
© Anatomical Travelogue/Photo Researchers, Inc.

If any one of these components is missing or not functioning properly, infertility may result. Usually a physician will make a diagnosis of infertility after a couple has been trying to conceive for about one year with no success. Although the problem in Brian and Laura's case is not clear, they seem to have an infertility problem. They are not alone. In the United States, the number of infertile couples has increased greatly over the last 10 years, to about 1 in 6 couples. In 40% of infertility cases, the woman is infertile, in 40% the male is infertile, and in 20% the cause is unexplained.

Infertility also becomes more common with increasing age; about 28% of couples in their late 30s are infertile, as shown in the graph below. Some of the reasons for age-related infertility are discussed in Chapter 3.

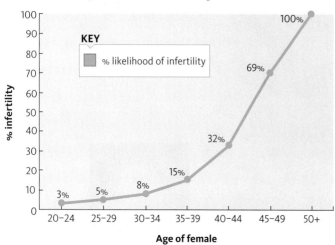

There are two types of infertility: primary and secondary. If a couple cannot become pregnant the first time they try, they are said to have primary infertility. But if a couple has had one child and has trouble conceiving a second, the problem is called secondary infertility. Some 3.3 million couples are thought to have secondary infertility. Its causes are unexplained, but age may be a factor in some cases. The discussion here will focus on primary infertility in both men and women.

What are the causes of infertility in women?

Three of the four things needed to conceive are associated with the female reproductive process: the egg, the fallopian tube, and the uterus (a functional sperm is the fourth). As you can see in the graph below, hormone regulation of ovulation is the most common cause of infertility in women. Ovulation depends on levels of the female hormones, estrogen and progesterone, and their interactions.

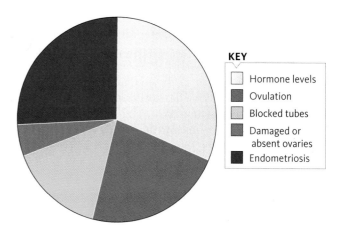

KEY

☐ Hormone levels
☐ Ovulation
☐ Blocked tubes
☐ Damaged or absent ovaries
☐ Endometriosis

Sperm can survive for several days in a woman's body, but an egg must be fertilized soon after ovulation. If a newly ovulated egg and sperm are not present in the fallopian tube at the same time, then conception will not occur. Therefore, ovulation and its timing are very important. Several problems can affect or prevent ovulation:

- **Hormone levels.** If there is no estrogen or too little estrogen, the ovary will not release an egg. Usually estrogen levels rise during the first part of a woman's cycle. When the estrogen level rises to a peak, it triggers ovulation and a rise in another hormone, called **luteinizing hormone** (LH) (home ovulation kits test for the presence of LH and can determine when a woman is ovulating). If estrogen levels are too low or are irregular, ovulation can be affected. Hormonal problems occur in about 50% of all cases in which ovulation does not occur.

- **Absent or damaged ovaries.** Surgical removal of the ovaries or damage to the ovaries by other surgeries, inflammation, radiation, or infection are causes of infertility. Ovarian cysts or infection can also prevent egg maturation or release. If there are no ovaries, then no eggs can be produced. Some genetic conditions cause a female child to be born without ovaries.

- **Premature menopause.** Although it is rare, some women stop menstruation and enter menopause at an early age. Some researchers think this happens more in women who are extremely athletic and who have had low body weight for many years. This might be why Maria Patino's condition was not detected earlier in her life (see Chapter 1). Recall that she would have been infertile because she had complete androgen insensitivity (CAI).

A related but reversible condition, **secondary amenorrhea,** is a lack of menstruation caused by several factors, including stress, low body weight, nutrition, and excessive physical conditioning.

Fallopian tube blockage, another cause of infertility, occurs in about 15% of all cases. If this tube is blocked, then the sperm and egg cannot connect.

Sexually transmitted bacterial and viral infections cause inflammation, scarring, and closing of the fallopian tubes. Seemingly unrelated conditions, such as appendicitis or a bowel problem called colitis, can cause inflammation in the abdomen that results in blocked fallopian tubes. Previous surgeries can cause formation of scar tissue, called adhesions, that can prevent eggs from moving through the fallopian tubes and contribute to infertility.

Also shown in the graph on the previous page, uterine problems can be a cause of infertility. If a woman has no uterus because it was removed or she was born without one, she will not be able to carry a child. About 10% of infertility cases are caused by *endometriosis,* a problem with the inner lining of the uterus. In addition, if the innermost layers of the endometrium are not properly formed at each menstrual cycle, an embryo cannot implant, or the fetus detaches from the uterine wall. This problem causes many miscarriages.

What are the causes of infertility in men?

In the male reproductive system, problems with sperm formation and with the sperm themselves are the main causes of infertility. As shown in the graph at right, low sperm count is the most common cause of infertility in men. Male infertility can be caused by any number of structural and functional problems:

- **Low sperm count.** Too few sperm (less than 20 million per ejaculation) make it difficult for the necessary number of sperm to swim up the fallopian tubes and meet

---

BY THE NUMBERS

6.1 million
The number of men and women who are infertile in the United States.

1 in 6
The number of couples who are infertile in the United States.

850 feet
The length of the seminiferous tubules in human testes.

Several hundred million
Number of maturing sperm in the testes on one day.

Less than 20 million sperm per ejaculate
The sperm level that indicates infertility among men.

70%
The percentage of women using IVF age 30–35 who become pregnant.

62%
The percentage of women using IVF age 40–44 who become pregnant.

14%
The percentage of women using IVF over age 44 who became pregnant.

30%
The percentage of surrogates who become pregnant.

---

the egg. Low sperm count can be caused by exposure to chemicals or radiation in the environment, tight underwear, drugs (including marijuana), mumps and other diseases (such as diabetes), lead exposure, pesticide exposure, obesity, and alcohol consumption. In addition, injury to the testes and undescended testes can also be the cause of infertility.

- **Low sperm motility.** When the sperm move too slowly, they cannot get to the egg in time for fertilization. Sometimes sperm have malformed or extra tails or do not swim toward the egg in an organized way. The results are similar to those associated with low sperm numbers.

- **Impotence.** If the penis cannot get enough blood supply, it cannot become erect, resulting in **erectile dysfunction** (ED). The causes of impotence can be emotional, physical (hormonal), or drug related. Certain conditions such as high blood pressure and diabetes can also cause erectile problems.

- **No sperm formation.** The causes of **aspermia** can include surgery (vasectomy), injury, drugs, and birth defects such as undescended testes or being born with no testes.

PHOTO GALLERY
A vasectomy is one of the causes of infertility. It can be reversed.
© SPL/Photo Researchers, Inc.

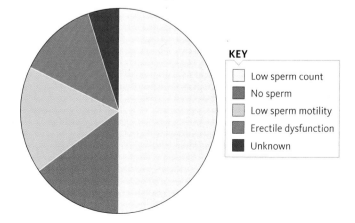

KEY
- Low sperm count
- No sperm
- Low sperm motility
- Erectile dysfunction
- Unknown

What other factors influence fertility in men and women?

Both personal habits and environmental factors can affect a couple's chances of having a child. For example, smoking lowers sperm count in men and increases the risk of miscarriage and low-birth-weight babies in women. Overall, for each menstrual cycle, smoking by men or women reduces the chances of conceiving by one-third. Women who are significantly overweight or underweight often have difficulty becoming pregnant.

Environmental factors, such as exposure to chemicals, pesticides, and radiation also affect fertility. For example, in one study, women living on farms who had mixed or applied pesticides in the two years before trying to conceive were 27 times more likely to be infertile than women who were not exposed to these chemicals. Other environmental agents such as lead, ethylene oxide (used to sterilize surgical instruments and medical supplies), and radiation have been shown to negatively affect fertility.

The relationship between age and infertility has taken on new importance as many people in the United States and other industrialized countries postpone marriage and/or pregnancy until educational or career goals are reached. Unfortunately, this delay means that the woman is older when she decides to have a child, and age is associated with decreased fertility. As we will discuss in Chapter 3, increased maternal age is also associated with an increased risk of having a child with a chromosomal abnormality. It is estimated that once a woman has celebrated her 42nd birthday, she has a less than 10% chance of having a baby with her own eggs.

During the "sexual revolution" of the 1960s and 1970s, there were higher incidences of infertility caused by untreated venereal diseases such as gonorrhea and chlamydia. This trend seems to have reversed since the appearance of AIDS in the 1980s; with it brought common use of barrier methods of contraception (condoms), which prevent HIV infection and most venereal diseases.

Another social factor, the increasing difficulty many couples encounter in finding a child to adopt (a result of improved birth control and the availability of legal abortion), has increased the demand for medical answers to infertility regardless of their complexity and high cost. We have also become more aware of our own genetics. Couples are more determined than ever to have their own genetic children.

PHOTO GALLERY
In artificial insemination, a doctor places sperm from a donor or from the woman's male partner into the woman's uterus.
© Mauro Fermariello/Photo Researchers, Inc.

2.4 Essentials

- **If there are problems with egg or sperm formation or with the uterus, infertility may result.**
- **Infertility is a serious problem in the United States.**
- **Either the male or the female of a couple can be infertile.**
- **Infertility in women increases with age.**
- **The most common form of female infertility is hormone imbalance.**

2.5 How Do Assisted Reproductive Technologies (ART) Help with Infertility Problems?

Because of the increase in infertility in the last 10 years, a medical specialty that treats infertile couples has been created. Many major medical centers offer fertility clinics, and there are independent programs in most cities.

This chapter will look closely at three of the most common methods used in assisted reproductive technologies (ART):

- donation of gametes
- *in vitro* fertilization (IVF)
- surrogacy

Most fertility specialists begin by analyzing the woman's and the man's hormone levels. If levels are low, then the physician may give the woman hormones (estrogen) to stimulate ovulation and/or give the man testosterone to stimulate sperm production.

If hormone treatments do not help with conception, more complicated procedures may be used. One of these is gamete donation. If a couple is not producing sperm, eggs, or both, they may look to a sperm and/or egg donor.

Sperm donation, one of the first methods of ART, has been widely used for almost 50 years. Sperm is donated by known or unknown donors and may be used immediately or frozen in liquid nitrogen for later use. Many sperm banks in the United States and all over the world offer sperm from qualified donors. Web sites offer couples, physicians, or women a chance to pick the traits of the donor they want and have the frozen sperm of their choice sent to them for artificial

PHOTO GALLERY
An IVF lab with incubators and lab equipment.
© AJP/Hopamerican/Photo Researchers, Inc.

insemination (■ Figure 2.1). The sperm is placed into the woman's uterus at ovulation, and if an egg is fertilized, a pregnancy may result. Sperm donation is used by those with low sperm count, low sperm mobility, or aspermia.

For men with low sperm counts, sperm from several ejaculations can be pooled and concentrated to increase the chances of *in vitro* fertilization (IVF). For men with blocked ducts, sperm can be retrieved from the epididymis or the testis using microsurgery.

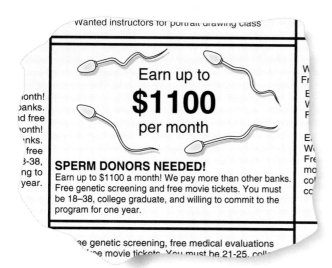

■ **Figure 2.1 Ad for Sperm Donors**
This ad on the Web site www.cryobank.com, offers special gifts for donors who successfully donate sperm.

■ **Figure 2.2 Ad for Egg Donors**
This ad ran in the Tufts University daily newspaper in November 1999. The price paid for the egg donors has increased tremendously since then.

A woman who is not producing eggs can opt to use an egg donor (■ Figure 2.2). Egg donation is more difficult because the procedure of egg retrieval is time-dependent and involves surgery. In addition, because of problems related to freezing eggs, they are generally used for IVF immediately after retrieval.

A donor who provides eggs to an infertile couple must undergo hormone shots to stimulate ovulation and a surgical procedure (**laparoscopy**) to remove the eggs. Using a microscope, a technician sorts the eggs to remove those that are too young or too old to be used in fertilization. Eggs that are not going to be fertilized immediately may be stored in liquid nitrogen for later use or for donation to another woman. Although freezing eggs has been somewhat successful, it is not performed regularly. However, it is an option in some cases. For example, women about to undergo chemotherapy who may want children in the future may freeze their own eggs for later use.

After retrieval, the eggs are used for *in vitro* fertilization (■ Figure 2.3), which can be done with donor eggs, donor sperm, or the sperm and egg of the couple themselves.

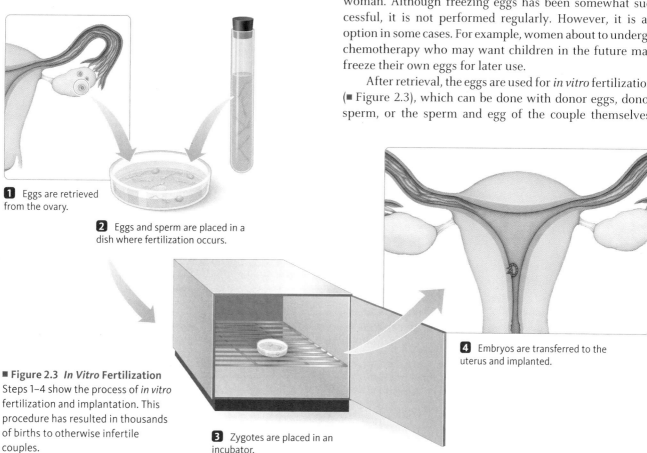

❶ Eggs are retrieved from the ovary.

❷ Eggs and sperm are placed in a dish where fertilization occurs.

❸ Zygotes are placed in an incubator.

❹ Embryos are transferred to the uterus and implanted.

■ **Figure 2.3 *In Vitro* Fertilization**
Steps 1–4 show the process of *in vitro* fertilization and implantation. This procedure has resulted in thousands of births to otherwise infertile couples.

Using sterile techniques, the egg and sperm are placed in a dish in the laboratory. Technicians watch the sperm swim to the eggs (usually more than one) and fertilize them. After fertilization, the dish containing the newly formed zygotes is placed in an incubator until they are ready to place in the woman's uterus. Often, up to three or four embryos are transferred to the uterus to increase the chances of implantation.

If a woman cannot carry a fetus to term because she has no uterus or her uterus does not function correctly, a surrogate mother can be used. Because of the barrier set up by the placenta and uterine wall, any woman can carry a genetically unrelated fetus without damage to mother or fetus.

There are two types of **surrogacy.** In egg donor surrogacy, the surrogate contributes an egg, which is fertilized by artificial insemination using the sperm of the infertile couple's male partner, and the surrogate carries the embryo to term. In gestational surrogacy, the surrogate carries the embryo to term but the egg and sperm are from two other people, who may be unrelated to the infertile couple.

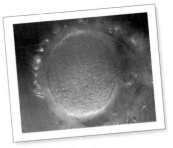

PHOTO GALLERY
In IVF, the sperm travel to the eggs in a dish, which reduces the distance the sperm have to travel.
© Lennart Nilsson

Often a relative or friend volunteers to carry the baby for an infertile couple and give the baby to them after birth. In other instances, the surrogate is a paid stranger (■ Figure 2.4). In such cases, a contract is signed (see Section 2.6) and the woman gives the baby to the couple after birth with any stipulations made clear in the contract.

As you can see in Table 2.3, a number of combinations of ART can be used. In one legal case, Buzzanca v. Buzzanca, a combination of techniques produced a girl with five parents. A contract was made among a sperm donor, an egg donor (both anonymous), and a surrogate (gestational) mother. The couple who contracted for the little girl, the egg donor, the sperm donor, and the surrogate mother are the five parents. This case went to court when the couple who contracted for the child decided to divorce.

2.5 Essentials

- *In vitro* fertilization (IVF) is a commonly used infertility treatment in which sperm and eggs are combined in a laboratory dish and embryos result.

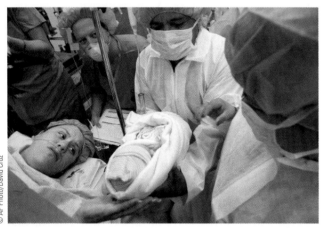

■ **Figure 2.4   A Surrogate Mother**
Teresa Anderson, a surrogate mother, is shown just after giving birth to one of the quintuplets she carried for Luisa and Enrique Gonzalez. The quints, who are all boys, were born on April 25, 2005. Their names are Enrique, Jorge, Gabriel, Victor, and Javier.

© AP Photo/David Cruz

CASE A QUESTIONS

Now that we know more about infertility and its treatments, let's address the problem that Brian and Laura are facing.

1. What tests should Dr. Franco perform on Brian and Laura?

2. List four things that Brian and Laura might do to solve their problem.

3. What do you think Brian and Laura should do? Why?

4. Give three reasons for using these new techniques.

5. Name three credible reasons to adopt.

6. Do you know anyone who has experienced infertility? Do you know more than one couple with infertility problems? If so, how many? See if they will tell you their story; if so, write a paragraph about it.

7. Give two reasons why ART has become so important in the last 10 years.

Table 2.3 New Ways to Make Babies

Artificial Insemination and Embryo Transfer	*In Vitro* Fertilization (IVF)
1. Father is infertile. Mother is inseminated by donor and carries child.	1. Mother is fertile but unable to conceive. Egg from mother and sperm from father are combined in laboratory. Embryo is placed in mother's uterus.
2. Mother is infertile but able to carry child. Donor egg is inseminated by father via IVF. Embryo is transferred and mother carries child. 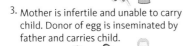	2. Mother is infertile but able to carry child. Egg from donor is combined with sperm from father and implanted in mother.
3. Mother is infertile and unable to carry child. Donor of egg is inseminated by father and carries child.	3. Father is infertile and mother is fertile but unable to conceive. Egg from mother is combined with sperm from donor.
4. Both parents are infertile, but mother is able to carry child. Donor of egg is inseminated by sperm donor. Then embryo is transferred and mother carries child.	4. Both parents are infertile, but mother is able to carry child. Egg and sperm from donors are combined in laboratory (also see number 4, column at left).
	5. Mother is infertile and unable to carry child. Egg of donor is combined with sperm from father. Embryo is transferred to donor (also see number 2, column at left).
	6. Both parents are fertile, but mother is unable to carry child. Egg from mother and sperm from father are combined. Embryo is transferred to donor.
	7. Father is infertile. Mother is fertile but unable to carry child. Egg from mother is combined with sperm from donor. Embryo is transferred to surrogate mother.

KEY

- Sperm from father
- Egg from mother
- Baby born of mother
- Sperm from donor
- Egg from donor
- Baby born of donor (Surrogate)

CASE B Single Mom Needs Dad

Jan Moppet wanted a child very badly but never found anyone she wanted to marry. So when she was about 35, her doctor told her that she had better think seriously about having a child before she got any older.

Dr. Ingram suggested that she have artificial insemination using sperm from a nearby sperm bank. In this procedure, donor sperm are placed inside the woman; in a large percentage of women, pregnancy results. This sounded just like what Jan wanted.

She went through the procedure later that year and gave birth to a baby girl she called Alex. Lately, though, Jan has been thinking what the future might bring.

1. Should Jan have chosen certain characteristics when picking her sperm donor?
2. Should Jan tell Alex about her father being a sperm donor?
   a. If yes, why and when?
   b. If no, why not?
3. When Alex reaches age 18, should she be allowed to find her father if she wants to do so?
4. You are Alex's attorney in her quest to find her father. Give five arguments you would make on her behalf.
5. If you were the sperm donor, list three reasons why you would not want Alex to find out your identity.

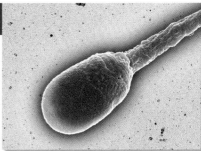

6. When Alex enters school, what might happen to her?
7. What is known about Alex's father's genetic makeup? Might she inherit a genetic disorder from him or carry a gene that she will pass on to her children?

2.6 What Are the Legal and Ethical Issues Associated with ART?

Many issues have come to light with the use of new ART procedures. A few of them will be discussed here.

During *in vitro* fertilization, a number of embryos can be produced. These embryos are placed in the woman's uterus or frozen for later use. Successful pregnancies have resulted from embryos frozen as long as 10 years. The big question is, what should be done with frozen embryos that no one wants?

PHOTO GALLERY
Sperm and embryos are stored at −180°C under liquid nitrogen in a cryogenic tank.
© Hank Morgan/Photo Researchers, Inc.

When several embryos are placed in the uterus at one time, the chance of a multiple birth increases. The number of multiple births has increased dramatically in the last 10 years as a result of ART.

In situations involving sperm and egg donors, questions often arise about the paternity or maternity rights of the donors; a number of legal cases have addressed this issue.

The following table lists some of the ethical and legal questions raised by these procedures.

Question	How are these questions decided?	Related case or legal issue
What should be done with leftover embryos?	This is usually decided by the fertility clinic and the parents of the embryos.	In the United Kingdom, a law demands that unclaimed embryos be destroyed after five years.
Should sperm and egg donors have parental rights to their children?	This is usually decided by courts if the donor wants to press the issue.	Individual cases have gone to court in the United States and the United Kingdom. For the most part, the parents who have raised the child retain the rights.
If a couple divorces, who should get custody of frozen embryos?	This is usually decided by courts if the couple wants to press the issue (see "Spotlight on Law: Davis v. Davis").	Most cases have used Davis v. Davis (see "Spotlight on Law") as a model and decided that no person can be forced to be a parent. Some have decided to destroy the embryos.
Should sperm and egg donors be paid for their contributions?	They are paid in the United States. Sperm donors can donate many times and are paid each time. Egg donors are given more compensation because of the complex process of egg retrieval.	No court case has developed from this issue, but each sperm bank and fertility clinic handles it differently.
Is a contract with a surrogate mother legal?	This legal question is dealt with on a state-by-state basis. One case in New Jersey, In re Baby M, decided the answer was no. New Jersey still does not accept these contracts, but other states do.	In a number of cases, surrogates have wanted to keep the babies they carried, but courts have upheld the contracts, especially in gestational surrogacy.
Should surrogates be compensated?	If the amount of money is indicated in the contract, this has been held to be legal.	Some feel that paying surrogates is the same as baby selling. Recently, however, the number of women being gestational surrogates has increased.

Davis v. Davis (842 S.W. 2d 588
(Tennessee Ct. of Appeals 1993))

During their marriage, Junior and Mary Sue Davis had gone to an infertility clinic to have *in vitro* fertilization. Mary Sue's eggs and Junior's sperm were fertilized in the laboratory, producing nine embryos. Two of these were implanted in Mary Sue's uterus, and seven were frozen. The two implanted embryos did not result in pregnancy. The Davises were planning to come back and implant additional embryos, but before that could happen, they developed marital difficulties and decided to divorce. Junior and Mary Sue asked the divorce court (they lived in Tennessee) to make a judgment no other court had previously decided: Each was asking for the embryos as part of the divorce settlement.

Junior wanted to destroy the embryos; Mary Sue wanted to implant the embryos and have children. In defense of her right to proceed with the implantation, she argued that the embryos were examples of potential human life, not typical property. But if they were to be regarded as property, she should have a say in their disposition under Tennessee divorce law. Mary Sue also entered a counterclaim that Junior be ordered to pay child support if she bore a child.

Junior's attorney argued that he should not be forced to be a parent; that this was his right under the Constitution; and that an embryo is not a person and therefore cannot be considered as a child.

The court needed to decide whether the embryos were property, children, or neither. If they were considered property, they would be split, as in most divorce law. If they were considered children, custody would have to be awarded.

The trial court awarded custody of the embryos to Mary Sue, calling them "children *in vitro*," and directed that she be allowed to implant them. Junior appealed the ruling to the appellate court of Tennessee.

RESULTS:

Trial Court:

The decision to award "custody" stated that Mary Sue should be "permitted the opportunity to bring these children to term through implantation." The court also decided that embryos were "instances of human life" and, therefore, had a right to life. The court used terminology that referred to the frozen embryos as "pre-embryos."

Appellate Court:

The state appellate court rejected the "right to life" rationale. Its decision was based on the fact that the trial court's decision was inconsistent with Tennessee law allowing abortion. It also stated that Junior had a right *not* to be a father and awarded control of the embryos to Junior and Mary Sue jointly. The court held that the Davises shared an interest in the frozen embryos.

Supreme Court of Tennessee:

The State Supreme Court concluded that frozen embryos are neither "persons" nor "property" but "occupy a position of special respect because of their potential for human life." It felt that the main issue is the individual's constitutional right to privacy. If a person has a right to procreate, then he or she also has a right *not* to procreate. If Mary Sue were allowed to implant the embryos, Junior would have no control over his parental status. Being an unwilling parent would place financial, emotional, and legal burdens on Junior. Therefore, the court awarded the embryos to Junior.

QUESTIONS:

1. In your opinion, to whom should the court have awarded the embryos? Why?
2. If neither spouse had been awarded the embryos, what should have happened to them? Argue both sides.
3. Which attorney's argument do you agree with? Give three reasons.
4. In Australia, two frozen embryos were left when a couple died in a plane crash. The couple had millions of dollars, and many women wanted to be implanted with these embryos because they thought the resulting children would inherit the money. The Australian court decided that no one would get the embryos. Do you agree with this decision? Why or why not?
5. Is there a duty to protect human embryos from harm? If so, who should protect them? If not, why not?
6. What should be done with unclaimed frozen embryos?
7. What should be included in a contract for a couple going through IVF? List some possible choices.
8. After many tries to get pregnant, what could the Davises have tried besides IVF? Research possible options.

✪ THE ESSENTIAL TEN

1. Sperm are produced by meiosis in the testes of the male. [Section 2.1]
   ▮ For every sperm-forming cell, four sperm are formed.

2. Eggs are produced by meiosis in the ovaries of the female. [Section 2.1]
   ▮ For every egg-forming cell, one egg is formed.

3. The major parts of the male reproductive system are the testes, urethra, and penis. [Section 2.2]
   ▮ The testes and penis are external, while the urethra is internal.

4. The major parts of the female reproductive system are the ovaries, fallopian tubes, and uterus. [Section 2.3]
   ▮ All major parts of the female reproductive system are internal.

5. If there are problems with egg or sperm formation or with the uterus, infertility may result. [Section 2.4]
   ▮ Eggs and sperm are needed to form a fetus; if either are missing or incomplete, the man or woman cannot reproduce without medical help.

6. Infertility is a serious problem in the United States. [Section 2.4]
   ▮ There are many causes of infertility, both environmental and genetic.

7. Either the male or the female of a couple can be infertile. [Section 2.4]
   ▮ Many statistics exist about the number of infertile couples. They show that the causes can be determined by medical tests.

8. Infertility in women increases with age. [Section 2.4]
   ▮ As a woman ages, her ovaries begin to shut down and she has trouble producing viable eggs.

9. The most common form of female infertility is hormone imbalance. [Section 2.4]
   ▮ Estrogen in females and testosterone in males can cause problems in egg and sperm production.

10. *In vitro* fertilization (IVF) is a commonly used infertility treatment in which sperm and eggs are combined in a laboratory dish and embryos result. [Section 2.5]
    ▮ *In vitro* fertilization was first done in the United Kingdom and has been practiced for almost 30 years.

KEY TERMS

androgens (p. 27)
aspermia (p. 31)
bulbourethral glands (p. 28)
cervix (p. 29)
egg (p. 27)
endometrium (p. 29)
epididymis (p. 28)
erectile dysfunction (p. 31)
estrogens (p. 27)
fallopian tubes (p. 28)
follicles (p. 29)
fructose (p. 28)
in vitro fertilization (IVF) (p. 32)
laparoscopy (p. 33)
luteinizing hormone (p. 30)
meiosis (p. 27)
myometrium (p. 29)
oocytes (p. 27)

oogenesis (p. 28)
ovaries (p. 27)
prostaglandins (p. 28)
prostate gland (p. 28)
secondary amenorrhea (p. 31)
seminal vesicles (p. 28)
seminiferous tubules (p. 28)
sperm (p. 27)
spermatids (p. 28)
spermatocytes (p. 28)
spermatogenesis (p. 28)
surrogacy (p. 34)
testes (p. 27)
urethra (p. 28)
uterus (p. 29)
vagina (p. 29)
vas deferens (p. 28)
vasectomy (p. 28)

ONLINE RESOURCES

Preparing for an exam? Log on at academic.cengage.com/login for a pre-test, a personalized learning plan, and a post-test in CengageNOW's Study Tools to help you assess your understanding.

If assigned by your instructor, the Case A and Spotlight on Law activities for this chapter, "Fertility Treatment" and "Custody of IVF Embryos," will be available in CengageNOW's Assignments.

What Would You Do If . . . ?

1. You needed to pick someone to be your surrogate?

2. A friend asked you to donate sperm for her baby?

3. A friend asked you to donate an egg so that she and her husband could have a baby?

4. You were a legislator asked to vote on a law to destroy all abandoned frozen embryos?

5. You were a fertility specialist and were counseling a couple carrying quadruplets?

REVIEW QUESTIONS

1. Fill in the following table by listing all of the infertility problems in women and the causes of each.

Infertility Problem	Cause 1	Cause 2

2. Fill in the following table by listing all of the infertility problems in men and the causes of each.

Infertility Problem	Cause 1	Cause 2

3. Look back at the preceding tables and discuss one problem from each, describing which assisted reproduction technique has been successfully used in treating it.
4. Of the two types of surrogacy, which might have more physical problems?
5. After a vasectomy, a man may decide that he wants to have children. What can be done? Research this topic.
6. What is it called when a man has few or no sperm?
7. Why is treating infertility with hormones the first choice of doctors?
8. What are the three main things needed to have a baby?
9. List three additional factors that can cause infertility.
10. Which causes of infertility are environmental?

APPLICATION QUESTIONS

1. Do you think that more doctors in the future will specialize in assisted reproduction? Why or why not?
2. If assisted reproduction increases in the future, what effect do you think this will have on adoptions? On abortions?
3. Of the two types of surrogacy, which might have more legal problems?
4. Research some other methods of assisted reproduction. Write them up and discuss how often they are used.
5. Expand the "By the Numbers" feature in this chapter. Research how successful the following types of ART are: IVF, surrogacy, sperm donor use, egg donor use. *Successful* means a baby has been born.
6. Research has shown that the age of the egg is more important than the age of the female, and that older women can have children by implanting donated eggs fertilized by IVF, or by freezing their own eggs for later use. This allows women to extend their child-bearing years, even after menopause. What problems, if any, do you foresee if older women (over 55 years of age) have children? List them.
7. Do some research about men who have fathered children after age 60. Why do you think it is more common for men to have children at an older age than women at the same age? Why does our society accept this?
8. Ovarian transplants are one way to treat female infertility. As a judge, how would you handle a case in which a woman receives an ovary transplant, has a child, and is faced with a claim by the donor that she is the genetic mother of the child and wants custody or visitation rights?

In The Media

Jodie Foster's Baby

Steve Sailer

Hollywood actor Jodie Foster searched for the perfect sperm donor in 1998, and proudly announced that after a long hunt, she had been impregnated with the gametes of a tall, dark, handsome scientist with an IQ of 160. Delighted with the result, she went back to the same clinic for her second child in 2002.

To access this article online, go to academic.cengage.com/biology/yashon/hgs1.

QUESTIONS:

1. How do you think being a celebrity affects a question such as sperm donation?
2. Research and find some other celebrities who have used sperm or egg donation.

New York Times, **November 20, 2005**

Hello, I'm Your Sister. Our Father Is Donor 150.

Amy Harmon

The Donor Sibling Registry is a Web site recently set up by a woman who wanted her children who had resulted from sperm donation to know their biological father. When she could not find his identity, an idea hit her: she would find the other children fathered by this donor. This Web site has located four siblings for her children and has helped others find their siblings.

To access this article online, go to academic.cengage.com/biology/yashon/hgs1.

QUESTIONS:

1. One boy used the Web site to locate his father. Do you think this is right? Why or why not?
2. What problems might result from these children finding each other?

The Sunday Times, **February 18, 2007**

Hello, I'm Donor 150. You're My Daughter.

Tony Allen-Mills

Jeffrey Harrison, or California Cryobank's Donor 150, had found the Donor Sibling Registry and contacted three of his genetic offspring nearly 20 years after describing himself glowingly as a writer with a philosophy degree. Ryann McQuilton, one of his "children," waited nervously to meet him, not knowing that he was now living out of a van with his dogs.

To access this article online, go to academic.cengage.com/biology/yashon/hgs1.

QUESTIONS:

1. How do you feel about this break in confidentiality?
2. If you were Ryann, would you have gone to meet the sperm donor? Why or why not?

Pink or Blue: Is It a Boy or Girl?

Acu-Gen Biolab launches *Baby Gender Mentor* marketing and information site

Acu-Gen, based in Boston, has developed a home test for the sex of a fetus using the mother's blood. A lancet and filter paper are included in the kit; the woman pricks her skin and the blood is sucked into the filter paper, which is sent back to the company for analysis. Acu-Gen claims a 99.9% accuracy rate. Parents can pay a small fee to find out the sex of their child without waiting for the ultrasound.

To learn more about Acu-Gen's home test product online, go to academic.cengage.com/biology/yashon/hgs1.

QUESTIONS:

1. Do you think many people will use this as a form of sex selection? Why or why not?
2. This is probably the first of many home tests for women who want to find out information about their fetuses. What other things might an expectant mother want to test for? What might be the result of this on a large scale?

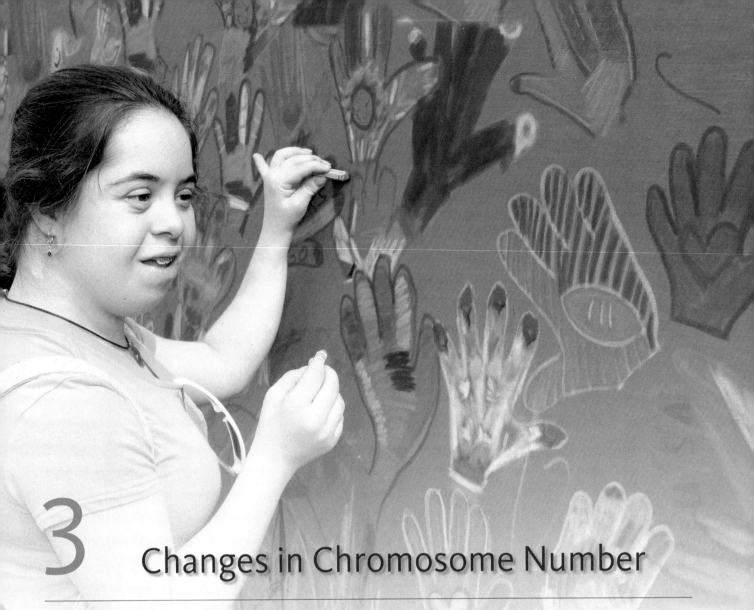

3 Changes in Chromosome Number

⊛ CENTRAL POINTS

▌ Chromosomes are made up of DNA combined with proteins.

▌ Chromosome numbers can vary among individuals. Most humans have 46 chromosomes.

▌ Several different tests can be used to determine the chromosome number in a fetus.

▌ Extra chromosomes can affect a fetus.

▌ Lawsuits can result from problems with genetic testing.

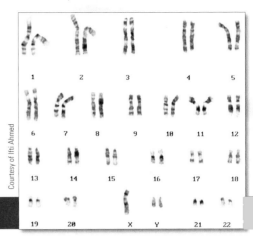

Courtesy of Ifti Ahmed

Results Worry Pregnant Woman

Martha Lawrence was nervous about going to the human genetics unit at the hospital. Her doctor had referred her because she was unexpectedly pregnant at age 41, and at that age she was at risk for having a child with a chromosomal abnormality, especially Down syndrome. She was 18 weeks pregnant, which would not leave much time if she wanted to have an abortion, considering it takes 7–14 days to get the results of an amniocentesis procedure. Women over the age of 35 have a much higher risk of having a child with a chromosomal abnormality than younger women.

Martha previously had two normal pregnancies; her children are now 13 and 17. The genetic counselor, Dr. Gould, suggested that Martha have amniocentesis (removal of fluid and cells from around the fetus) to examine the fetus's chromosomes.

The test showed that the fetus did not have an extra copy of chromosome 21 and therefore did not have Down syndrome. But the analysis did show the presence of an extra Y chromosome (XYY instead of XY), a condition called Jacobs syndrome. This condition is fairly common; each day in the United States, 5–10 boys are born with an XYY set of chromosomes.

Some questions come to mind when reading this case. Before we can address those questions, let's look at how mistakes in cell division can lead to chromosomal abnormalities.

3.1 What Is a Chromosome?

As discussed in "Biology Basics: Cells and Cell Structure," chromosomes are thread-like structures in the nucleus of a cell that carry genetic information (■ Figure 3.1).

Humans have 46 chromosomes, two of which are called the *sex chromosomes.* Remember from Chapter 1 that these are designated X and Y. As shown in Figure 3.1, a chromosome has three important parts: (1) the centromere, (2) the short arm (p arm), and (3) the long arm (q arm). At their ends, chromosomes have specialized regions called **telomeres.** Telomeres keep chromosomes from sticking to each other and may play a role in aging.

3.1 Essentials

- **The normal chromosome number in humans is 46.**

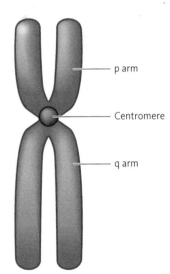

■ **Figure 3.1 Anatomy of a Chromosome**
The parts of a chromosome include the centromere and two arms (the short arm is the p arm, the long arm is the q arm).

3.2 What Can Cause Changes in the Number of Chromosomes?

Eggs and sperm (gametes) are produced in the ovaries and testes by a form of cell division called **meiosis.** Cells that begin meiosis contain two copies of each chromosome (one from each parent), for a total of 46 chromosomes. During meiosis, members of all chromosome pairs separate from each other to produce *haploid* cells that each contain 23 chromosomes. Meiosis accomplishes this reduction in chromosome number by two rounds of cell division called meiosis I and meiosis II, which are shown below.

MEIOSIS I

Before cells begin meiosis, the chromosomes duplicate. As meiosis begins, chromosomes coil and shorten, and become visible in the microscope. Each chromosome has a matching partner and the two chromosomes may exchange parts (cross over) during this stage, called **prophase I.**

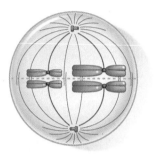

The chromosome pairs line up along the middle of the cell, and spindle fibers attach to the *centromere* of each pair. This stage is called **metaphase I.**

Members of each homologous pair separate and move toward opposite sides of the cell. This stage is called **anaphase I.**

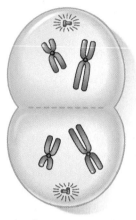

The chromosomes reach opposite poles of the cell, and the nuclei begin to re-form. This stage is called **telophase I.** The cytoplasm divides, and two cells are formed. These cells have half the number of chromosomes of the original cells and are called *haploid* cells.

MEIOSIS II

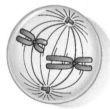

Two cells formed during meiosis I. In **prophase II**, the chromosomes of these cells become coiled, and move toward the center of the cell.

The 23 chromosomes in each cell attach to spindle fibers at their centromeres. This stage is called **metaphase II.**

Each centromere divides, and the newly formed chromosomes (also called sister chromatids) move to opposite ends of the cell. This stage is called **anaphase II.**

Finally, the chromosomes uncoil and the nuclear membrane re-forms. This stage is called **telophase II.** After the cytoplasm divides, the result is four cells, each with the haploid number of chromosomes. Meiosis is now completed.

The cell division in meiosis I separates members of a **homologous chromosome pair.** In meiosis II, the centromeres split, and the sister chromatids become chromosomes that are distributed to daughter cells.

Table 3.1 below summarizes the events of meiosis.

Table 3.1	Meiosis
Stage	**Characteristics**
Interphase I	Chromosome replication.
Prophase I	Chromosomes become visible, homologous chromosomes pair, chromatids undergo crossing over (recombination).
Metaphase I	Paired chromosomes align at equator of cell.
Anaphase I	Members of each chromosome pair move to opposite poles of the cell.
Telophase I	Chromosomes uncoil, nuclei form.
Cytokinesis	Cytoplasm divides, forming two cells.
Prophase II	Chromosomes re-coil.
Metaphase II	Unpaired chromosomes align at the equator of the cell.
Anaphase II	Centromeres split, daughter chromosomes pull apart.
Telophase II	Chromosomes uncoil, nuclear membrane forms, meiosis ends.
Cytokinesis	Cytoplasm divides, forming four cells.

Normally, meiosis results in gametes that contain 23 chromosomes. However, in a small percentage of cases, chromosomes fail to separate properly during one of these two divisions (■ Figure 3.2). This event, called **nondisjunction,** results in some gametes with two copies of a given chromosome, and some gametes with no copies of that same chromosome. If gametes with an abnormal number of chromosomes are involved in fertilization, the resulting zygote will have an abnormal number of chromosomes. Variations in chromosome number that involve one or a small number of chromosomes are called **aneuploidy.** Depending on which chromosomes are involved, the effects of aneuploidy can range from non-life-threatening physical symptoms to devastating and lethal effects. An error during meiosis that resulted in a sperm carrying two copies of a Y chromosome is probably what happened to Martha Lawrence's fetus. Aneuploidy is a major cause of early miscarriages; about 5% of all pregnancies are aneuploid. In addition, aneuploidy is the leading cause of mental retardation and developmental disabilities and places a significant social, emotional, and financial burden on millions of families.

3.2 Essentials

- Meiosis is involved in formation of the sperm and egg.
- In meiosis, members of a chromosome pair separate from each other, forming gametes that contain the haploid chromosome number.
- Most aneuploidy is the result of mistakes in meiosis.

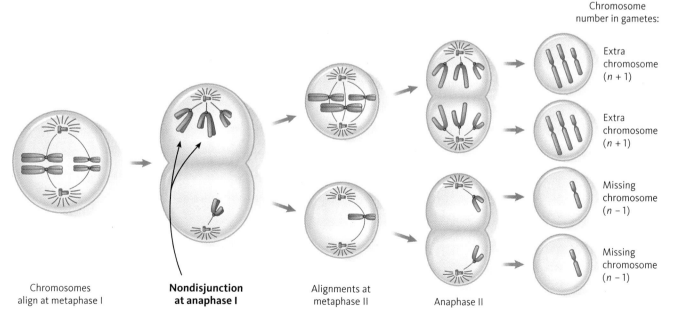

Chromosome number in gametes:

Extra chromosome ($n + 1$)

Extra chromosome ($n + 1$)

Missing chromosome ($n - 1$)

Missing chromosome ($n - 1$)

Chromosomes align at metaphase I

Nondisjunction at anaphase I

Alignments at metaphase II

Anaphase II

■ **Figure 3.2 Nondisjunction**
Nondisjunction can cause abnormal chromosome numbers in gametes (eggs or sperm).

3.3 Can We Identify Chromosomal Abnormalities in a Fetus?

Because of Martha Lawrence's age, her genetic counselor suggested that she have a procedure called **amniocentesis** to analyze the fetus's chromosomes and detect any abnormalities. The drawing below shows a developing fetus in the uterus and the use of amniocentesis to obtain a sample for analysis.

Removal of about 20 ml of amniotic fluid containing suspended cells that were sloughed off from the fetus

Biochemical analysis of the amniotic fluid after the fetal cells are separated out

Centrifugation

Analysis of fetal cells to determine sex

Fetal cells are removed from the solution

Cells are grown in an incubator

Karyotype analysis

How does amniocentesis work?

Amniocentesis is routinely used to collect fetal cells and amniotic fluid for analysis. Aneuploidy, other chromosomal abnormalities, and biochemical disorders can be detected by amniocentesis, and the sex of the fetus can also be determined.

During amniocentesis, the fetus and placenta are first located by ultrasound. Ultrasound machines use sound waves to create a picture of the fetus so the doctors can locate it in the uterus (shown above). A needle is inserted through the abdominal and uterine walls (avoiding the placenta and fetus) and into the sac surrounding the fetus. Some of the fluid surrounding the fetus, called **amniotic fluid,** is collected. This fluid consists of fetal urine, water, and cells shed from the fetus. The fluid and the cells it contains therefore reflect the genetic makeup of the fetus.

After the cells are removed from the fluid, they are grown in the laboratory for several days, and then cells are broken open and the chromosomes spread onto a microscope slide and stained with one or more dyes. A camera attached to a microscope transmits the chromosome images

to a computer, where they are digitized and processed to make a **karyotype.** In a karyotype, as shown below, chromosomes 1–22 (the autosomes) are arranged in pairs. The sex chromosomes are usually placed on the karyotype separately from the autosomes.

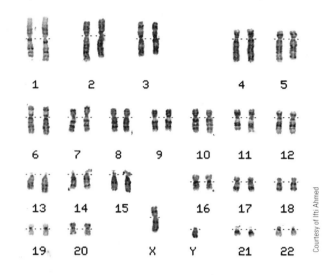

Courtesy of Ifti Ahmed

Amniocentesis is usually performed at or after the 16th week of pregnancy. Processing and analyzing the results usually takes about 7–14 days. As with all medical procedures, some risks are associated with amniocentesis. Because there is a 0.06% (1 in 1,600) risk of miscarriage associated with amniocentesis, the procedure is normally only used under certain conditions:

- When the mother is over age 35. The risk for having children with chromosomal abnormalities increases dramatically for women in this age group. At age 41, Martha Lawrence is a candidate for this test.
- When the mother has already had a child with a chromosomal aberration. The recurrence risk in such cases is 1–2%.
- When either parent has one or more structurally abnormal chromosomes. This situation may cause an abnormal number of chromosomes in the fetus.
- When the mother is a carrier of a genetic disorder caused by a gene on the X chromosome.
- When the parents have experienced unexplained infertility or a number of previous miscarriages.

Are there other tests for aneuploidy?

Another method of prenatal chromosome analysis, called **chorionic villus sampling (CVS),** can be performed earlier in the pregnancy (10 to 12 weeks, compared with 16 weeks for amniocentesis). Because the cells removed by CVS are rapidly dividing, test results are available within a few hours or a few days. With amniocentesis, fetal cells must be grown in the laboratory for several days before test results can be read. A drawing of the CVS procedure is shown on the next page.

In CVS, a flexible catheter is inserted through the vagina or abdomen into the uterus, guided by ultrasound images.

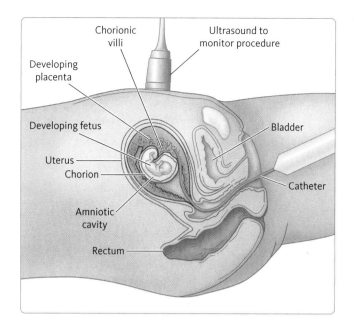

Cells of the chorionic villi (a fetal tissue that forms part of the placenta) are removed by suction and analyzed immediately. This gives CVS a distinct benefit over amniocentesis because the results can be determined in the first trimester of pregnancy. Both CVS and amniocentesis can be coupled with recombinant DNA technology for prenatal diagnosis. The use of recombinant DNA techniques for prenatal diagnosis of genetic disorders is discussed in Chapter 6.

As with amniocentesis, some risk is associated with CVS. Women who have had this procedure may experience cramps, infection, and miscarriage. Studies indicate that the risk for CVS-related miscarriage is 1–2% (1 in 100 to 1 in 50).

Are there other chromosomal variations?

In addition to the XYY condition present in Martha Lawrence's fetus, other birth defects are caused by changes in chromosome number or structure. These changes may involve additional copies of all the chromosomes in a cell (a cell with 69 or 92 chromosomes instead of 46), a condition called **polyploidy;** the gain or loss of individual chromosomes (aneuploidy); or structural changes within individual chromosomes.

The figure in the next column shows some of the structural changes found in chromosomes.

The diploid number (2*n* or 46) of chromosomes in body cells and the haploid number (*n* or 23) in gametes are called the normal, or **euploid,** condition. The related term *aneuploidy* means without the normal condition (*a*- means "without"). The simplest form of aneuploidy involves the gain or loss of a single chromosome. The gain of one chromosome is known as **trisomy** (it is written as 2*n* + 1, or 47 chromosomes); the loss of a single chromosome is known as **monosomy** (written as 2*n* − 1, or 45 chromosomes).

The most common cause of both trisomy and monosomy in humans is *nondisjunction,* the failure of chromosomes to separate properly during meiosis. There are two cell divisions in meiosis, and nondisjunction can occur in either the first or second division.

STRUCTURAL CHANGES IN CHROMOSOMES

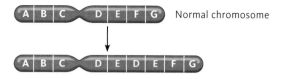

Normal chromosome

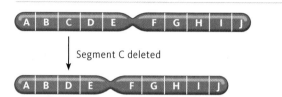

One segment repeated three times

Duplication of parts of a chromosome, in which sections of the chromosome are repeated.

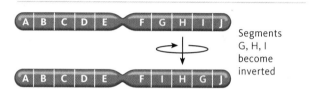

Segment C deleted

Deletion of parts of a chromosome, in which sections of the chromosome are missing.

Segments G, H, I become inverted

Inversion of parts of a chromosome, in which sections of the chromosome are in reverse order.

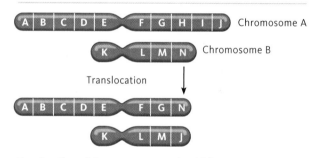

Chromosome A

Chromosome B

Translocation

Translocation of chromosome parts, in which parts of a chromosome are exchanged.

3.3 Essentials

- Two tests that can detect aneuploidy during pregnancy are amniocentesis and chorionic villus sampling (CVS).

- An extra chromosome, a condition called trisomy, can be fatal to the fetus.

- A missing chromosome, a condition called monosomy, may be fatal to the fetus.

3.4 What Are the Effects of Monosomy and Trisomy?

In the following sections, we will examine some of the symptoms of monosomy and trisomy involving chromosomes 1–22, the autosomes. Then we will consider monosomy and trisomy involving the sex chromosomes (X and Y).

Monosomy involving any autosome is fatal and results in miscarriage early in pregnancy. This happens because when a chromosome is missing, *all* the genes on that chromosome are also missing. Trisomies involving the majority of autosomes also cause the death of the fetus early in development. Only a few autosomal trisomies result in live births (trisomy 8, 13, 18, and 21). *Trisomy 21* (Down syndrome) is the only autosomal trisomy in which survival into adulthood is possible.

The following chart shows the most commonly formed trisomies and their karyotypes:

AUTOSOMAL TRISOMIES

Karyotype

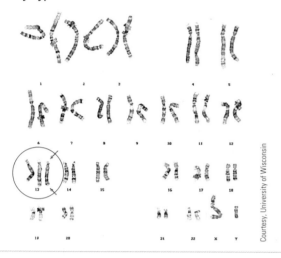

Courtesy, University of Wisconsin

Name
Trisomy 13: Patau syndrome (47,+13)

Number of Chromosomes	**How Common?**	**Survival**
47, extra #13	1 in 15,000 live births	1–2 months

Symptoms
Trisomy 13 involves facial malformations, eye defects, extra fingers or toes, and feet with large protruding heels. Internally, there are usually severe malformations of the brain and nervous system, as well as congenital heart defects.

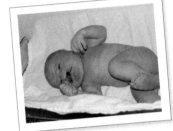

PHOTO GALLERY
Trisomy 13
Courtesy, Mike and Judie Grabski

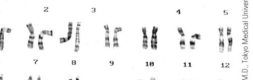

Courtesy of Ifti Ahmed

Name
Trisomy 18: Edwards syndrome (47,+18)

Number of Chromosomes	**How Common?**	**Survival**
47, extra #18	1 in 11,000 live births; 80% are females	2–4 months

Symptoms
Infants are small at birth, grow very slowly, and have mental disabilities. Clenched fists, with the second and fifth fingers overlapping the third and fourth fingers, and malformed feet are also characteristic. Heart malformations are almost always present, and heart failure or pneumonia usually causes death.

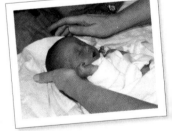

PHOTO GALLERY
Trisomy 18
Courtesy, Mike & Judy Grabski

© 1997 Hironao Numabe, M.D., Tokyo Medical University

Name
Trisomy 21: Down syndrome (47,+21)

Number of Chromosomes	**How Common?**	**Survival**
47, extra #21	1 in 800 live births; (changes with age of mother)	Up to age 50

Symptoms
Down syndrome is a leading cause of childhood mental retardation and heart defects in the United States. Affected individuals usually have a wide, flat skull; folds in the corner of the eyelids; and spots on the iris. They may have furrowed, large tongues that cause the mouth to remain partially open. Physical growth, behavior, and mental development are retarded, and approximately 40% of all children with Down syndrome have congenital heart defects. In spite of these handicaps, many individuals with Down syndrome lead rich, productive lives.

PHOTO GALLERY
A child with Down syndrome
© Lauren Shear/Photo Researchers, Inc.

What is the leading risk factor for trisomy?

The only known risk factor for Down syndrome and other forms of autosomal aneuploidy is maternal age (■ Figure 3.3). Young mothers have a lower probability of having a child with Down syndrome than older mothers. However, the risk increases rapidly after age 35, which is why the genetic counselor recommended amniocentesis for Martha Lawrence. At age 20, a mother's risk of having a child with Down syndrome is 1 in 2000 (0.05%); by age 35, the risk has climbed to 1 in 111 (0.9%); and at age 45, 1 in 33 (3%) of all newborns have trisomy 21. Paternal age has also been proposed as a factor in autosomal trisomy, but no clear link has been found.

Why is maternal age a risk factor?

No one knows for certain why the age of the mother increases the risk for aneuploidy. One idea is that older eggs are more prone to nondisjunction. The eggs of a female fetus begin meiosis well before the child is born. These eggs enter meiosis I and then stop. Under hormonal stimulation that begins at puberty, one egg per month completes meiosis I and enters meiosis II. The final steps in meiosis are not completed until the egg is fertilized. Younger women have a low risk of having an aneuploid child, and their eggs have been in meiosis I for 15–25 years. Eggs ovulated at age 40 have been in meiosis I for more than 40 years. During this time, events in the cell or environmental agents might damage the egg, making nondisjunction more likely.

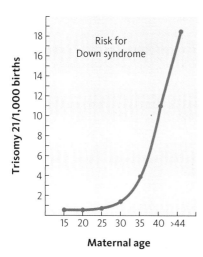

■ **Figure 3.3 Maternal Age and Down Syndrome** The percentage of births in which the child has Down syndrome increases with the age of the mother.

Another explanation focuses on a process called *maternal selection.* According to this idea, interactions between the uterus and an implanting embryo normally lead to spontaneous abortion of embryos with an abnormal number of chromosomes. As women age, maternal selection may become less effective, allowing more chromosomally abnormal embryos to implant and develop. More research is needed to clarify the underlying mechanisms.

PHOTO GALLERY
An adult with Down syndrome
© Lauren Shear/Photo Researchers, Inc.

Does aneuploidy affect the sex chromosomes?

Aneuploidy involving the X and Y chromosomes is more common than autosomal aneuploidy. Monosomy involving autosomes is always fatal, but this is not necessarily true for monosomy of the sex chromosomes. Monosomy involving the X chromosome, a condition called Turner syndrome (45,X), occurs in about 1 in 10,000 female births. Although 1 in 10,000 female births are 45,X, this represents only a small fraction of the 45,X embryos produced; most of these miscarry. On the other hand, monosomy for the Y chromosome (45,Y) always results in miscarriage; no live births for this condition have been reported.

Other aneuploid conditions that involve the sex chromosomes include Klinefelter syndrome (47,XXY), and Jacobs syndrome (47,XYY), the condition present in Martha Lawrence's fetus. These conditions are summarized in the chart on the next page.

3.4 Essentials

- Individuals with Down syndrome (trisomy 21) can survive to adulthood.

Sex Chromosome Trisomies

Karyotype

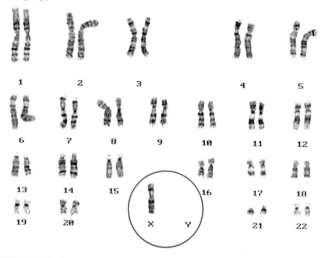

1	2	3	4	5
6	7	8	9	10	11	12
13	14	15		16	17	18
19	20	X	Y	21	22

Name
Turner syndrome (45,X)

Number of Chromosomes
45, missing one X

Survival
To adulthood

Symptoms
Babies born with Turner syndrome are always female and live to adulthood. They are short and wide-chested with undeveloped ovaries. At birth, puffiness of the hands and feet is prominent, but this disappears in infancy. Many also have narrowing of the aorta. They have normal intelligence.

How Common?
It is estimated that 1% of all conceptions are 45,X and that 95–99% of all 45,X embryos die before birth. Turner syndrome occurs with a frequency of 1 in 10,000 female births.

PHOTO GALLERY
A child with Turner syndrome
UNC Medical Illustration and Photography

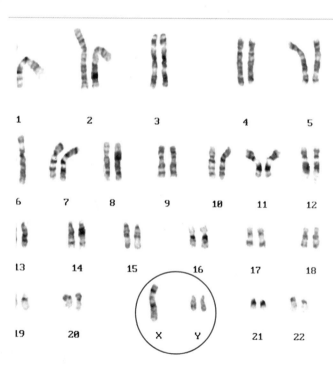

1	2	3	4	5
6	7	8	9	10	11	12
13	14	15		16	17	18
19	20	X	Y	21	22

Name
Klinefelter syndrome (47,XXY)

Number of Chromosomes
47, an extra X

Survival
To adulthood

Symptoms
The features of this syndrome do not develop until puberty. Affected individuals are male but are usually sterile. Some men with Klinefelter syndrome have learning disabilities or mild retardation. Many men with Klinefelter are *mosaics*. This means some cells have the XY chromosome combination and others have XXY. In these men, the symptoms are less severe.

How Common?
1 in 1,000 males

PHOTO GALLERY
A child with Klinefelter syndrome (XXY)
Stefan Schwarz.

1	2	3	4	5
6	7	8	9	10	11	12
13	14	15		16	17	18
19	20	X	Y	21	22

Name
XYY syndrome (47,XYY)

Number of Chromosomes
47, with an extra Y

Survival
To adulthood

Symptoms
These individuals are above average in height and thin, have personality disorders, and may have some form of retardation. Some have learning disabilities; some have very mild symptoms. They often experience severe acne during adolescence and may have some behavioral problems.

How Common?
1 in 1,000 male births
(about 0.1% of the males
in the general population)

3.5 Is There a Way to Evaluate Genetic Risks?

Understanding the results of a genetic test and what it means for a family and its present and future children can be an overwhelming responsibility. Fortunately, a health professional called a **genetic counselor** can help individuals or families to understand what the results of a genetic test mean. Some of the information provided by a genetic counselor may include the following:

- medical facts, including the diagnosis, progression, management, and any available treatment for a genetic disorder
- how heredity contributes to the disorder and the risk of having children with this disorder
- alternatives for dealing with the risk of recurrence
- how to adjust to the disorder in an affected family member or manage the risk of recurrence

Genetic counselors achieve these goals in a nondirective way. They provide all the information available to individuals or family members, so the person or family can make the decisions best suited to them based on their own cultural, religious, and moral beliefs.

Who are genetic counselors?

Genetic counselors are health care professionals with specialized training in medical genetics, psychology, and counseling. They usually work as part of a health care team, providing information and support to people and families who have genetic disorders or may be at risk for an inherited disorder. Genetic counselors identify families at risk, investigate the condition in the family, interpret information about the disorder, analyze inheritance patterns and risks of recurrence, and review available options with the family.

Why do people go to genetic counselors?

People ask to see a genetic counselor if they have a family history of a genetic disorder, cancer, a birth defect, or a developmental disability. Women over age 35 and members of specific ethnic groups in which particular genetic conditions occur more frequently are often seen by genetic counselors who teach them about their increased risk for genetic or chromosomal disorders and the genetic tests available to them.

Counseling is especially recommended for several groups of people:

- women who are pregnant, or are planning to become pregnant, after age 35
- couples who already have a child with mental retardation, a genetic disorder, or a birth defect
- couples who would like testing or information about genetic disorders that occur more frequently in their ethnic group
- couples who are first cousins or other close blood relatives
- individuals who are concerned that their job, lifestyle, or medical history may pose a risk to a pregnancy, including exposure to radiation, medications, chemicals, infection, or drugs
- women who have had two or more miscarriages or babies who died in infancy
- couples whose infant has a genetic disease diagnosed by routine newborn screening; and
- people who have, or are concerned that they might have, an inherited disorder or birth defect.

How does genetic counseling work?

Most people see a genetic counselor after a prenatal test or after the birth of a child with a genetic disorder. Individuals may also seek genetic counseling to determine their risk of being affected by a condition they may already have, or their risk of being a carrier. The counselor usually begins by constructing a detailed family and medical history and

CASE A QUESTIONS

Now that we understand some basic information about chromosomal abnormalities and how they arise, let's look at some of the issues raised in Martha Lawrence's situation. The uncertainty surrounding the impact of an XYY child illustrates Dr. Gould's dilemma about Martha's case.

1. Based on the karyotype that Dr. Gould obtained by amniocentesis, list five options that Martha has.

2. Which do you think is the best option in this situation?

3. Give three reasons why someone might disagree with your choice.

4. Knowing what you do about the different aneuploid conditions, if the diagnosis were Down syndrome, would your decision change? Why or why not?

While analyzing the results of Martha's amniocentesis, Dr. Gould had some questions herself:

5. Keeping in mind that Martha was primarily concerned about her age and the chances of having a child with trisomy 21, should Dr. Gould tell her about the abnormal XYY karyotype? Why or why not?

6. Dr. Gould considered telling Martha's husband about the test results before she told Martha. Should she?

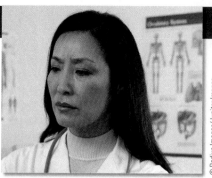

Martha Lawrence's case really had Dr. Gould thinking. She remembered a patient back in 1996 who taught her quite a lot about the legal aspects of her practice. This patient was named Donna Slotin. She gave birth to a son named A.J. At the time, Donna was 31 and was not considered a high-risk pregnancy; however, A.J. was born with a genetic condition that causes serious mental retardation and other physical disabilities. After the birth and diagnosis, an attorney called Dr. Gould and said that the Slotins were going to sue her because Donna was not given an amniocentesis.

1. What is the first thing Dr. Gould should do?
2. What would the Slotins' attorney argue?
3. What would Dr. Gould's attorney argue?
4. What do you think could be done to avoid a lawsuit?

a pedigree. A pedigree shows the family's genetic history as a diagram using symbols to represent family members (pedigrees will be discussed in Chapter 4). Then he or she explains the basic concepts of biology and how the condition is inherited.

When couples seek information before they have a child, blood tests of the couple and their families can be used along with pedigree analysis to help determine what, if any, risks are present. If a trait is found to be genetically determined, the counselor constructs a risk assessment profile that explains the risk of having a child who has the condition.

To help both couples and individuals understand how genes and gene products are related to a genetic disorder, genetic counselors use a number of teaching techniques. They share information that allows an individual or a couple to make informed decisions.

What is the future of genetic counseling?

As information from the Human Genome Project (HGP) changes medical care, the number of genetic disorders that can be diagnosed by genetic testing is increasing dramatically, and the role of the genetic counselor will become more important. Results from the HGP are already changing the focus of genetic counseling from reproductive risks to adult-onset conditions such as cancer, diabetes, and cardiovascular diseases. The information provided by the counselor allows at-risk individuals to adopt lifestyles that may reduce the impact of the disorder, to make decisions about whether to have children, and to plan for medical care that they or their affected children may require later in life.

3.5 Essentials

- As a woman ages, the chances that she will have a child with aneuploidy increases and she can get help from genetic counselors.

3.6 What Are the Legal and Ethical Issues Associated with Chromosomal Abnormalities?

In the U.S. legal system, Martha Lawrence would have an option to sue Dr. Gould if she were not offered amniocentesis and later gave birth to a child with a genetic disorder. There are two possible legal scenarios: a **wrongful-birth suit** or a **wrongful-life suit.** Both types of lawsuits are based on the idea that the birth or the life of the child is wrong because of either the action or the inaction of the physician or genetic counselor.

The chart on page 53 explains the differences between the two legal actions.

The outcomes in these types of cases are based on two questions:

1. Could a diagnosis of this condition have been made in time to have an abortion?
2. Was the condition serious enough that a reasonable person would have had an abortion?

Most courts of law allow wrongful-birth suits for two reasons: Roe v. Wade gave a woman an alternative to birth; and doctors have extensive medical malpractice insurance, so financial compensation is often not a problem. However, only five states allow wrongful-life suits because courts are uncomfortable declaring that someone should never have been born. An example of a wrongful-life suit is examined in "Spotlight on Law: Becker v. Schwartz."

Wrongful-life and wrongful-birth suits are different from malpractice suits because in these cases the doctor has done nothing to *cause* the injury (the birth defect). In malpractice suits, the doctor is accused of *mal*practice ("bad" practice) because he or she caused the damage to the patient.

In a Wrongful-Birth Case	In a Wrongful-Life Case
The parents sue the doctor.	The child sues the doctor.
The parents argue that because the doctor's inaction did not make it possible for them to take a prenatal test, they could not make an informed decision as to whether they should have the baby.	The child argues that because of inadequate advice by the doctor, he or she was born into a life of pain and suffering.
This type of case relies on Roe v. Wade, the case that legally allowed abortion in the United States, because the mother *must* say that if she had known that the child had this condition, she would have aborted it.	This type of case also relies on Roe v. Wade for similar reasons.
Reasons for such suits include the following: The doctor neglected to tell the parents of the test or the risk of having such a child; the doctor or lab mixed up the results, did the test incorrectly, or did the wrong test; or the doctor deliberately chose not to tell the parents.	Reasons for such suits are the same as for wrongful-birth suits.

Some other types of cases have evolved from these types of lawsuits:

- Wrongful conception: when birth control methods (surgical or chemical) don't work.
- Wrongful pregnancy: when an abortion isn't completed and the mother has medical problems.

Martha Lawrence may want to read about some of the interesting legal aspects of the XYY karyotype. Early investigations found that people with the XYY karyotype have a tendency toward aggressive behavior associated with the presence of an extra Y chromosome. In effect, this may mean that some forms of violent behavior are genetically determined. In fact, the XYY karyotype has been used on several occasions as a legal defense (unsuccessfully, so far) in criminal trials (see, for example, People v. Tanner [1970], 13 Cal.App. 3d).

The question asked in these trials and the subsequent studies of children with the XYY karyotype is: Can we find a direct link between the XYY condition and criminal behavior? Using today's science, there doesn't seem to be strong evidence to support such a link. In fact, most males with the XYY karyotype lead socially normal lives.

In the United States, long-term studies of the relationship between antisocial behavior and the XYY karyotype were discontinued. Researchers feared that identifying children with potential behavioral problems might lead parents to treat them differently and result in behavioral problems as a self-fulfilling prophecy. The uncertainty surrounding the XYY condition illustrates Dr. Gould's dilemma about Martha Lawrence's case, and may make Martha's decision all the more difficult.

3.6 Essentials

- **Wrongful-life and wrongful-birth lawsuits can result from incorrect prenatal diagnosis.**

Becker v. Schwartz
(386 N.E.2d 807 (NY 1978))

The following case is an example of the type of lawsuit Ms. Lawrence's child might file if Dr. Gould fails to tell Ms. Lawrence about the XYY condition: a wrongful-life case.

Delores Becker got pregnant at age 37. Her physician, who treated her for her entire pregnancy, did not tell her about amniocentesis and her increased risk of having a child with Down syndrome. Her son was born with Down syndrome. Even though it was 1978, amniocentesis was being used.

The Beckers filed a wrongful-life suit on behalf of their son, and the resulting case, Becker v. Schwartz, was eventually heard by the New York Court of Appeals—the highest judicial body in New York. Delores Becker testified that if she had been informed of her son's condition, she would have had an abortion. As a result of her doctor's failures, she did not learn of her son's mental disability until after he was born.

RESULTS:
The court dismissed the complaint, holding that no one has the right to be born free of disease. A $2,500 settlement was reached, and the baby was given up for adoption. Some attorneys analyzing the case later thought that the Beckers could have received more money if they had gone to an even higher court.

QUESTIONS:
1. What was the basis for the suit?
2. Do you think that the court found correctly?
3. How much money do you think the jury might have given for the care and "loss" of their son?
4. Knowing what you do about Down syndrome, do you think the parents should have raised the child and not sued? Why or why not?

✪ THE ESSENTIAL TEN

1. The normal chromosome number in humans is 46. [Section 3.1]
   ▍ There are 23 pairs of chromosomes, including the sex chromosomes (XX or XY).

2. Meiosis is involved in formation of the sperm and egg. [Section 3.1]
   ▍ The egg and sperm have the haploid number of chromosomes.

3. In meiosis, members of a chromosome pair separate from each other, forming gametes that contain the haploid chromosome number. [Section 3.2]
   ▍ If the chromosomes don't separate correctly, aneuploidy can occur.

4. Most aneuploidy is the result of mistakes in meiosis. [Section 3.2]
   ▍ Aneuploidy can result in monosomy or trisomy.

5. Two tests that can detect aneuploidy during pregnancy are amniocentesis and chorionic villus sampling (CVS). [Section 3.3]
   ▍ Each one of these tests has associated risks and benefits.

6. An extra chromosome, a condition called trisomy, can be fatal to the fetus. [Section 3.3]
   ▍ Risk to the fetus depends on which chromosome is involved.

7. A missing chromosome, a condition called monosomy, may be fatal to the fetus. [Section 3.3]
   ▍ When a chromosome is missing, all genes on that chromosome are also missing.

8. Individuals with Down syndrome (trisomy 21) can survive to adulthood. [Section 3.4]
   ▍ People with Down syndrome have many serious symptoms.

9. As a woman ages, the chances that she will have a child with aneuploidy increases and she can get help from genetic counselors. [Section 3.5]
   ▍ Women over 35 years of age are recommended to have amniocentesis when pregnant.

10. Wrongful-life and wrongful-birth lawsuits can result from incorrect prenatal testing. [Section 3.6]
    ▍ Testing problems can be due to problems in laboratories.

KEY TERMS

amniocentesis (p. 46)
amniotic fluid (p. 46)
anaphase I (p. 44)
anaphase II (p. 44)
aneuploidy (p. 45)
chorionic villus sampling (CVS) (p. 46)
euploid (p. 47)
genetic counselor (p. 51)
homologous chromosome pair (p. 45)
karyotype (p. 46)
meiosis (p. 44)
metaphase I (p. 44)
metaphase II (p. 44)
monosomy (p. 47)
nondisjunction (p. 45)
polyploidy (p. 47)
prophase I (p. 44)
prophase II (p. 44)
telomeres (p. 43)
telophase I (p. 44)
telophase II (p. 44)
trisomy (p. 47)
wrongful-birth suit (p. 52)
wrongful-life suit (p. 52)

ONLINE RESOURCES

Preparing for an exam? Log on at academic.cengage.com/login for a pre-test, a personalized learning plan, and a post-test in CengageNOW's Study Tools to help you assess your understanding.

If assigned by your instructor, the Case A and Spotlight on Law activities for this chapter, "Results Worry Pregnant Woman" and "Becker v. Schwartz," will be available in CengageNOW's Assignments.

REVIEW QUESTIONS

1. At which stages of meiosis is a chromosomal aberration most likely to occur?
2. Why is meiosis so important?
3. Karyotypes on newborns are not done very often. Why do you think that is the case?
4. Would trisomy involving a large chromosome necessarily be more serious than one involving a small chromosome? Why or why not?
5. Do some research on how often trisomies in other chromosomes occur and report your findings.
6. What are the stages of meiosis?
7. Why does nondisjunction occur?
8. What are the risk factors in amniocentesis?
9. List three types of trisomy.
10. Who might need a genetic counselor?

What Would You Do If . . . ?

1. You were Martha Lawrence? (See Case A.)

2. Your daughter, who has Down syndrome, wanted to have a child?

3. You were a lawyer who had to defend a doctor accused of wrongful birth? What arguments would you use?

4. You were Dr. Gould and thought that Martha Lawrence was mentally unstable? (See Case A.)

5. You were a member of a legislature that was voting on a bill that required every pregnant woman to be tested by amniocentesis?

6. You were a doctor whose patient refused amniocentesis even though she was 42 years old?

7. You were a pregnant woman considering amniocentesis, and you knew that the chance of miscarriage with an amniocentesis is 0.01%?

8. You were a doctor considering whether to tell your patients that the chance of miscarriage with an amniocentesis is 0.01%?

APPLICATION QUESTIONS

1. If a woman is younger than age 35, physicians don't usually recommend prenatal testing, because the maternal risk factor for genetic disorders is low. Do some research on this topic and discuss why this practice may change in the future.

2. During amniocentesis, a very long needle is used to extract the amniotic fluid. Find a photo of an amniocentesis needle and discuss how the test can be discussed with a woman without frightening her.

3. Occasionally a baby is harmed during prenatal diagnosis. Does a doctor need to discuss this possibility with the patient? Why or why not?

4. Obviously, some chromosomal aberrations are more serious than others. What are the differences between monosomy of the X chromosome and monosomy for any of the other chromosomes?

5. It seems pretty clear that a doctor who diagnoses a chromosomal aberration did not cause the condition. Yet some are sued by the parents of these children. Is this ethical?

6. There is a chromosomal abnormality called triploidy that occurs when the fertilized egg contains three copies of each chromosome (69 chromosomes). Where do you think the extra set of chromosomes comes from? *Hint:* Many triploid embryos have an XYY sex chromosome set. Do you think these embryos live? Why or why not?

7. Amniocentesis is not done until about the 16th week of pregnancy. One reason physicians wait is to make sure that the cells recovered are from the fetus and not from the mother. Clearly, if a karyotype of the recovered cells shows a Y chromosome, the cells are from the fetus. Why?

8. A pregnant mother comes to you and insists on having the procedure at week 12. The result shows cells that have two X chromosomes. How could you test to find out if the recovered cells belong to the mother or the fetus?

9. Why would someone want an amniocentesis as early as 12 weeks?

10. Expand on the "By The Numbers" chart. Fill in the chart below by doing research.

How many amniocenteses are done every year?	
How many Down syndrome babies were born in one recent year?	
How many Down syndrome babies were born 10 years ago?	

KETV.com, May 2007

Study: Down Syndrome Testing Good for Moms of All Ages

Two specialized tests, a nuchal translucency ultrasound* and a test on blood taken from the mother, can test for Down syndrome. There is a movement to test all pregnant women regardless of age by using these tests. The nuchal translucency screening test uses ultrasound to measure the clear space in the tissue at the back of a developing baby's neck and the blood tests shows the levels of two proteins.

The two tests are used together in a computer program to calculate whether an amniocentesis should be done to determine if a fetus has Down syndrome.

To access this article online, go to academic.cengage.com/biology/yashon/hgs1.

QUESTIONS:
1. If more women are tested in this way, what problems do you foresee?
2. Ultrasound tests are already done routinely in prenatal checkups, so would physicians have to be specially trained to do these new tests?

Science, **January 12, 1973**

Behavioral Implications of the XYY Genotype

One genetic study was carried out by Harvard child psychiatrist Stanley Walzer and Harvard Medical School geneticist Park Gerald. By 1968, they were screening all newborn males at Boston Hospital for Women and following up by studying the development of those with abnormal karyotypes such as XYY or XXY. The research was funded by a grant from the Centers for Studies of Crime and Delinquency of the National Institute for Mental Health.

Source: Ernest B. Cook, "Behavioral Implications of the XYY Genotype," *Science*, vol. 179, pp. 139–150, January 12, 1973.

QUESTIONS:
1. What problems do you see with this study?
2. Do some research on this study and find out why it was stopped.

Law & Order: SVU (USA Network)
"Competence," May 10, 2002
"Clock," September 26, 2006

Law and Order Raises Questions

In the first of these episodes, a woman with Down syndrome who was raped wants to keep her baby, but her mother objects. In the second, a woman with Turner syndrome who looks young for her age fakes a kidnapping to run away with her boyfriend.

To access episode summaries and downloads online, go to academic.cengage.com/biology/yashon/hgs1.

QUESTIONS:
1. Should these topics be covered in television shows?
2. Both of these episodes are about competency of individuals with genetic conditions. Who should determine if a person is competent?

4 How Genes Are Transmitted from Generation to Generation

⊛ CENTRAL POINTS

▪ Genes are transmitted from generation to generation.

▪ Traits are inherited according to predictable rules.

▪ Dominant, recessive, X-linked traits, and diseases follow these rules.

© catchlightisual Service./Alamy

CASE A A Family's Dilemma

Alan Franklin, 17, has only vague memories of his mother; she died before he was 5 years old. But his older brother, John, told him about her. She had been sick for a very long time; she stumbled around and walked with jerky movements, and finally was in bed all the time. John said she died of **Huntington disease (HD).**

Huntington disease is caused by a mutant gene. A person who carries just one copy of this mutant gene will eventually have the disease. Most people with HD begin to show symptoms between ages 30 and 50. Muscle control and mental capacity slowly disintegrate until the person is bedridden and dependent on others for care. The disease is fatal, but its progression can take as long as 10–20 years.

Alan is a good student and wants to be a pilot in the Air Force; he wonders if he is at risk for developing HD. Recently in his biology class he learned that HD is inherited

and that a test is available for people at risk for this disorder.

There are three children in the Franklin family. Mary Franklin, Alan's older sister, is engaged to be married to Bob Wilson next June. John's wife, Emily (Alan's sister-in-law), is pregnant, and they already have a 2-year-old son, Matthew.

Alan asks his biology teacher for more information about the test and she gives him the name of Marta Wright, a genetic counselor at a nearby medical center. She suggests that he ask his father's permission to visit the counselor.

Alan's father encourages him to see the genetic counselor. During their first appointment, Ms. Wright diagrams Alan's personal family history:

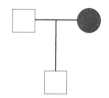

She explains that the top of this diagram contains a square, symbolizing his father, connected by a line to a circle symbolizing his mother. Her symbol is filled, because she had Huntington disease. Below these symbols is a square representing Alan. The counselor tells Alan that because his mother had HD, there is a 50% chance that he also carries the mutant gene. If he does, he will begin to show symptoms in middle age.

Ms. Wright tells Alan that if he takes the test for HD and the results show that he carries the mutant gene, he probably will not get into the Air Force, and his dreams of being a pilot will not come true.

Some questions come to mind when reading about Alan's case. Before we address those questions, let's look at how genetic traits, including Huntington disease, are passed from generation to generation.

4.1 How Are Genes Transmitted from Parents to Offspring?

Thanks to the work of Gregor Mendel, a European monk in the mid-1800s, we understand how genes are transmitted from parents to offspring in all plants and animals, including humans. In his experiments with pea plants, Mendel chose plants that each had a different, distinguishing characteristic, called a *trait*. These traits included plant height and flower color.

In one experiment, Mendel crossed tall pea plants and short pea plants to see how height was passed from parents to offspring. Geneticists use letters as symbols to identify each trait: in this case, *T* for tall, and *t* for short. Mendel discovered specific patterns in the way traits are passed from parent to offspring. Some traits, such as shortness, disappeared in the first generation of offspring, but reappeared in the next generation in about ¼ of the offspring (a 3:1 ratio of tall to short). He called the trait that was present in the first generation offspring a **dominant** trait, and the trait that was absent but reappeared in the next generation a **recessive** trait.

We now know that Mendel's conclusions about how traits are inherited are correct and apply to humans as well as pea plants. He predicted that traits are passed from generation to generation by "factors" (we call them *genes*) transmitted from parent to offspring. Each parent carries a pair of genes for a given trait, but contributes only one of these genes to its offspring. This separation of members of a gene pair occurs during meiosis when members of a chromosome pair separate from each other.

Members of a gene pair can differ. In our example, each parent carries a pair of genes specifying plant height, but this gene has two versions; one specifies tall plants, the other short plants (■ Figure 4.1). We call these variations of a gene **alleles.** Organisms that carry two identical alleles of a gene (*TT* or *tt*) are **homozygous;** those that carry nonidentical alleles (*Tt*) are **heterozygous.**

So, using letters to symbolize alleles in Mendel's experiment, the first cross looked like this:

$$TT \ \times \ tt \text{ (both parents are homozygous)}$$

$$Tt \text{ (all the offspring plants are heterozygous)}$$

In addition, Mendel concluded that the appearance of an organism (tall or short) can be different from its genetic makeup. What an organism looks like, or what we can observe about it, is called its **phenotype,** and the genetic makeup of an organism is called its **genotype.** Therefore, *TT, Tt,* and *tt* are genotypes, and *tall* and *short* are phenotypes. Organisms can have identical phenotypes but different genotypes (■ Figure 4.2). For example, because *tall* is a dominant trait, tall pea plants can be homozygous (*TT*) or heterozygous (*Tt*).

■ **Figure 4.2 Sorting of Alleles**
When Mendel crossed two first generation tall pea plants, three tall plants resulted for every short one. Where did the short plants come from?

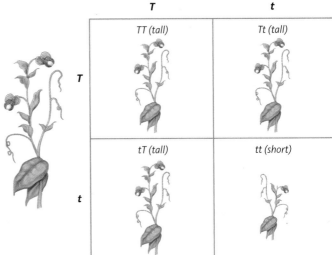

	T	**t**
T	*TT (tall)*	*Tt (tall)*
t	*tT (tall)*	*tt (short)*

a. **b.**

■ **Figure 4.1 Different Plant Heights**
These two pea plants are similar to what Mendel used in his experiments. One plant (a) is tall, while the other (b) is short.

Mendel is still remembered because two of his conclusions became the foundation of genetics:

1. Two copies of each gene separate from each other during the formation of egg and sperm (which takes place during meiosis). As a result, only one copy of each gene is present in the sperm or egg and is contributed to the offspring. This idea is called the **law of segregation.**
2. Members of a gene pair segregate into gametes independently of other gene pairs, so that gametes can have different combinations of parental genes. This idea is called the **law of independent assortment.**

Mendel's methods are still used to study the genetics of experimental organisms such as fruit flies, rats, and plants. These same laws are applied every day by genetic counselors such as Ms. Wright to help their clients understand their family's genetics.

How do Mendel's findings apply to human traits?

To illustrate that Mendel's two laws apply to human traits, let's follow the inheritance of a trait called *albinism*. Below is a photo of a person with albinism.

We can use the letters *A* and *a* to represent the alleles in this case: *A* is the dominant allele for **pigment** formation (color), and *a* is the recessive allele for lack of pigment formation. People with albinism carry two copies of the recessive allele (*aa*) and cannot make pigment; they have pale white skin, white hair, and colorless eyes (see the photo at left).

Anyone carrying at least one dominant allele (*Aa* or *AA*) can make enough pigment to have colored skin, hair, and eyes.

The discussion of how Mendel's laws apply to humans will start with parents who are heterozygotes (*Aa*) with normal pigmentation (■ Figure 4.3).

Aa × *Aa*

During meiosis, the dominant (*A*) and recessive (*a*) alleles carried by each parent separate from each other and end up in different gametes. Because each parent can produce two different types of gametes (one with *A* and another with *a*), the random union of gametes at fertilization will produce four possible combinations of alleles in the offspring:

AA, Aa, aA, aa

Mendel experimented with plants and could do many different crosses over a short period of time. Obviously, we cannot do this with humans. However, if the parents in this family have enough children (say 20 or 30), we will see something close to the predicted ratio of 3 pigmented to 1 albino that Mendel's laws predict.

Important: This means that if both parents are heterozygotes (*Aa*), each child has a 75% chance of having normal pigmentation and a 25% chance of being albino (a 3:1 ratio).

How is inheritance of traits studied in human genetics?

To study the inheritance of human traits, genetic counselors take a detailed family history. This history is used to construct a diagram called a **pedigree,** which shows all family

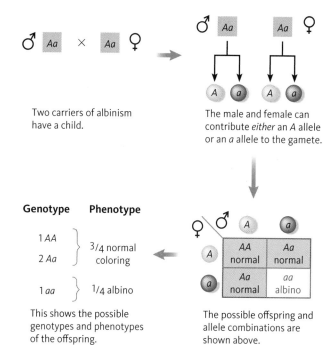

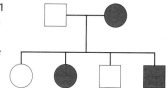

■ **Figure 4.3 Segregation of the Albino Allele in Humans**

members and identifies those affected with the genetic disorder in question as well as unaffected family members (■ Figure 4.4). In addition to interviews, counselors can also use letters, diaries, photographs, family records, and medical records to assemble the family history and construct the pedigree.

■ **Figure 4.4 Sample Pedigree 1**
This sample pedigree shows two generations. Those colored red have the genetic disorder. Those blank are unaffected.

A numbering system (■ Figure 4.5) is used to indicate generations:

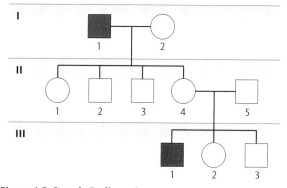

■ **Figure 4.5 Sample Pedigree 2**
This sample pedigree shows three generations (numbered I, II, and III).

The following symbols are used in pedigrees:

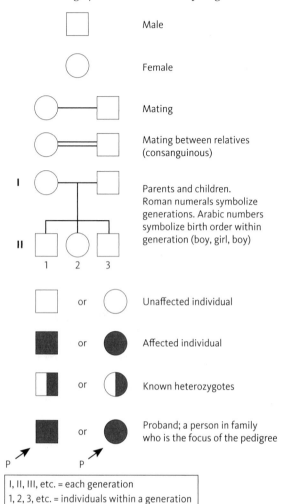

	Male
	Female
	Mating
	Mating between relatives (consanguinous)
	Parents and children. Roman numerals symbolize generations. Arabic numbers symbolize birth order within generation (boy, girl, boy)
or	Unaffected individual
or	Affected individual
or	Known heterozygotes
or	Proband; a person in family who is the focus of the pedigree

I, II, III, etc. = each generation
1, 2, 3, etc. = individuals within a generation

Pedigrees are often constructed after a family member who has a genetic disorder has been identified, or when a family member requests counseling (■ Figure 4.6). In Alan's case, he requested that Ms. Wright construct the pedigree. When the family pedigree is constructed, Alan will be identified as the **proband,** the person who is the focus of the pedigree. He will be indicated on the pedigree by an arrow and the letter *P*:

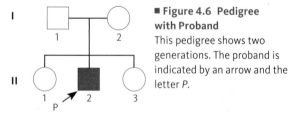

■ Figure 4.6 Pedigree with Proband
This pedigree shows two generations. The proband is indicated by an arrow and the letter *P*.

Other symbols used in pedigree construction are shown below:

| | Offspring of unknown sex |
| | Questionable whether individual has trait |

Collection of pedigree information is not always easy because knowledge about distant relatives is often incomplete, recollections about medical conditions can be forgotten over time, and family members are sometimes reluctant to discuss relatives who had abnormalities.

Look at the two pedigrees below. One is Alan's (made by Ms. Wright, the genetic counselor); the other is a family with albinism. What differences do you see?

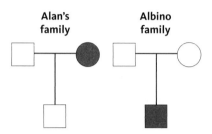

4.1 Essentials

- Members of a gene pair segregate into gametes independently of other gene pairs, so that all gametes have different combinations of genes.
- Gregor Mendel, a monk who lived in the 1800s, was the father of genetics.
- Mendel did his experiments with plants, but his important laws of genetics are still used today to study human genetics.

4.2 What Can We Learn from Examining Human Pedigrees?

Analysis of a pedigree can determine whether a trait present in a family has a dominant or recessive inheritance pattern. Pedigrees can also be used to predict genetic risk in several situations, including the following:

- the risk in a fetus
- the risk of having an adult-onset disorder
- risks in future offspring

Based on Mendel's conclusions, we know that traits in humans can be inherited in several ways, each with a distinctive pattern of inheritance. To appreciate the differences in inheritance and the resulting differences in genetic risk, we will discuss three possible patterns of inheritance: **autosomal recessive, autosomal dominant,** and **X-linked recessive.** Autosomal traits are carried on chromosomes 1–22, and X-linked traits are carried on the X chromosome.

How are autosomal recessive traits inherited?

A pedigree showing affected and unaffected members over several generations can determine whether a trait has a recessive pattern of inheritance. The pedigrees of recessive traits, such as albinism, have several distinguishing characteristics:

- Unaffected parents can have affected children:

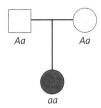

- All children of affected parents are affected.

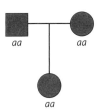

- For two heterozygotes (carriers), the risk of having an unaffected child is 75% and the risk of having an affected child is 25%.
- Male and female children are affected in roughly equal numbers. In addition, both the male and female parent must transmit the gene for a child to be affected.
- Unaffected (heterozygous) parents of an affected (homozygous) individual may be related to each other as shown in the pedigree below. This is called **consanguinity** and is indicated by a double line between the parents.

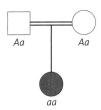

A pedigree illustrating the inheritance of an autosomal recessive trait is shown in ■ Figure 4.7. Table 4.1 lists several autosomal recessive genetic disorders.

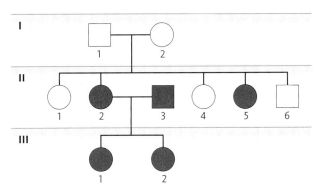

■ **Figure 4.7 Autosomal Recessive Pedigree**
This pedigree shows three generations of a family with an autosomal recessive trait.

Table 4.1 Examples of Autosomal Recessive Genetic Disorders

Trait	Phenotype
Cystic fibrosis (CF)	Mucus production that blocks ducts of certain glands and lung passages
Phenylketonuria (PKU)	Excess accumulation of phenylalanine in blood, mental retardation
Sickle cell anemia (SCA)	Abnormal hemoglobin, blood vessel blockage
Xeroderma pigmentosum (XP)	Lack of DNA repair enzymes, sensitivity to light, skin cancer
Tay-Sachs disease (TSD)	Improper metabolism in nerve cells, death in childhood

Some recessive traits have relatively moderate effects and show up in traits such as hair color and eye color. Others can be life-threatening or even fatal disorders, including cystic fibrosis and sickle cell anemia (discussed shortly). Genetic conditions such as the ones listed below are caused when an important protein in the body is produced incorrectly or is missing altogether. In albinism, for example, the affected protein is involved in making melanin or pigment.

1. **Albinism** (A = normal coloring; a = albinism): Albinism is a group of genetic conditions associated with a lack of pigmentation (melanin) in the skin, hair, and/or eyes. In normal individuals, melanin is found in pigment granules inside cells (called melanocytes) present in the skin, hair, and eyes. In people with albinism, melanocytes are present but contain no pigment because the individual cannot make melanin.

   One common type of albinism is called *oculocutaneous albinism type I* (OCA1). Affected individuals have no pigment in their hair, skin, and eyes, and also have reduced eyesight, sensitivity to bright lights, and involuntary eye movements.

2. **Cystic fibrosis** (C = normal condition; c = cystic fibrosis): Cystic fibrosis (CF) is another recessive trait. CF affects the glands that produce mucus and digestive enzymes; it has far-reaching effects because these glands perform a number of vital functions. CF causes the production of thick mucus in the lungs that blocks airways; most people with cystic fibrosis develop obstructive lung diseases and infections that lead to premature death.

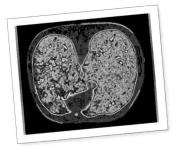

PHOTO GALLERY
The lungs of an individual with CF; they show a definite clogging of passageways
© James Cavallini/Photo Researchers, Inc.

The CF gene and its protein product (**CFTR**) have been identified and studied in detail and used to develop new methods of treatment, including gene therapy, a procedure that will be discussed in Chapter 9.

3. **Sickle cell anemia** (S = normal red blood cells; s = sickle cell): People with ancestors from parts of West Africa, the lowlands around the Mediterranean Sea, or some regions of the Indian subcontinent have a high frequency of a genetic disorder called sickle cell anemia (SCA) (■ Figure 4.8). SCA is inherited as a recessive disorder that causes production of abnormal hemoglobin, a protein found in red blood cells. Hemoglobin transports oxygen from the lungs to the tissues of the body. In SCA, abnormal hemoglobin molecules

PHOTO GALLERY
The red blood cells of a person with sickle cell anemia; notice the varying shapes of the cells
© Eye of Science/Photo Researchers, Inc.

group together to form rods that cause red blood cells to become crescent- or sickle-shaped (■ Figure 4.9).

These deformed cells are fragile and break open as they circulate through the body. New red blood cells are not produced fast enough to replace those that are lost, causing anemia. This also reduces the oxygen-carrying capacity of the blood.

People with sickle cell anemia tire easily and can develop heart failure caused by an increased load on the circulatory system. The deformed blood cells clog small blood vessels and capillaries, further reducing oxygen transport. As oxygen levels fall in the body, more red blood cells become deformed, causing intense pain as

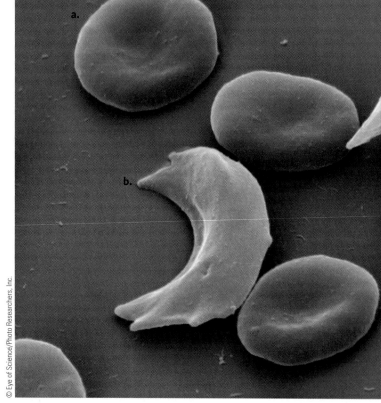

© Eye of Science/Photo Researchers, Inc.

■ **Figure 4.9 Normal and Sickled Cells**
A normal red blood cell (a) and a sickled red blood cell (b).

blood vessels are blocked. In some affected areas, ulcers and sores appear on the skin. Blockage of blood vessels in the brain can also cause strokes and paralysis.

How are autosomal dominant traits inherited?

In dominantly inherited disorders, anyone who carries one copy of a mutant allele is a heterozygote (Aa) and, therefore, has the condition. Only in rare cases are dominant genetic disorders present in a homozygous condition (AA). Because it results in a double dose of the incorrect protein, being homozygous is fatal early in life. Unaffected individuals, on the other hand, carry two recessive alleles (aa).

Pedigrees of dominantly inherited traits have several distinctive characteristics:

• Every affected individual should have at least one affected parent:

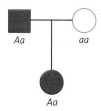

• Because most affected individuals are heterozygotes (Aa) and have a homozygous recessive (unaffected) partner (aa), each child has a 50% chance of being affected:

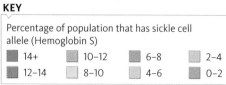

KEY

Percentage of population that has sickle cell allele (Hemoglobin S)

■ 14+	10–12	6–8	2–4
■ 12–14	8–10	4–6	0–2

■ **Figure 4.8 Geography of Sickle Cell Anemia**
This map shows the distribution of sickle cell anemia in Africa and the Middle East by percentages.

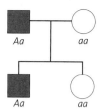

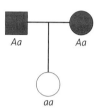

- The numbers of affected males and females are roughly equal.
- Two affected individuals may have unaffected children, again because most affected individuals are heterozygous:

- For most dominantly inherited disorders, homozygous dominant individuals (*AA*) are more severely affected than heterozygous individuals and often die before birth or in childhood. This is probably because only the defective protein is produced in the body.

A pedigree showing the inheritance of an autosomal dominant trait is shown in ■ Figure 4.10.

Table 4.2 lists several autosomal dominant genetic disorders.

Genetic conditions such as those listed below are inherited as dominant traits.

1. **Neurofibromatosis** (*N* = Neurofibromatosis 1; *n* = normal): A common autosomal dominant trait is type I neurofibromatosis (NF). This mutant gene produces several different phenotypes. Some affected individuals (*Nn*) have only pigmented spots on their skin, called *café-au-lait spots* (■ Figure 4.11). Others have noncancerous tumors in the nervous system (■ Figure 4.12). In some cases, these tumors can be large and press on nerves, causing blindness or paralysis in body parts. In a small number of cases, tumors can cause deformities of the face or other body parts.

Table 4.2	Other Examples of Autosomal Dominant Genetic Disorders
Trait	**Phenotype**
Achondroplasia	Dwarfism associated with abnormalities in growth
Brachydactyly	Malformed hands with shortened fingers
Familial hypercholesterolemia	Elevated cholesterol levels, cardiovascular disease
Marfan syndrome	Connective tissue disorder, possible aortic aneurysm
Porphyria	Inability to metabolize porphyrin, episodes of mental disturbance, such as hallucinations or paranoia

■ **Figure 4.11 Neurofibromatosis and Café-au-Lait Spots**
This child has marks on the skin called *café-au-lait* spots, one of the symptoms of neurofibromatosis.

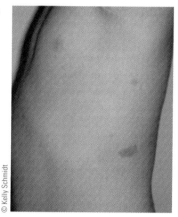

© Kelly Schmidt

■ **Figure 4.12 Neurofibromatosis and Tumors**
This woman has neurofibromatosis. In this case, tumors have grown on her skin.

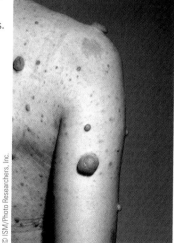

© ISM/Photo Researchers, Inc.

■ **Figure 4.10 Autosomal Dominant Pedigree**
This pedigree shows four generations of a family with an autosomal dominant trait. Symbols with diagonal slashes represent deceased family members.

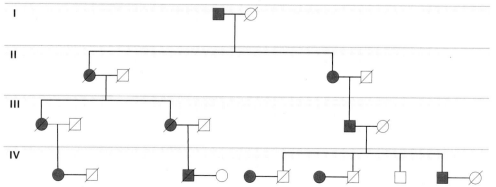

Although NF is an autosomal dominant condition, it often appears in the children of normal parents. This happens because the gene for neurofibromatosis, *NF1*, mutates spontaneously from the normal to the mutant allele. The *NF1* gene has a very high mutation rate. When this happens, one parent's normal allele is changed, causing NF in the child. The causes of the high mutation rate in this gene are under study.

2. **Huntington disease** (*H* = Huntington disease; *h* = normal): Huntington disease (HD) causes destruction of cells in certain areas of the brain. Symptoms begin slowly, usually between ages 30 and 50. Because HD occurs in midlife, affected individuals like Alan's mother may have already had children, who in turn may inherit this disorder.

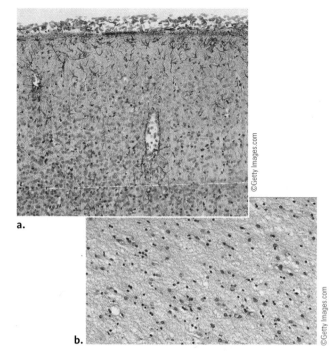

PHOTO GALLERY
This individual with Huntington disease has difficulty in controlling movement
© Conor Caffrey/Photo Researchers, Inc.

Early signs of HD include irritability or apathy. Later, uncontrollable rapid arm and leg movements and a lack of coordination become more frequent. The disorder becomes progressively worse, and individuals with HD develop slurred speech and have difficulty eating or swallowing. Soon they are mentally and physically unable to care for themselves. There is no treatment for HD, and most affected individuals die within 10–25 years after symptoms begin, from complications such as pneumonia, falls, or choking.

As the disease progresses, cells in certain regions of the brain die because large amounts of a defective protein called **huntingtin** accumulate in cells (■ Figure 4.13). Why the protein accumulates, and how it kills the cells is still unknown.

Because HD is inherited as a dominant trait, only one copy of the mutant gene is needed to cause this disorder (*H* is the Huntington allele; *h* is the normal allele). Most affected individuals are heterozygous (*Hh*), and each of their children have a 50% chance of having the disease.

■ **Figure 4.13 The Brain Cells of a Person with Huntington Disease** Microscopic photographs of human brain cells. (a) Normal brain, showing dense clusters of cells (nuclei are stained blue). (b) Brain of a person with Huntington disease. There are fewer cells because of cell death caused by the accumulation of the defective protein, huntingtin.

Pedigree Analysis and Adult-Onset Disorders

Many of the genetic disorders we've discussed in this chapter have phenotypes expressed at or shortly after birth. However, some disorders, such as Huntington disease (HD), develop only later in life, and are called **adult-onset disorders.** In HD, symptoms appear between the ages of 30–50 years. Affected individuals may have already had children, who will each have a 50/50 chance of inheriting the condition.

Other genetic disorders also develop later in adult life. One of these is adult polycystic kidney disease (ADPKD), a very common disease found in 1 in 600 to 1 in 1,000 individuals. ADPKD is inherited as an autosomal dominant trait like HD and symptoms first appear in mid-life and kidney failure and death may develop by age 60. Cysts form in the

CASE A QUESTIONS

Now that we understand how Huntington disease and other disorders are inherited, let's look at some of the issues raised in Alan's case and the rest of the Franklin family.

1. Now that you know about Huntington disease, is Alan at a high enough risk to be tested?

2. If Alan is tested, must the genetic counselor or the doctor tell his siblings the results?

3. Even though Alan is 17, should he be able to have the test without his father's consent? Give three reasons pro or con.

4. What might happen if Alan's father says no?

5. Now that you know when the symptoms of HD first appear, would the results of a test affect your decision whether Alan should be admitted to the Air Force Academy? Why or why not?

kidney, but at first there are very few symptoms, often just headaches and elevated blood pressure.

Adult-onset disorders can present a problem in pedigree analysis. At the time a pedigree is constructed and analyzed, some family members that carry the mutant gene may be too young to show any symptoms. They are placed on the pedigree and indicated as normal. However, if they carry the mutant allele, they will develop symptoms later in life. In order to calculate the risk to family members in these disorders, genetic tests might have to be done.

4.2 Essentials

- Pedigrees are used to study family history and the distinctive patterns of inheritance seen in genetic disorders.

- From a pedigree, one can often determine how a trait is inherited.

- In recessive traits such as albinism, two unaffected parents may have a child with the condition. Two copies of the mutant gene are required for the condition to be expressed.

- In dominant traits, a heterozygote (*Aa*), who carries one copy of a mutant dominant allele, has the condition.

4.3 How Are X-Linked Recessive Traits Inherited?

Because females have two X chromosomes, females can be heterozygous (*Aa*) or homozygous (*aa*) for any gene carried on the X chromosome. Males, on the other hand, carry only one copy of the X chromosome and also carry a Y chromosome. The Y chromosome does not contain the same genes as the X chromosome and carries very few genes of its own. Genes on the X chromosome are called **X-linked,** and genes on the Y chromosome are called **Y-linked.**

This chromosomal difference means that males carrying a mutant gene on the X chromosome that causes a recessive disorder cannot carry a normal, dominant allele of the gene to mask expression of the recessive allele as occurs in females.

In other words, males cannot be heterozygous for any gene on the X chromosome. Because X-linked recessive genes are always expressed in males, they are affected by X-linked recessive genetic disorders far more often than females.

X^*X: females

X^*Y: males

X^*: recessive mutant allele

Because males cannot be homozygous or heterozygous for genes on the X chromosome, they are said to be **hemizygous** for all genes on the X chromosome.

Males give an X chromosome to their daughters and a Y chromosome to their sons. They never give an X to their sons. Females give an X chromosome to each of their children (review the inheritance of X and Y chromosomes in Chapter 1).

A distinctive pattern of inheritance for X-linked recessive traits shows up in pedigrees:

- Hemizygous (X^*Y) males and homozygous (X^*X^*) females are affected.
- These conditions are much more common in males than in females. In the case of rare traits, males are almost exclusively affected.
- Affected males who get the mutant allele from their mothers and transmit it to all of their daughters but not to any sons:

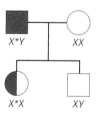

- Daughters of affected males are usually heterozygous, or carriers, and therefore unaffected. However, sons of heterozygous females each have a 50% chance of receiving the recessive gene:

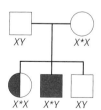

Table 4.3	Examples of X-Linked Recessive Genetic Disorders
Trait	**Phenotype**
Red-green colorblindness	Inability to see green and red colors
Hemophilia A	Inability to form blood clots
Lesch-Nyhan syndrome	Lack of HGPRT protein, mental retardation, self-mutilation
Muscular dystrophy	Duchenne-type, progressive condition with muscle wasting

Courtesy of Dr. Poh-San Lai (both)

a.   b.

■ **Figure 4.14 Cells of a Person with Muscular Dystrophy**
Dystrophin (the protein coded for in DMD). (a) Normal cells with dystrophin in the cell membrane. (b) Cells from a child with DMD. No dystrophin is present in the membranes.

Geneticists have identified more than 850 X-linked recessive traits to date, including color blindness, muscular dystrophy, and hemophilia. Table 4.3 lists several X-linked recessive genetic disorders. The following pedigree shows the inheritance of an X-linked disorder:

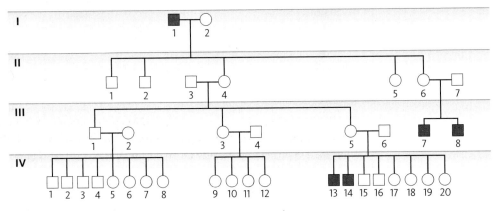

Genetic conditions such as the ones listed below are inherited as a sex linked trait.

1. **Muscular dystrophy** (X^M = normal; X^m = muscular dystrophy): The most common form of muscular dystrophy is Duchenne muscular dystrophy (DMD), which affects about 1 in 3500 males. Infant boys with DMD appear healthy at birth, but begin to develop symptoms between ages 1 and 6. Progressive muscle weakness is one of the first signs of DMD. The disease progresses rapidly, and by age 12 affected individuals usually must use a wheelchair because of muscle degeneration. Death usually occurs by age 20 because of respiratory infection or cardiac failure.

PHOTO GALLERY
This boy has Duchenne muscular dystrophy
© Index Stock Imagery/Photolibrary

The DMD gene is located near one end of the X chromosome and encodes a protein called *dystrophin*.

Dystrophin works inside the muscle cell to prevent the plasma membrane from breaking during muscle contraction (■ Figure 4.14). When dystrophin is absent or defective, these cells are torn apart and eventually die.

There are two forms of X-linked muscular dystrophy. In DMD, which is the more serious form, no detectable amounts of dystrophin are present in muscle tissue. However, in the second, less serious form, called *Becker muscular dystrophy (BMD)*, a shortened, partially functional form of dystrophin is made. As a result, boys with BMD develop symptoms at a later age, have milder symptoms, and live longer than those with DMD. These two diseases are caused by different mutations in the same gene.

2. **Hemophilia** (X^H = normal; X^h = hemophilia): One X-linked genetic disorder associated with lack of blood clotting is a recessively inherited disorder known as *hemophilia A*. In hemophilia A, a clotting factor called **factor VIII** is absent or present in reduced amounts. This genetic disorder causes a failure to form blood clots. As a result, affected individuals are in danger of bleeding to death from minor cuts or from hemorrhages caused by a bruise. They often require hospitalization to treat bleeding.

Hemophilia A is the most common form of X-linked hemophilia. Females are affected only if they are homozygous for this recessive gene (X^hX^h).* Very few females have hemophilia A because it can occur only when a heterozygous female and a male with hemophilia have a girl. Even then the chances of the girl having hemophilia A is only 50%. Until recently, very few males with hemophilia survived to reproductive age, a fact that further lowers the chances that a female will have hemophilia.

*X^h = hemophilia gene.

Treatment for hemophilia formerly involved injections of concentrated clotting factor made from blood serum collected from a large number of donors. In the early to mid-1980s, some clotting factor preparations were contaminated with HIV, the virus associated with AIDS. As a result, about half of all people (males) with hemophilia became infected with HIV. Many developed AIDS and died as a result of their treatment with contaminated blood products.

Today, recombinant DNA technology is used to make clotting factors that are free from contamination with viruses and other disease agents (■ Figure 4.15).

4.3 Essentials

- In X-linked traits such as hemophilia, the mutant gene is carried on the X chromosome and males are more often affected than females.
- Many traits are inherited through dominant, recessive, and X-linked genetics.

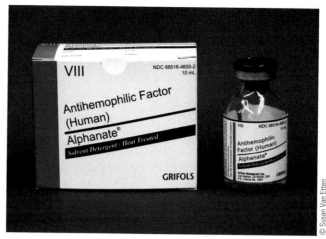

■ **Figure 4.15 Vials of Factor VIII, a drug used to treat hemophilia** The clotting factor (Factor VIII or IX) that people with hemophilia need so that their blood can form clots. These factors usually made through recombinant DNA technology (see Chapter 6).

CASE B The Franklins Find Out More

Alan is now very concerned about his future. Because the genetic counselor has told him that there is a test for HD, he wonders whether he will need written permission from his father to be tested.

Alan decides to talk to his older siblings. He talks to Mary first. She says she wants to talk to her fiancé about testing. Bob, Mary's fiancé, thinks the test is a good idea but suggests that if the results are positive, she sign a prenuptial agreement stating that they will not have any children.

John says he would take the test but wants his wife's fetus tested too. He believes that the baby should be aborted if it carries the Huntington gene—he still remembers his mom. His wife agrees.

The genetic counselor then completes the family's pedigree:

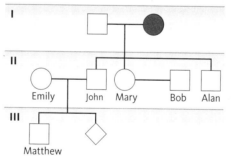

Alan wonders if he should even have the HD test.

1. Do you think Alan should have the test? Give three reasons for your answer.
2. What might happen if Alan tests positive? List a few possibilities.
3. Now that you know how serious HD is, how do you think this might affect health insurance coverage for this family?

4. Remember Martha Lawrence's case from Chapter 3? Her doctor had questions about whether she should tell Martha the results of her amniocentesis. Here the test results affect not only Alan but his entire family. If the test results were given to Alan but not the rest of the family, what might happen? List three issues that might arise from this situation.

4.4 What are the Legal and Ethical Issues Associated with Dominant and Recessive Genetic Conditions?

One of the questions a person might ask after reading Alan's family's dilemma is "if the family takes the Huntington test, who can see the test results?" Alan wants to have a test to find out his genetic status, but could his future employers know his status?

Legally, medical records and test results are confidential and doctor/patient confidentiality is one of the legal principles that allows patients to trust doctors with information. Because of this confidentiality, patients feel safe telling their doctors private medical concerns. A law was passed in 1996 that mandated such privacy rights. This law, called The Health Insurance Portability and Accountability Act, or HIPAA, created national standards to protect individuals' medical records and other personal health information.

HIPAA:

- gives patients the ability to control their health information.
- allows only certain uses of health records.
- protects the privacy of health information.
- establishes civil and criminal penalties if patients' privacy rights are violated.

- allows for certain exceptions—for example, to protect public health.

If Alan discusses his test results with his physician, the family's insurance company might be able to see the results. When applying for health insurance, patients are asked to sign a waiver allowing the insurance company to see their medical records. In our case, Alan's family must have signed such a waiver.

The reasons that insurance companies and employers can access medical records vary, but some are:

- to keep statistics on certain conditions
- to get information for medical studies that will help the medical population as a whole.
- to find information that will protect employees from illness
- to find information that might lower health care costs
- to decide whether to employ or cover certain individuals
- to decide the cost of insurance for an individual or company.

On some levels this seems to be the opposite of confidentiality, but insurance companies need to see patient's medical conditions before, during, and after they work with them.

The chart below addresses some of the ethical and legal questions regarding privacy of medical information:

Question	How are these questions decided?	Related case or legal issue
Is a minor protected under the HIPPA laws?	State laws set guidelines for the age that allows coverage under HIPAA. However, teens in many states are allowed privacy.	In many states, a pregnant woman, no matter what her age, is protected under HIPAA and can make her own medical decisions.
Can someone sue if their medical records are released without their permission?	Yes, law suits were common before the passage of the HIPAA laws, but now there are fewer. Often, these questions are settled by arbitration with the employer and a mediator.	Some states limit the amount of money awarded in such law suits.
Can an employer mandate that employees or potential employees be tested for a condition such as Huntington disease?	If the condition is tied directly into one's employment, a test could be mandated. In the case of Huntington disease, this would probably not be applicable.	In EEOC v. Burlington N. Santa Fe Railway Co., discussed in Chapter 5, the company tried to test its employees for an inherited form of *carpal tunnel syndrome*. Read Spotlight on Law, Chapter 5.
Can an insurance company drop your insurance if you have a positive genetic test for a condition such as Huntington disease?	The Kennedy/Castlebaum Law specifically states that this could not happen. In addition, in 2008, the Genetic Information Non-discrimination Act (GINA) passed in the Senate and House of Representatives. This bill will prevent discrimination.	On May 21, 2008, President George W. Bush signed GINA into law. Hopefully, this will result in greater privacy in genetic testing.

When doctors have genetic information that applies to other members of the family, such as Alan's, are they obligated to reveal this information? The question of whether medical records are confidential is a serious issue in the legal profession. We sign privacy agreements every time we go to the doctor based on HIPAA laws. The following case directly applies to this question.

Dr. Lawrence Moore, a psychologist employed by the University of California hospitals, was treating Prosenjit Poddar as an outpatient. During a session in 1969, Poddar confided that he intended to kill his ex-girlfriend, Tatiana Tarasoff, when she returned from her summer vacation.

Dr. Moore wasn't sure that Poddar would actually kill Tatiana, but he wanted to be safe so he notified the campus police. The police detained Poddar but released him because he seemed rational and denied that he was going to kill his ex-girlfriend. Soon after, he stopped his treatment with Dr. Moore.

Three months later, Poddar killed Tatiana.

Tatiana's parents sued the University of California for malpractice, claiming that Dr. Moore had a "duty to warn" Tatiana and that it could have saved her life.

RESULTS:

Trial Court:

The trial court found for the University of California (defendant), stating that no duty to warn exists; the Tarasoff family appealed the decision.

Supreme Court of California:

The state supreme court reversed the decision of the lower court, stating that therapists have a duty to take actions to protect third parties from violent patients, even if this breaches confidentiality. The court said that Dr. Moore could not escape liability just because Tatiana was not his patient; he still should have warned her.

Also, when a therapist determines that a patient presents a serious danger to another, he or she has an obligation to use reasonable care to protect the intended victim from danger. However, the doctor must have a special relationship with either the person who might cause harm or the potential victim before he or she can breach confidentiality. The potential victim must be identifiable and the harm must be foreseeable and serious. The court stated, "Privilege ends where public peril begins."

Legislation:

Some states have enacted laws based on the Tarasoff decision that require disclosure when certain dangers exist. They state that the therapist or doctor has an obligation to use reasonable care to protect a potential victim.

QUESTIONS:

1. Do you agree with the California Supreme Court's decision? Why or why not?
2. The Tarasoffs were asking for monetary damages to compensate them for the loss of their daughter. Do you think this is a good argument? Why or why not?
3. Did Dr. Moore do enough to protect Tatiana?
4. If not, what more should he have done?
5. What could this mean for the future of medicine?
6. This decision has been cited in cases in which a physician was questioning whether to tell a spouse that his or her partner had HIV. Does Tarasoff apply in this situation?
7. Does the Tarasoff case apply in Alan's situation? How?
8. If you were the attorney for the Tarasoffs, would you want a jury trial or a judge-only trial? Why?
9. A woman has a genetic test and finds that she carries the gene for cystic fibrosis, a recessive trait. Should the doctor tell the patient's fiancé? Why or why not?
10. How could the decisions in the Tarasoff case be used in the argument in the case in question 9 above?

✪ THE ESSENTIAL TEN

1. Only one copy of each gene is present in the sperm or egg and is contributed to the offspring. [Section 4.1]
   ▮ When the sperm and egg combine, the number of chromosomes is 46.

2. Members of a gene pair segregate into gametes independently of other gene pairs, so that all gametes have different combinations of genes. [Section 4.1]
   ▮ Only one member of each gene pair goes into an egg or sperm.

3. Gregor Mendel, a monk who lived in the 1800s, was the father of genetics. [Section 4.1]
   ▮ Mendel's work was done with pea traits such as tall and short stems.

4. Mendel did his experiments with plants, but his important laws of genetics are still used today to study human genetics. [Section 4.1]
   ▮ Human traits are inherited in the same way as traits in pea plants.

5. Pedigrees are used to study family history and the distinctive patterns of inheritance seen in certain genetic conditions. [Section 4.2]
   ▮ Anyone can construct a pedigree of their family.

6. From a pedigree, one can determine how a trait is inherited. [Section 4.2]
   ▮ Using a pedigree, counselors can determine a couple's risk of certain conditions.

7. In recessive traits such as albinism, two unaffected parents may have a child with the condition. [Section 4.2]
   ▮ Two copies of the mutant gene are required for the condition to be expressed.

8. In dominant traits, a heterozygote (*Aa*), who carries one copy of a mutant dominant allele, has the condition. [Section 4.2]
   ▮ If a person has a dominant allele, their offspring have a 50% chance of inheriting it.

9. In X-linked traits such as hemophilia, the mutant gene is carried on the X chromosome and males are more often affected than females. [Section 4.3]
   ▮ One trait that is inherited this way is hemophilia.

10. Many traits are inherited through dominant, recessive, and X-linked genetics. [Section 4.3]
    ▮ These traits run in families and can be tracked by pedigrees.

KEY TERMS

adult-onset disorder (p. 66)
albinism (p. 63)
alleles (p. 60)
autosomal dominant (p. 62)
autosomal recessive (p. 62)
CFRT (p. 64)
consanguinity (p. 63)
cystic fibrosis (p. 63)
dominant (p. 60)
dystrophin (p. 68)
factor VIII (p. 68)
genotype (p. 60)
hemizygous (p. 67)
hemophilia (p. 68)
heterozygous (p. 60)
homozygous (p. 60)
huntingtin (p. 66)
Huntington disease (p. 59)
law of independent assortment (p. 60)
law of segregation (p. 60)
muscular dystrophy (p. 68)
neurofibromatosis (p. 65)
pedigree (p. 61)
phenotype (p. 60)
pigment (p. 61)
proband (p. 62)
recessive (p. 60)
sickle cell anemia (p. 64)
X-linked (p. 67)
X-linked recessive (p. 62)
Y-linked (p. 67)

ONLINE RESOURCES

Preparing for an exam? Log on at academic.cengage.com/login for a pre-test, a personalized learning plan, and a post-test in CengageNOW's Study Tools to help you assess your understanding.

If assigned by your instructor, the Case A and Spotlight on Law activities for this chapter, "A Family's Dilemma" and "Confidentiality of Genetic Information," will be available in CengageNOW's Assignments.

REVIEW QUESTIONS

1. When Mendel studied his pea plants, he knew nothing about genes or chromosomes. How did he figure out that these traits were passed on? Do physicians today use this technique? When?

2. The law of segregation describes what cellular activity?

3. With any recessive trait (in humans), the same rules apply as with Mendel's pea plants. Does this make it simpler or harder for parents of an affected child to understand a genetic counselor's explanation?

4. Are dominant traits more or less common than recessive traits? Explain.

5. Choose four conditions discussed in this chapter and complete the following chart.

Condition	Is the Condition Dominant, Recessive, or X-Linked?	On What Chromosome Is the Gene?	What Group Tends to Have the Condition?	Is There a Test for the Condition?

6. What does hemizygous mean?
7. How does dystrophin work in the muscles?
8. What goes wrong with the dystrophin gene to cause Duchenne muscular dystrophy?

9. In sex-linked traits, explain why more males are affected than females.
10. Why were hemophiliacs affected by HIV in the 80s?

APPLICATION QUESTIONS

1. Mendel is mentioned in this chapter, but other scientists were also important in the early study of genetics. Research two of them and write a short paragraph about their contributions.
2. Research the problems a person with albinism may have throughout his or her life and write a short report.
3. Look at the pedigrees of the royal houses of Europe and draw one. Why did certain traits (hemophilia and others) show up so frequently?
4. Make your own family pedigree and trace one trait through a few generations.
5. Research three conditions not discussed in this chapter and complete the following chart.

6. Research and consider how geneticists are studying Duchenne muscular dystrophy in the Amish population.
7. When you look at "By the Numbers," you can see how many carriers of CF there are in certain populations. How could you use this information to lower the numbers of babies born with CF?
8. Now that you know how tallness is inherited in pea plants, do you think this is the same in humans? Explain your answer.

Condition	Is the Condition Dominant, Recessive, or X-Linked?	On What Chromosome Is the Gene?	What Group Tends to Have the Condition?	Is There a Test for the Condition?

9. In the pedigree at right, determine how the gene is inherited. Is it dominant, recessive, or X-linked recessive? How can you tell?

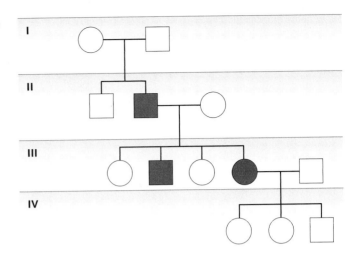

Learn by Writing

In Chapter 1 we suggested that you start a blog with members of your class or others who are interested in our topics. Now is a time to revisit your blog and consider some of the issues surrounding testing for sickle cell anemia. E-mail others you think might be interested and invite them to contribute.

Here are some ideas to address in your blog:

- Is a single gene defect such as sickle cell anemia more serious than an extra or missing chromosome?
- Ms. Lawrence's case involves early testing. Alan's family might have benefited from these procedures. How?
- Should the law actually get involved in any of these personal decisions?
- Address any other question or comment that came to mind as you read this chapter.

What Would You Do If . . . ?

1. You were Alan? (See Case A and Case B.)

2. You were Mary? (See Case B.)

3. You were on a committee deciding whether members of this family (Case A and Case B) could obtain health insurance?

4. You were the genetic counselor advising Emily (Case B) about her unborn fetus?

5. You were a member of a legislature voting on a bill that made it mandatory for all couples applying for a marriage license to be tested as carriers of cystic fibrosis? Why?

6. You were asked to counsel a person whose genetic test was positive for HD?

In The Media

New York Times, June 8, 2008

Albinos, Long Shunned, Face Deadly Threat in Tanzania

Jeffrey Gettleman

Throughout sub-Saharan Africa albinos have been discriminated against for many years. Recently, however, in Tanzania, this has become much more deadly.

Nineteen albinos, adults and children, have been killed this year and their body parts sold.

One in 3000 people in Tanzania are albinos and witch doctors are selling the skin, bone, and internal organs to make into potions that are supposed to help people become rich.

In addition, some Africans believe that albinos have magical powers and their hair has been sold on the market for high prices. Fishermen weave the hair into their fishnets in the belief that it will increase the catch.

To access this article and related video online, go to academic.cengage.com/biology/yashon/hgs1.

QUESTIONS:

1. How might you try to protect albinos who are in fear for their lives?

2. What ideas do you have to educate the population and make them understand how albinism is inherited?

Inter Press Service News Agency, July 2, 2006

Striking Down the Taboos about Albinism

The Nigerian government plans to establish an agency to coordinate the affairs of its citizens with albinism, who number about 600,000. The people with albinism in this country are having similar problems as those in Namibia and other sub-Saharan countries.

Faced with many challenges, this agency for people with albinism is conducting a variety of initiatives in a bid to improve matters.

It will visit church leaders, tribal leaders, and councilors to appeal to them to help people with albinism with hats, creams, and sunglasses. Joseph Ndinomupya, president of the Namibia Albinism Association Trust, a private company, thinks this will work.

To access this article online, go to academic.cengage.com/biology/yashon/hgs1.

QUESTIONS:

1. Is this type of action enough to change people's minds?
2. Does discrimination against people with albinism occur in the United States and other countries?
3. How does this story show progress in the protection of albinos?

Biology Basics: DNA

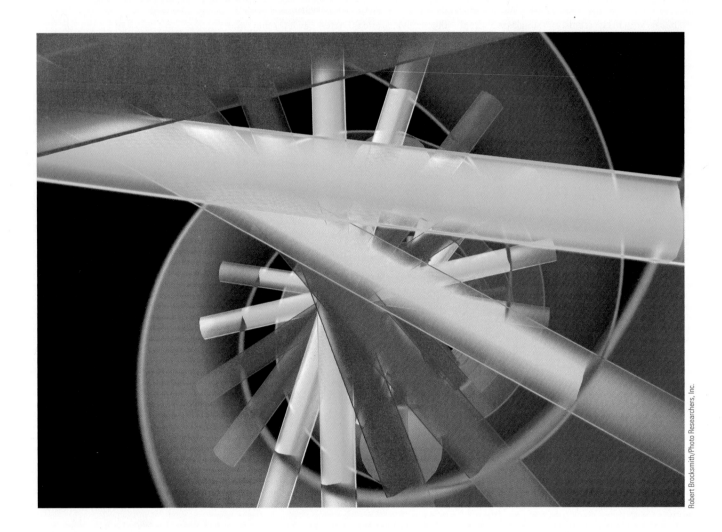

Robert Brocksmith/Photo Researchers, Inc.

BB2.1 What is the evidence that DNA carries genetic information?

In the early part of the 20th century, biologists identified chromosomes as the carriers of genetic information. When observed in the microscope, the behavior of chromosomes paralleled the movement of genes from generation to generation as described by Mendel. Next, they identified which chemical components carried this information. These components were identified as two major biomolecules: proteins and DNA (deoxyribonucleic acid).

The evidence for DNA as a carrier of genetic information came from an unexpected source: the study of an infectious disease. By the 1920s, it was known that one form of pneumonia was caused by the bacterium called *Streptococcus pneumoniae,* shown below:

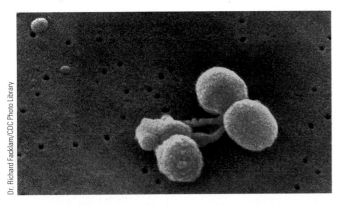

Frederick Griffith studied two different strains of this bacterium. One strain (the S strain) caused pneumonia when injected into mice. Cells of this strain were surrounded by a capsule. The other strain (the R strain) was not surrounded by a capsule and did not cause pneumonia when injected into mice. Griffith wondered if the capsule caused the pneumonia. So he devised a clever experiment. He killed the S cells by heating them and mixed them with the living R cells. When he injected this mixture into the mice, they developed pneumonia and died.

Griffith concluded that after injection into the mice, living R cells were able to form a capsule similar to the S cells. The hereditary information somehow passed from the dead S cells into the R cells, allowing them to make a capsule and, therefore, cause pneumonia. He called this process **transformation.**

In the 1940s, Oswald Avery, Colin MacLeod, and Maclyn McCarty removed and purified certain biomolecules (DNA, RNA, protein, etc.) from the heat-killed S cells. They tested them one at a time to see which one could allow the R cells to make a capsule. Only DNA was able to cause this transformation. The result of this experiment clearly established that DNA is the carrier of genetic information and that it could be transferred from one bacterial cell to another.

The discovery that DNA carries genetic information helped fuel efforts to understand the structure of DNA. From the mid-1940s through 1953, several laboratories made significant strides in unraveling the structure of DNA. This resulted in the model for DNA structure proposed by James Watson and Francis Crick in 1953. It is shown below:

Taken in 1953

Taken in 1993

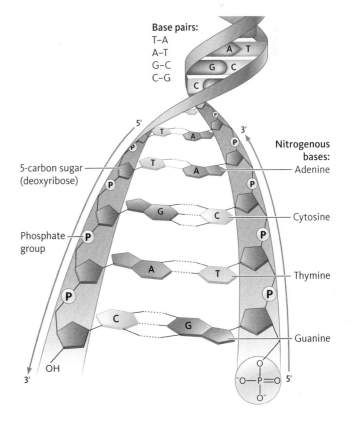

Base pairs:
T–A
A–T
G–C
C–G

Nitrogenous bases:

5-carbon sugar (deoxyribose)

Phosphate group

Adenine

Cytosine

Thymine

Guanine

BB2.2 What are the chemical subunits of DNA?

When Watson and Crick began their work, it was already known that organisms contain two types of nucleic acids: **deoxyribonucleic acid (DNA)** and **ribonucleic acid (RNA).** Both are made up of subunits known as **nucleotides.** DNA is found in the nucleus, and RNA is found in both the nucleus and the cytoplasm.

The drawings below show nucleotides of RNA and DNA linked together to form *polynucleotides.* A single nucleotide has three components:

1. A nitrogen-containing base (there are two types of bases: **purines** and **pyrimidines**). The purine bases **adenine** (A) and **guanine** (G) are found in both RNA and DNA. The pyrimidine bases are **thymine** (T), found only in DNA; **uracil** (U), found only in RNA; and **cytosine** (C), found in both RNA and DNA. RNA has four bases (A, G, U, C), and DNA has four bases (A, G, T, C).

2. A sugar (either **ribose**, found in RNA, or **deoxyribose**), found in DNA. The difference between the two sugars is a single oxygen atom, which is present in ribose and absent in deoxyribose.

3. A **phosphate group.** Phosphate groups are strongly acidic, which is why DNA and RNA are called acids.

The components of a nucleotide are assembled by chemically linking a base to a sugar, which in turn is linked to a phosphate group. As shown at left, nucleotides can be strung together in chains by linking the phosphate group of one nucleotide to the sugar of another nucleotide to form polynucleotides.

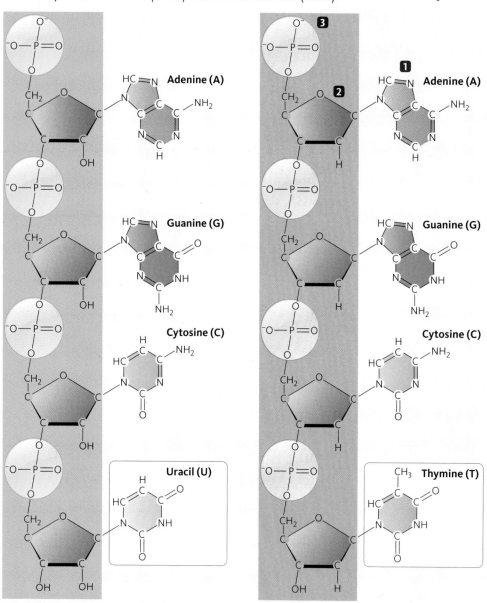

RNA

DNA

BB2.3 How are DNA molecules organized?

As shown in the drawing below, DNA contains two chains of nucleotides and is a ladder-like molecule. Within the DNA molecule, a purine base on one chain always pairs with a pyrimidine base on the other chain. More specifically, A in one chain always pairs with T in the opposite chain, and C always pairs with G, forming a base pair. This pairing makes the two chains of DNA *complementary* in base composition. This means that if one strand of nucleotides has the sequence A C G T C, the other strand must be T G C A G.

The Watson-Crick model has three important properties:

1. *Genetic information is stored in the sequence of bases in the DNA.*
2. *The model offers a molecular explanation for mutation.* Because genetic information can be stored as a chain of bases in DNA, any change in the order or number of bases in a gene can result in a mutation that may produce a genetic disorder.
3. *The complementary sequence of bases in the two strands of DNA explains how DNA copies itself before each cell division.* In copying the DNA, each strand can be used as a template to reconstruct the base sequence in the opposite strand.

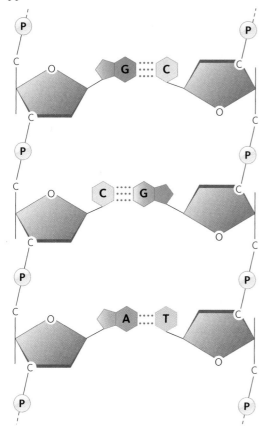

■ Figure BB2.1
Rosalind Franklin, who worked on the discovery of DNA structure.

To create their model, Watson and Crick used information supplied from an image of DNA created by Rosalind Franklin (Figure BB2.1) using **x-ray diffraction.** The photo below shows the image created by Franklin:

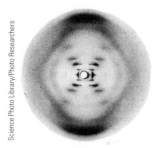

From observing Franklin's photograph Watson and Crick determined that the nucleotide strands of DNA are organized as a double helix as shown below:

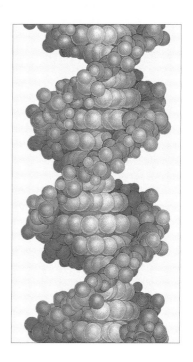

BB2.4 How is DNA organized in a chromosome?

One long DNA molecule is coiled and combined with proteins to form a chromosome. The total length of the DNA in all 46 human chromosomes is about 6 ft (1.8 m). To fit all this DNA inside the nucleus of a cell, the DNA has to be folded over and over. The diagram on the right shows how this is accomplished. Follow along with the description:

First, DNA is wound around the outside of a protein cluster (histone core) to form structures called **nucleosomes** (a). These nucleosomes are connected to each other by short threads of DNA (b). Next, nucleosomes are coiled to form a thin cylinder (c). Then the cylinders are coiled to form thicker and thicker fibers, eventually forming the loops and fuzzy-looking fibers we can see as part of chromosomes in the electron microscope (d). Two photomicrographs of replicated chromosomes are shown below:

Centromere

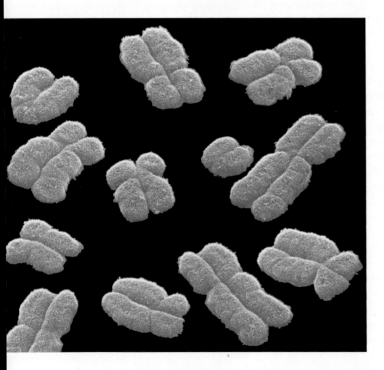

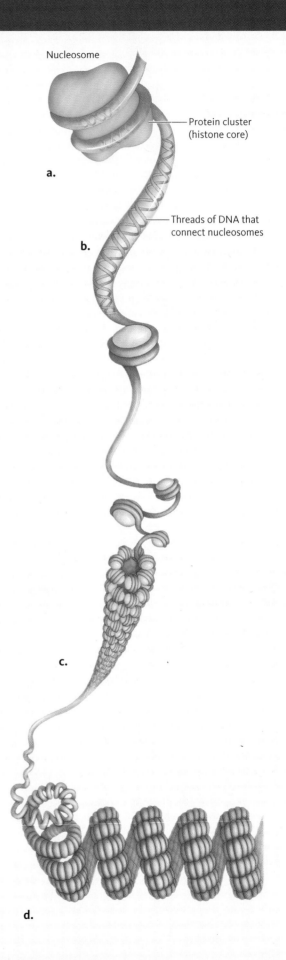

Nucleosome

a.

Protein cluster (histone core)

Threads of DNA that connect nucleosomes

b.

c.

d.

BB2.5 How are RNA molecules organized?

The second type of nucleic acid, RNA, is found in the nucleus *and* the cytoplasm. RNA transfers genetic information from the nucleus to the cytoplasm, participates in the formation of proteins, and is a component of ribosomes (see "Biology Basics: Cells and Cell Structure").

RNA (shown below) differs from DNA in two respects: the sugar in RNA is ribose (instead of deoxyribose in DNA), and the base uracil takes the place of the base thymine. Table BB2.1 shows the differences between RNA and DNA.

As shown below, RNA is usually single-stranded, and there is no complementary strand.

Table BB2.1	Differences between DNA and RNA	
	DNA	RNA
Sugar	Deoxyribose	Ribose
Bases	Adenine	Adenine
	Cytosine	Cytosine
	Guanine	Guanine
	Thymine	Uracil

BB2.6 How is genetic information carried in a DNA molecule?

DNA is found only in the nucleus. It functions as a storehouse of genetic information. Based on their model for the structure of DNA, Watson and Crick proposed that genetic information is encoded in the sequence of base pairs in a DNA molecule.

The amount of information contained in DNA ranges from a few thousand nucleotides in some viruses to more than 3 billion nucleotides in humans, and even more in some plants and amphibians.

A gene typically consists of hundreds or thousands of nucleotides.

Each gene has a beginning and end, marked by specific nucleotide sequences. There is one molecule of DNA in each chromosome which can contain thousands of genes. A change in any base pair in a gene can be considered a mutation.

BB2.7 What are the two main functions of DNA?

DNA has two major functions in the cell: (1) to replicate itself and (2) to carry the information for proteins and RNA molecules necessary for life. Here, we will explore how DNA replicates itself. In Chapter 5, we will discuss how DNA encodes information for proteins and the steps in forming these proteins. Later chapters will examine how an understanding of the structure and properties of DNA led to the development of recombinant DNA technology, its applications in medicine and forensics, and finally, the Human Genome Project.

BB2.8 How is DNA replicated?

Before cell division begins, all cells must copy their DNA so that each daughter cell will receive an exact copy of the genetic information in the chromosomes. When a DNA molecule unwinds, each half serves as a template or pattern for making a new, complementary strand.

DNA replication occurs in all living things from bacteria to humans. The steps are as follows (follow along with part C of the figure on the next page):

1. The double-stranded DNA molecule unwinds, opening the molecule to the action of an enzyme, **DNA polymerase.**
2. DNA polymerase is able to read the sequence of the template strand and link together nucleotides (already present in the nucleus) to form a complementary strand that attaches to the old strand.
3. Each newly replicated DNA molecule contains one old strand (the template strand that was copied) and one new strand (that is complementary to the old strand).

Thinking about this in terms of chromosomes, recall that each chromosome contains one single double-stranded DNA molecule, running from end to end. When DNA replication is complete, each chromosome consists of two DNA molecules. These two molecules are visible in the microscope as structures called *chromatids* that are joined at a common centromere. See the photomicrographs on page 80.

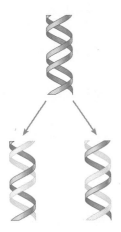

a. An unreplicated DNA molecule (top) and two replicated DNA molecules (bottom).

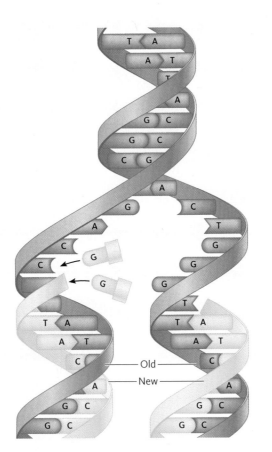

Old — New

b. During replication, the old strand (blue) serves as a template for the synthesis of a complementary new strand (gold).

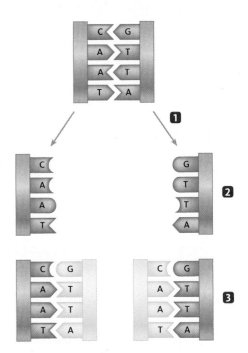

c. During replication, the DNA strands unwind (1), and each strand serves as a template (2) for replication. The sequence of the new strands (gold) is complementary to the template strands (blue) as shown in 3.

Each chromatid contains a DNA molecule composed of one old strand and one new strand.

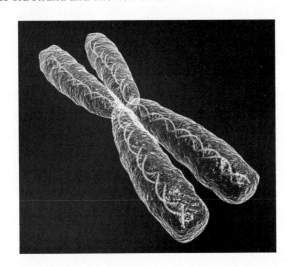

When the centromeres divide during mitosis, each chromatid becomes a separate chromosome.

Can DNA replication cause mutations?

DNA replication is not a perfect process. DNA polymerase moves quickly (about 20 bases per second in humans) and sometimes the wrong nucleotide is inserted into the newly made strand. In most cases, DNA polymerase detects the error and repairs the mistake. Sometimes, however, the error is undetected, creating a mutation.

DNA can be damaged when it is exposed to physical or chemical agents that cause mutations. For example, when DNA is exposed to ultraviolet light (UV), changes can occur in the base pairing. Usually, these errors can be detected and repaired by enzymes, as shown in the figure on the next page. If these errors are not detected or repaired properly, a mutation results.

What can happen if errors are not repaired?

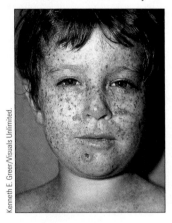

Kenneth E. Greer/Visuals Unlimited.

Cells have a number of DNA repair systems, but even with all these checkpoints, errors in replication or DNA damage can still produce mutations. This can happen when the genes for the repair systems are themselves damaged. One example of a condition that results from mutations in DNA repair systems is **xeroderma pigmentosum (XP).** People with XP are extremely sensitive to UV light, and develop skin cancer on sun-exposed areas of the body (see photo above). About half of all XP individuals develop skin cancer by 8 years of age, and also have a 20-fold higher risk of developing other cancers, including brain tumors.

Chapter 5 focuses on genes and how gene action is related to genetic traits, including genetic diseases. Chapter 6 describes recombinant DNA technology and how it is used in biotechnology. Chapter 7 considers how molecular biology has revolutionized genetic testing and prenatal diagnosis. Chapter 8 discusses the methods and applications of DNA forensics in several fields. Chapter 9 summarizes the history and findings of the Human Genome Project, and its future impact on medicine and society.

Glossary

adenine A nitrogen-containing purine base found in nucleic acids.

cytosine A nitrogen-containing pyrimidine base found in nucleic acids.

deoxyribonucleic acid (DNA) A molecule consisting of antiparallel strands of polynucleotides that is the primary carrier of genetic information.

deoxyribose A pentose sugar found in DNA.

DNA polymerase An enzyme that catalyzes the synthesis of DNA using a template DNA strand and nucleotides.

gene regulation The process by which genes are turned on and off.

guanine A nitrogen-containing purine base found in nucleic acids.

histone The protein that DNA coils around to form nucleosomes.

nucleosome Structures formed when DNA is coiled around proteins.

nucleotide The basic building block of DNA and RNA. Each nucleotide consists of a base, a phosphate, and a sugar.

phosphate group A compound containing phosphorus chemically bonded to four oxygen molecules.

purine A class of double-ringed organic bases found in nucleic acids.

pyrimidine A class of single-ringed organic bases found in nucleic acids.

ribonucleic acid (RNA) A nucleic acid molecule that contains the pyrimidine uracil and the sugar ribose. The several forms of RNA function in gene expression.

ribose A pentose sugar found in RNA.

Streptococcus pneumoniae A bacterial species that causes pneumonia.

thymine A nitrogen-containing pyrimidine base found in nucleic acids.

transformation A heritable change caused by DNA that originates outside the cell.

uracil A nitrogen-containing pyrimidine base found only in RNA.

xeroderma pigmentosum A skin condition that results from a problem with DNA repair.

x-ray diffraction A method of studying the structure of a crystal using x-rays.

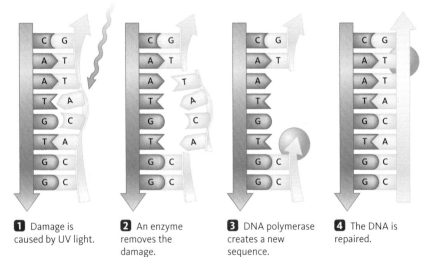

UV light

1 Damage is caused by UV light.

2 An enzyme removes the damage.

3 DNA polymerase creates a new sequence.

4 The DNA is repaired.

5 Genes as DNA: How Genes Encode Proteins

⊛ CENTRAL POINTS

- Genes are made of DNA that encode proteins.

- During transcription genetic information DNA is copied into messenger RNA.

- During translation the information is transferred to the amino acid sequence of proteins.

- Mutations are changes in a gene's DNA.

- Changes in DNA can produce changes in proteins.

CASE A Marcia Johnson's Surprising Test Results

Marcia Johnson thought she was going to her obstetrician for a routine checkup. She had just found out she was pregnant, so she expected to get her prenatal vitamins and go home. In the waiting room, a nurse approached her and asked if she would be willing to be part of a study and give a blood sample for a simple test. The test was for sickle cell anemia. The nurse explained that researchers were trying to establish how many people in northern Illinois were carriers of the mutant gene that causes the disorder. Marcia thought, "Why not?" She signed the consent form, gave blood, and promptly forgot about it.

Two weeks later her obstetrician called to tell her that she was indeed a carrier for sickle cell anemia. Marcia was surprised, but she had searched the Internet after she gave the blood sample and learned that carriers were common among those with West African ancestors. The doctor said not to worry, but that she should ask her husband, Mark, to be tested. Mark was tested and found out he was also a carrier.

At the visit when Mark received his results, they also had the first ultrasound of Marcia's pregnancy. She was carrying twins! Marcia and Mark were disturbed by what she had learned about sickle cell anemia on

the Internet, and they made an appointment with her obstetrician to have a prenatal test on the fetuses. A week later, a genetic counselor recommended by her doctor called to tell them the results.

Some questions come to mind when reading this case. Before we address these questions, let's look at what genes are, how they work, and why a mutant gene may result in a genetic disease.

In "Biology Basics: Cells and Cell Structure," you learned that DNA has two major functions. In this chapter we will look at one of these: how DNA codes for proteins.

5.1 How Do Genes Control Traits?

Each of the trillions of cells in the human body (except mature red blood cells) contains a nucleus. Inside the nucleus is a set of 46 chromosomes, carrying two copies of the 20,000–25,000 genes that make us human. Recall from Chapter 3 that we get one copy of our gene set from our mother and the other from our father. Either or both of these gene sets can be identical or can carry different forms of each gene (these different forms are called **alleles**).

Genes (our *genotype*) are composed of DNA. Most, but not all, genes contain the information to make a **protein.** Proteins, or more properly, their functions, contribute to the observable traits, or **phenotype,** that make us unique. One example of a phenotype is the symptoms associated with cystic fibrosis (■ Figure 5.1). The normal allele of this gene codes for a cell membrane protein that controls the flow of chloride ions into and out of a cell. The most common mutant allele of this gene produces a protein (CFTR) that cannot function. Those who carry two copies of the mutant allele lack this protein. They show the symptoms of cystic fibrosis, a life-threatening disorder that causes severe lung damage and nutritional deficiencies.

■ **Figure 5.1 A child with cystic fibrosis.**
This child is using a bronchodilator, similar to those used by asthma patients. The mist she is inhaling helps open her airways.

To make all the proteins our bodies need, we must have a supply of all 20 amino acids. Some of these can be made by the cells of our body, but others must be included in our diet; these are called **essential amino acids** (Table 5.1). A balanced diet is necessary to ensure that we have an adequate supply of all the essential amino acids. Vegetarians must be especially careful in planning meals because some plants contain low amounts of certain essential amino acids. Vegans, who consume no animal or dairy products, must select complementary protein sources to ensure that they eat a balanced diet that contains all the essential amino acids.

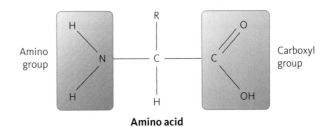

Amino acid

What is a protein?

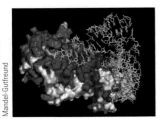

A three-dimensional model of a protein is shown at left. Proteins take many forms and have many functions. They provide structure (hair, nails), serve as enzymes (proteins that control chemical reactions in the cell), can be chemical messengers (hormones such as human growth hormone and insulin), act as receptors (for sight, smell, hearing), and are carrier molecules (hemoglobin, the molecule altered in sickle cell anemia).

Refer to ■ Figure 5.2. Proteins are composed of subunits called **amino acids.** Twenty different types of amino acids are found in proteins. What makes each amino acid different is a chemical group (called an R group) attached to the backbone of the molecule. Two other chemical groups are found at the ends of all amino acids. One end carries an amino group (NH_2), and the other has a carboxyl group ($COOH$). These amino acid subunits are chemically linked together by the cell to form proteins (■ Figure 5.3).

The diversity of proteins found in nature comes from the astronomical number of possible combinations of these 20 different amino acids. For example, in a protein composed of only five amino acids, 20^5 or 3,200,000 combinations are possible. Most proteins are composed of several hundred amino acids, so billions of combinations are possible.

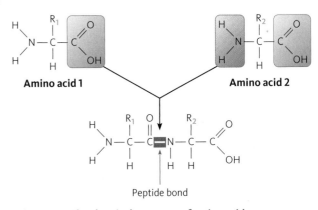

Amino acid 1   **Amino acid 2**

Peptide bond

■ **Figure 5.2 The chemical structure of amino acids.**
The twenty amino acids differ from each other in the chemical nature of the R groups. Peptide bonds are formed between the amino group of one amino acid and the carboxyl group of another amino acid.

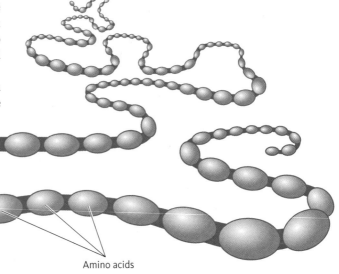

Amino acids

■ **Figure 5.3 Linked Amino Acids**
Amino acids linked together form a protein.

Table 5.1 Essential Amino Acids

At various stages in our lives, our bodies can produce 10 of the 20 amino acids. We must consume the other 10, which are listed here.

Amino Acid	Abbreviation
Arginine	Arg
Histidine	His
Isoleucine	Ile
Leucine	Leu
Lysine	Lys
Methionine	Met
Phenylalanine	Phe
Threonine	Thr
Tryptophan	Trp
Valine	Val

As we will see in a later discussion, after a protein is formed, it folds into a three-dimensional shape based on the order of amino acids. This functional protein then plays its role in producing a phenotype.

How does DNA carry the information to a protein?

At first glance, it seems difficult to envision how the information for billions of different combinations of 20 amino acids can be carried in DNA. If DNA contains only four different nucleotides (A, T, C, and G), and there are 20 different amino acids in proteins, how can only four nucleotides encode the information for all 20 amino acids? The answer is that a sequence of three nucleotides, called a **codon,** carries the information that specifies one (and only one) amino acid. That amino acid is inserted at a specific location in a protein. Because the order of nucleotides in mRNA is determined by the sequence in DNA, this means that the order of bases in DNA ultimately determines the order of amino acids in the protein a gene encodes.

Not all of the bases in DNA contain information for proteins. In addition to the information that encodes an amino acid sequence, DNA also carries information that controls when and how much of a specific protein is made. In humans and other organisms whose cells have a nucleus, almost all the cell's DNA is found in the nucleus (some is in the mitochondria), and almost all of the proteins are manufactured by ribosomes in the cytoplasm (see "Biology Basics: Cells and Cell Structure").

This means that the information encoded in DNA must pass through the membrane surrounding the nucleus and into the cytoplasm. Then it must move to the ribosomes (organelles that are the site of protein synthesis). The chromosomes remain in the nucleus, but an intermediate, called **messenger RNA** or **mRNA,** carries genetic information from the nucleus to the cytoplasm.

As shown on the next page, the process of information transfer from a gene (a DNA sequence) to a protein occurs in several steps:

1. DNA $\rightarrow$ mRNA (transcription)
2. mRNA $\rightarrow$ protein (translation)

5.1 Essentials

- Each gene encodes a gene product called a protein.
- Proteins give us the phenotype that makes us unique.
- Proteins are composed of one or more chains of amino acids.

5.2 What Happens in Transcription, the First Step of Information Transfer?

In step 1, called **transcription,** information encoded in the DNA sequence of a gene is copied into the sequence of bases in a messenger RNA (mRNA) molecule. Messenger RNA is similar to DNA (see "Biology Basics: DNA"), but its function is different.

Follow along with the figure on the next page while reading the following text:

Transcription begins when the DNA in a chromosome at the location of a gene unwinds and one strand is used as a template to make an mRNA molecule. Then:

1. **RNA polymerase** (an enzyme) binds to a specific nucleotide sequence (called a **promoter**) at the beginning of a gene.
2. Using the DNA sequence of the strand being copied as a template, RNA polymerase moves along the template, linking RNA nucleotides together, forming an mRNA molecule.
3. The mRNA molecules produced by transcription are not exact copies of the gene's nucleotide sequence, but because of the rules of base pairing in nucleic acid molecules, they are called *complementary copies.* The rules of base pairing in DNA transcription are the same as in DNA replication, with one exception: an A on the DNA template ends up as a U in the mRNA. For example, if the nucleotide sequence in the DNA strand being copied is C G G A T C A T, the mRNA will have the complementary sequence G C C U A G U A.
4. As the RNA polymerase moves along the DNA template making mRNA, it eventually reaches the end of the gene, marked by a specific nucleotide sequence (called a **termination sequence**). When the RNA polymerase reaches the end of the gene (termination), the polymerase is removed from the DNA strand, the mRNA molecule is released, and the separated DNA strand reforms as a double helix.

After it is made, mRNA is processed to remove sequences that do not code for amino acids in the protein product.

Once processed, the mRNA moves through pores in the nuclear membrane into the cytoplasm, where the protein

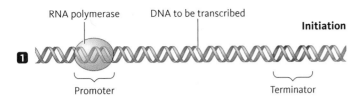

RNA polymerase
DNA to be transcribed
Initiation

1

Promoter
Terminator

Elongation

2

mRNA transcript

Termination

3

mRNA

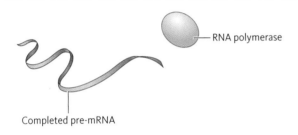

RNA polymerase

Completed pre-mRNA

will be assembled based on the information contained in the mRNA. This second step is called **translation.**

5.2 Essentials

- The DNA sequence of a gene is copied into the nucleotide sequence of messenger RNA (mRNA).

5.3 What Happens in Translation, the Second Step of Information Transfer?

Once in the cytoplasm, the nucleotide sequence in the mRNA molecule is translated into the order of amino acids in the protein. Each codon in the mRNA codes for one amino acid in the protein. However, mRNA does not directly assemble amino acids into proteins.

For this, another nucleic acid molecule, called **transfer RNA (tRNA),** is needed.

Transfer RNA (tRNA) molecules, shown in the next column, perform two important tasks in translation: (1) they recognize and bind to one specific amino acid, and (2) they recognize the mRNA codon for that amino acid. At one end, each tRNA molecule has a region that binds a specific amino

acid, and at the other end, each has a set of three nucleotides (called an **anticodon**) that can pair with the mRNA codon for that amino acid.

During protein synthesis, amino acids specified by the mRNA codons are linked together on ribosomes by the formation of chemical bonds.

Follow along with the figure below while reading about the process of translation.

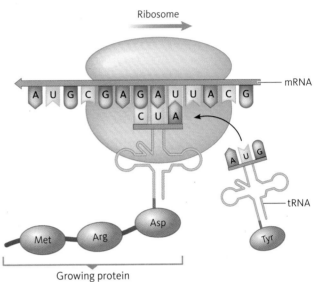

Ribosome

mRNA

tRNA

Met   Arg   Asp   Tyr

Growing protein

1. An mRNA molecule is loaded onto ribosomes. Recall that ribosomes are cellular organelles that can float free in the cytoplasm or are attached to the outer membrane of the endoplasmic reticulum (ER) (see "Biology Basics: Cells and Cell Structure"). Protein synthesis occurs on the ribosomes. Translation begins at a specific end of the mRNA molecule. The first codon is always AUG and is called the **start codon** (Table 5.2). It encodes an amino acid called *methionine*, which is the first amino acid in all human proteins.

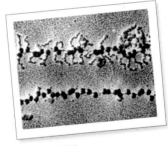

PHOTO GALLERY
DNA translation. © Dr. Kiseleva and Dr. Donald Fawcett/ Visuals Unlimited

2. When the second amino acid is in position, an enzyme within the ribosome forms a chemical bond (called a **peptide bond**) between the two amino acids. After this bond is formed, the tRNA for the first amino acid is released and moves out of the ribosome.

3. Next, the ribosome moves down the mRNA to the next codon, and the process repeats itself, adding amino acids to the growing protein.

4. Protein synthesis continues until the ribosome reaches a **stop codon.** Stop codons (UAA, UAG, and UGA) do not code for amino acids. When the stop codon is reached, the ribosome detaches from the mRNA, the completed amino acid chain is freed, and the amino acid chain folds into a three-dimensional structure we call a protein.

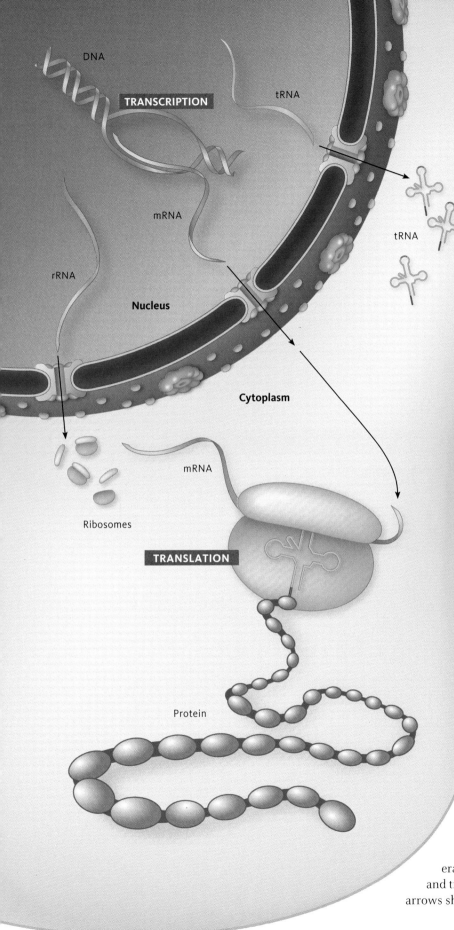

DNA

TRANSCRIPTION

tRNA

mRNA

Nucleus

rRNA

tRNA

tRNA

Cytoplasm

Ribosomes

mRNA

TRANSLATION

Protein

• In the cytoplasm, each amino acid specified by an mRNA codon is linked to other amino acids to make a protein.

5.4 Does a Cell Have Ways of Turning Genes On and Off?

Each of our chromosomes carries hundreds of the 20,000 genes that make us what we are, but not all those genes are active in every cell. In fact, almost all genes are switched off most of the time. Liver cells, for example, do not express the genes for eye color and cells of the brain do not make proteins that digest food. In any given cell, only about 5%–10% of the genes are active. The turning on and off of genes is called **gene regulation.**

One of the most important ways of regulating gene expression is by controlling access to the promoter region at the beginning of a gene. Recall that in a chromosome, DNA is wound around histone molecules. If the DNA is wound tightly around the histones, any genes in that region cannot be expressed. If the histones are modified, or the DNA is partly unwound, the gene becomes accessible and can be expressed.

Cells receive and process outside signals including hormones, nutrients, and other chemicals from their environment. Often, cells respond to these signals by changing the number of active genes and the type and number of proteins produced. Several types of DNA sequences that are not genes are involved in regulating gene expression. As you recall from reading the section on transcription, promoter sequences are DNA regions that allow proteins such as RNA polymerase to bind and switch on genes. In addition, regulatory proteins can attach to other sequences, called **enhancers,** and cause an increase in protein production.

Control over gene expression occurs at several stages before, during, and after transcription and translation as shown in the figure on page 90. The arrows show the stages where these events can occur.

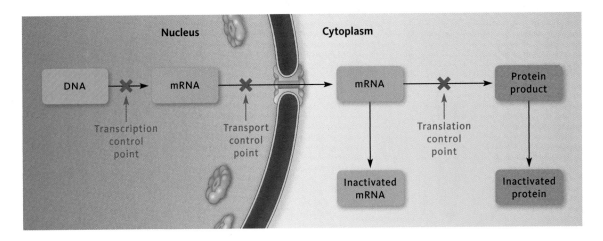

Nucleus   Cytoplasm

DNA → ✖ → mRNA → ✖ ⟶ mRNA → ✖ → Protein product

Transcription control point

Transport control point

Translation control point

Inactivated mRNA

Inactivated protein

5.4 Essentials
- Promoter sequences turn genes on or off.

5.5 What Is a Mutation and How Does a Gene Mutate?

Genes can be altered by a process called **mutation.** Mutations are changes in a DNA sequence. They can occur in any cell of the body, including those that give rise to sperm and eggs. Mutated genes produce either an abnormal, nonfunctional protein; a partially functional protein; or, in some cases, no protein at all. Other mutations can affect the timing and level of gene expression, while some may result in no change, and are essentially invisible.

Mutations occur in cells as a result of mistakes made during DNA replication and as a by-product of normal

Table 5.2 The Genetic Code for Amino Acids

The messenger RNA codon for each amino acid.

Amino Acid	Abbreviation	mRNA codons
Alanine	Ala	GCA GCC GCG GCU
Arginine	Arg	AGA AGG CGA CGC CGG CGU
Asparagine	Asn	AAC AAU
Aspartic acid	Asp	GAC GAU
Cysteine	Cys	UGC UGU
Glutamic acid	Glu	GAA GAG
Glutamine	Gln	CAA CAG
Glycine	Gly	GGA GGC GGG GGU
Histidine	His	CAC CAU
Isoleucine	Ile	AUA AUC AUU
Leucine	Leu	CUA CUC CUG CUU UUA UUG
Lysine	Lys	AAA AAG
Methionine*	Met	AUG
Phenylalanine	Phe	UUC UUU
Proline	Pro	CCA CCC CCG CCU
Serine	Ser	AGC AGU UCA UCC UCG UCU
Threonine	Thr	ACA ACC ACG ACU
Tryptophan	Trp	UGG
Tyrosine	Tyr	UAC UAU
Valine	Val	GUA GUC GUG GUU
Stop codons		UAA UAG UGA

Codon letters: A = adenine, C = cytosine, G = guanine, U = uracil.
*AUG signals "start" of translation when it occurs at the beginning of a gene.

cellular functions. Environmental agents, also called **mutagens,** including radiation and chemicals, can also cause mutations.

How do we know when a mutation has occurred?

If someone with a genetic disorder is born into a family with no history of that disorder, the question is whether the cause is genetic or nongenetic. For example, a woman who drinks alcohol while pregnant may have a child with facial defects, growth problems, and nervous system abnormalities. This condition could be fetal alcohol syndrome, a nongenetic disorder. However, these symptoms are similar to those seen in several genetic disorders, including Williams syndrome and Noonan syndrome. To determine whether a condition is caused by genetic or nongenetic factors, a family history should be taken, a pedigree constructed, and a genetic test done, if available.

If pedigree analysis shows that a genetic disorder has appeared in a family with no history of the condition, it is often assumed that a mutation has occurred. If, however, an autosomal recessive trait suddenly appears in a family, it is usually very difficult to identify when the mutation first occurred. Recessive alleles can be carried through several generations before becoming visible. Mutations leading to new alleles produce the genetic diversity we see in those around us. This includes disorders such as sickle cell anemia, the condition that Marcia Johnson is so worried about, as well as other traits that make up our phenotype.

BY THE NUMBERS

4 million
Number of base pairs in a bacterial cell.

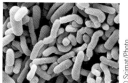
© Scimat/Photo Researchers, Inc.

3.2 billion
Number of base pairs of DNA in a human cell.

146
Number of amino acids in the beta globin molecule.

1
Number of mistakes in a gene that can cause sickle cell anemia.

As we will see later, this change created the mutant sickle cell anemia gene carried by Mark and Marcia Johnson.

How else can altered proteins cause a genetic disease?

During and after synthesis, a protein folds into a three-dimensional shape that is determined by its amino acid sequence. This three-dimensional shape, ultimately determined by the order of bases in DNA, can alter a protein's function (■ Figure 5.4).

Some mutations alter the three-dimensional shape of a protein, causing a genetic disorder. For example, cystic fibrosis and a genetic form of Alzheimer disease are associated with defects in folding. In these disorders, defective folding prevents the protein from functioning normally. In other disorders, such as mad cow disease and its human equivalent, Creutzfeldt-Jakob disease (CJD), defective protein folding can cause other proteins in the body to malfunction, producing an abnormal phenotype.

Which protein is altered in sickle cell anemia?

Remember from Chapter 4 that sickle cell anemia is inherited as an autosomal recessive trait. Affected individuals can have a wide range of symptoms, including weakness, abdominal pain, kidney failure, and heart failure (■ Figure 5.5), that can lead to early death if left untreated. This painful and disabling condition (■ Figure 5.6) is caused by a mutation in the gene that encodes **hemoglobin,** the oxygen-carrying protein in the blood.

5.5 Essentials

- Genes can be changed by mutations, which change the function of the proteins they produce.
- A gene mutation can cause one amino acid of a protein to be substituted for another amino acid.

5.6 How Does an Altered Gene Cause a Genetic Disorder?

Remember from Table 5.2 how similar the sequences of many codons are? A mutation in DNA that changes only one base pair in a gene can change a single amino acid in a protein, which in turn, may cause a genetic disorder. For example, Table 5.2 shows that one of the codons for glutamic acid is GAG. If a mutation in DNA changed the mRNA codon to GUG, the mutated codon would specify insertion of the amino acid valine into the protein in place of glutamic acid.

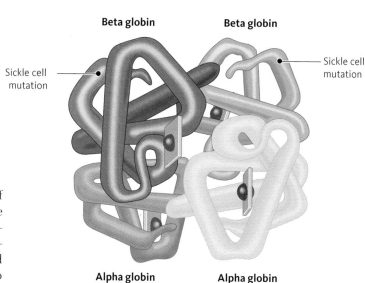

Beta globin   **Beta globin**

Sickle cell mutation

Sickle cell mutation

Alpha globin   **Alpha globin**

■ **Figure 5.4 Hemoglobin Molecule**
A three-dimensional representation of a hemoglobin molecule.

**A PERSON WITH TWO MUTATED GENES FOR THE
BETA CHAIN OF HEMOGLOBIN HAS SICKLE CELL ANEMIA**

Abnormal hemoglobin

Sickling of red blood cells

Rapid destruction of sickle cells

Clumping of cells and interference with blood circulation

Collection of sickle cells in the spleen

Anemia

Local failures in blood supply

Overactivity of bone marrow

Heart damage

Muscle and joint damage

Gastrointestinal tract damage

Increase in amount of bone marrow

Dilation of heart

Lung damage

Brain damage

Kidney damage

Weakness and fatigue

Poor physical development

Pneumonia

Paralysis

Kidney failure

Skull deformation

Impaired mental function

Heart failure

Rheumatism

Abdominal pain

Enlargement, then fibrosis of spleen

■ **Figure 5.5 Symptoms of Sickle Cell Anemia**
Sickle cell anemia begins with a mutation in DNA that results in abnormal hemoglobin. The effects of the mutation at the molecular, cellular, and organ levels appear in many different ways that can cause illness.

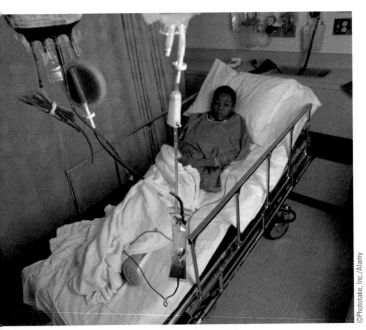

■ **Figure 5.6 A Treatment for Sickle Cell Anemia**
This child is being treated for sickle cell anemia using a blood transfusion.

In adults, each hemoglobin molecule (called HbA) is composed of two different proteins: **alpha globin** and **beta globin.** The mutation in sickle cell anemia affects the beta globin component of hemoglobin. The difference between normal beta globin and the beta globin in sickle cell anemia is a change in only one of the 146 amino acids in the protein.

After oxygen has been transported from the lungs and unloaded into cells in the body, hemoglobin molecules containing the mutant beta globin stick together, forming long tubular structures inside the cell.

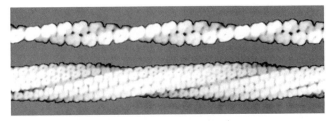

©Phototake, Inc./Alamy

The tubes (shown above) distort and harden the membrane of the red blood cell, twisting the cell into a sickle shape (■ Figure 5.7). The deformed blood cells easily break down and no longer function. Losing red blood cells reduces the oxygen-carrying capacity of the blood and results in anemia.

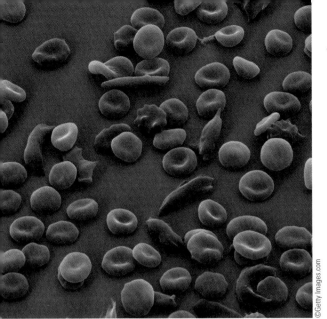

■ Figure 5.7 Red Blood Cells
Compare the normal red blood cells with the sickled one.

The sickled cells also clog capillaries and small blood vessels, producing pain and tissue damage.

What kind of mutation is responsible for sickle cell anemia?

The mRNA for the normal form of beta globin contains the codon GAG (for the amino acid glutamine) at amino acid number 6 (■ Figure 5.8). A single nucleotide change in DNA, a **point mutation,** changes the mRNA codon from GAG to GUG. Instead of glutamic acid at position 6, the mutant form of beta globin has valine at this position (■ Figure 5.9). This single amino acid substitution changes the way beta globin interacts with other hemoglobin molecules, causing them to clump together to form fibers.

To put it simply, all the symptoms of sickle cell anemia come from a change in *one* amino acid out of the 146 found in the beta globin gene. This is caused by a change in *one* base pair in the DNA of the beta globin gene.

Normal hemoglobin A		1	2	3	4	5	6	7	8
	DNA	CAC	GTG	GAC	TGA	GGA	CTC	CTC	TTC
	mRNA	GUG	CAC	CUG	ACU	CCU	GAG	GAG	AAG
	Amino acid	Val	His	Leu	Thr	Pro	Glu	Glu	Lys
Hemoglobin in sickle cell anemia		1	2	3	4	5	6	7	8
	DNA	CAC	GTG	GAC	TGA	GGA	CAC	CTC	TTC
	mRNA	GUG	CAC	CUG	ACU	CCU	GUG	GAG	AAG
	Amino acid	Val	His	Leu	Thr	Pro	Val	Glu	Lys

■ Figure 5.8 mRNA for Beta Globin
The DNA code, the mRNA codon, and the corresponding amino acid for a small section of the hemoglobin molecule. The top sequence is for normal hemoglobin; the bottom sequence illustrates sickle cell anemia.

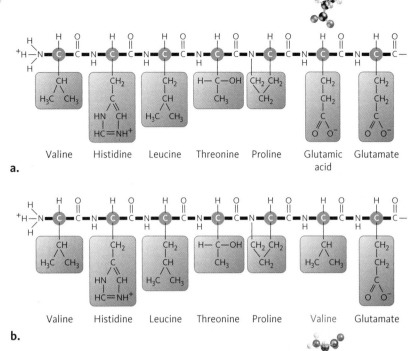

Valine   Histidine   Leucine   Threonine   Proline   Glutamic acid   Glutamate

a.

Valine   Histidine   Leucine   Threonine   Proline   Valine   Glutamate

b.

■ Figure 5.9 Amino Acid Chains for Hemoglobin Molecules
The top sequence shows part of the amino acid chain for normal hemoglobin. Below it is the same sequence for a person with sickle cell anemia.

5.6 Essentials

- **A protein with an incorrect amino acid sequence can produce disease symptoms, as in sickle cell anemia.**

What can be done to treat sickle cell anemia?

The genetic counselor should tell the Johnsons that treatment for sickle cell anemia is available. First, she will show them a drawing of the hemoglobin molecule similar to Figure 5.4. Then, she will explain that there is more than one gene that contributes to the formation of beta globin, the mutant protein in sickle cell anemia.

Everyone carries three slightly different versions of the gene for beta globin. Early in development, the embryonic version of the beta globin gene is active. This beta globin combines with alpha globin to form embryonic hemoglobin (HbE). When fetal development begins, the embryonic beta globin gene is switched off, and a fetal version of the beta globin gene is switched on, forming fetal hemoglobin (HbF). Just before birth, the fetal beta globin gene is switched off, and the adult beta globin gene becomes active, producing adult beta globin which unites with alpha globin to form adult hemoglobin (HbA). These changes in gene activity in the beta globin genes is called **gene switching.** The mutation causing sickle cell anemia is found only in the adult version of the beta globin gene, and is not present in the embryonic or fetal beta globin genes, even in those with sickle cell anemia.

CASE A QUESTIONS

Now that we understand how proteins are made and how mutations can result in genetic disorders (specifically sickle cell anemia), let's consider some of the questions raised in Marcia and Mark Johnson's case.

1. What should Marcia do before she calls her physician?

2. As we discussed in the case involving Ms. Lawrence and Dr. Gould (Chapter 3), we know that problems can arise with the doctor's role in cases involving genetic testing. Does the doctor have to tell Marcia she is a carrier (heterozy-gous) for sickle cell anemia? Why or why not?

3. Does the doctor have to tell Mark that he is a carrier?

4. Sickle cell anemia does not affect an entire chromosome. How will this impact Ms. Johnson's decision?

The Johnsons met with the genetic counselor, who revealed the results of the tests on the twin fetuses. Both twins, being identical girls, have sickle cell anemia. Now there are even more questions.

5. What are Mr. and Ms. Johnson's options?

6. Do some research on how sickle cell anemia affects individuals and their families. Should this information affect the Johnsons' decision? Why or why not?

7. Should all pregnant women have their fetuses tested for sickle cell anemia? Why or why not?

8. Should everyone be tested to determine whether he or she is a carrier for sickle cell anemia before marrying or having children? Why or why not?

Recently, in India and Saudi Arabia, it has been discovered that some people with sickle cell anemia also inherit a genetic condition that keeps the fetal beta globin (HbF) gene active after birth. These individuals have a very mild form of sickle cell anemia. This finding sparked the idea that if the fetal version of the beta globin gene could be switched on in sickle cell patients, it might be used as a treatment to decrease the symptoms of the disease.

The first breakthrough came from using a drug called *hydroxyurea* to treat leukemia and other blood disorders. In patients treated with hydroxyurea, the fetal beta globin gene is active, and produces HbF. This finding led to studies with hydroxyurea as a treatment for sickle cell anemia and eventually the U.S. Federal Drug Administration (FDA) approved it as the first drug to be used in treating sickle cell anemia.

Unfortunately, about one-third of those with sickle cell anemia do not respond to treatment with hydroxyurea, and in cases where the fetal beta globin gene is active, the drug only reduces, but does not eliminate the painful and life-threatening symptoms. Evidence shows that the higher the levels of HbF in a sickle cell patient, the more the symptoms are reduced. Researchers are now working to find drugs that raise the levels of HbF high enough to eliminate the clinical symptoms of this disorder.

5.7 How Are Other Single-Gene Defects Caused?

Another condition we discussed in Chapter 4 and earlier in this chapter is cystic fibrosis (CF). The most common mutation in the CF gene causes changes in the protein product after it is formed and begins to fold. This mistake in protein folding causes the protein to be destroyed before it can be inserted in the plasma membrane.

In Huntington disease (HD), also discussed in Chapter 4, the situation is different. The mutation adds nucleotides to the gene, called *trinucleotide repeats*. The repeat in Huntington disease is CAG. The more CAG repeats are present in the gene, the earlier the symptoms seem to appear. The gene and the protein (huntingtin) that cause HD have been isolated, but the exact function of the protein has not been determined.

5.7 Essentials

- In sickle cell anemia and other genetic disorders, a mistake in *one* base pair in the DNA can cause an abnormal protein and an abnormal phenotpye.

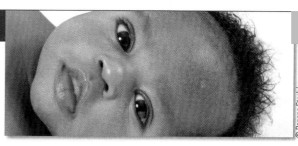

© Banana Stock/
Jupiter Images

Now the twins are 15 months old. Having known before the birth that the twins would have sickle cell anemia, the Johnsons and their doctors were able to plan specialized medical care from the moment they were born. The twins are healthy, but the Johnsons continue to monitor their condition carefully. They have made quite a few visits to the doctor and to the hospital for tests, treatment, and blood transfusions in the first 15 months. The Johnsons' insurance covered the most of the cost of the medical care, but it was still expensive.

Now Marcia thinks she may be pregnant again. She bought a home pregnancy test, and the result was positive—she *is* pregnant. Marcia and Mark are worried. What if this baby also has sickle cell anemia?

1. What is the first thing Marcia should do?

2. What would you advise her to do if you were her doctor?

3. What laws will protect Marcia's decision on what to do with her unborn fetus?

4. Go back to Chapter 3 and list what tests Marcia should have done.

5. What might happen if her doctor says, "Don't worry—everything will be all right"?

5.8 What Are the Legal and Ethical Issues?

In the Law and Ethics section of Chapter 4 we discuss the confidentiality issue and how the law addresses it. More than 10 years ago, a law was passed (Health Insurance Portability & Accountability Act of 1996 or HIPAA) that required not only that a patient's records be kept confidential, but that each patient be informed of this in writing.

The HIPAA law applies to Marcia Johnson's case as well. It is clear that she signed an informed consent form that allowed for the testing but did not address what might happen to these test results. The nurse who collected her blood was obviously collecting samples from many people, each of whom had to give their consent. The form that they signed was required by the HIPAA law to disclose how the results of the test were to be used.

The next step in our case was to test Marcia's husband, Mark. What were his privacy rights? A consent form prob-

ably wasn't necessary here because his doctor would do the testing, not the group doing the study. Private tests fall under the HIPAA rules.

But, an interesting problem might arise here. If Marcia didn't want to share her results with anyone, then the lab, the study, or the doctor could not release them. This would apply to Mark or anyone else. Privacy is an important part of our medical treatment because most doctors feel that in order for patients to trust them, they must keep their confidential information out of the hands of anyone, even the patient's relatives.

Employers might also want health and genetic information. If Marcia and Mark's carrier status were known by insurance companies and employers, they might be discriminated against. As we discussed in Chapter 1, this can be a serious problem and is against the law.

The chart on the next page addresses some of the ethical and legal questions regarding discrimination.

Question	How are these questions decided?	Related case or legal issue
If someone collects your blood for a genetic test, can they use the results in a study?	Information on how one's blood tests are used should be spelled out in any consent form they sign.	We can volunteer our information but even then HIPAA requires the results be kept anonymous.
Can the information used in a study be made available to anyone who wants it?	The FDA and drug companies are required to keep the results of large scale studies private. This includes individual test results.	Most law suits involving clinical trials are about side effects and deaths. The rules of such studies require that patient's information be kept private.
What if Marcia didn't want Mark to know her results?	Marcia's result is her private business. Doctors cannot give this information to anyone unless the patient is in danger.	In Chapter 4 we discussed the case of Tarasoff v. the Regents of CA. In this case, it was held that doctors should only share information if the third party is in imminent danger.

SPOTLIGHT ON LAW

Norman-Bloodsaw v. Lawrence Berkeley Laboratory (135 F.3d 1260, 1269 (9th Cir. 1998))

AND

EEOC v. Burlington N. Santa Fe Railway Co., No. 02-C-0456 (E.D. Wis. 2002)

In 2000, President Clinton signed Executive Order 13145 to prohibit discrimination in federal employment based on genetic information, and in 2005, the Genetic Information Non-discrimination Act was introduced that expanded this to all employers. It was signed into law on May 21, 2008, by President George W. Bush.

This action was a result of a number of cases in which genetic testing was used to discriminate, including Norman-Bloodsaw v. Lawrence Berkeley Laboratory (135 F.3d 1260, 1269 (9th Cir. 1998)) and, more recently, EEOC v. Burlington N. Santa Fe Railway Co., No. 02-C-0456 (E.D. Wis. 2002).

Lawrence Berkeley National Laboratory is the oldest national research laboratory in the United States. Operated by the University of California, it conducts research on human genetics, breast cancer, astrophysics, and nuclear science, among other subjects. For years, hundreds of employees gave blood samples as part of hiring and employment.

One day, Vertis Ellis, an administrative assistant, looked at her medical records, which were delivered to her office by mistake. Inside she found the results of syphilis tests taken without her knowledge in medical exams during her 29 years of employment. When she asked about these tests, laboratory officials acknowledged testing the blood and urine samples of its employees. They said the tests, for syphilis, the sickle cell trait, and pregnancy, were done under the guidance and approval of the U.S. Department of Energy. They claimed the testing had stopped. Neither Ellis nor any other employee had either given permission for such tests or received results.

In September 1995, Marya S. Norman-Bloodsaw, Ellis, and six other employees filed a class-action suit naming the laboratory, the secretary of the U.S. Department of Energy, and the regents of the University of California as defendants. The employees claimed that the testing occurred without their knowledge or consent and without any later notification that the tests had been conducted. The defense argued that the employees had consented to a complete medical exam, and the blood tests were part of that.

RESULTS:

The first court to hear the case, the district court in San Francisco, dismissed it, stating, "The three tests in question were administered as part of a comprehensive medical examination to which plaintiffs had consented."

Next, the employees appealed to the U.S. Court of Appeals for the Ninth Circuit in San Francisco. The court of appeals looked at three facts in the case:

1. Only black employees were tested for sickle cell anemia.
2. Only female employees were tested for pregnancy.
3. No safeguards were taken to keep the results of these tests private.

The Ninth Circuit Court of Appeals asked: What was the government's interest in taking these tests? It found that the government had no interest in the results of the tests because they had no effect on the job performance of the employees. Then the court found that because only women and black employees were given some of the tests (pregnancy and sickle cell anemia), this was discrimination based on sex and race.

Our second case is <u>EEOC v. Burlington N. Santa Fe Railway Co.</u>, No. 02-C-0456 (E.D. Wis. 2002).

Gary Avery, an employee of the Burlington Northern and Santa Fe Railway, became unable to work because of a severe case of carpal tunnel syndrome. In this condition, a wrist nerve is compressed, which causes numbness, tingling, and muscle weakness in the hand. After having surgery and returning to work, Gary was asked to supply seven vials of blood for tests. He was curious what these tests were and asked his wife, a nurse, what she thought. She wondered why they needed so much blood. In discussions with his friends at work, he discovered that other employees with carpal tunnel syndrome had also been asked for large amounts of blood.

When asked to come back for another test, Gary, worrying, did not go. The next day, he received notice that he would be fired if he failed to submit a blood sample. Gary feared for his job and filed a complaint with the Equal Employment Opportunity Commission (EEOC) and his union. The EEOC regulates the Americans with Disabilities Act (ADA), which prohibits employers from discriminating against people with disabilities.

When the EEOC began to investigate the blood tests, they found that the company was testing for a hereditary form of carpal tunnel syndrome called hereditary neuropathy with liability to pressure palsies (HNLPP). This inherited condition, whose gene is found on chromosome 17, is an autosomal dominant disorder that affects approximately 16 out of 100,000 people and puts those who are affected at a predisposition for carpal tunnel syndrome. Two weeks after finishing its investigation, the EEOC filed a lawsuit on behalf of four employees, including Gary, to end genetic testing on employees.

The EEOC argued that the company based employment decisions on the genetic testing, and therefore violated the ADA. Employers may require employees to submit to a medical examination only if the examination is job related and consistent with business necessity. Any test that may predict future disabilities is unlikely to be relevant to the employee's present ability to perform his or her job.

Burlington argued that the incidence of carpal tunnel syndrome among its employees had recently increased. The company was testing employees to find out whether this increase was due to a genetic condition. It claimed that it was not using the information to screen potential employees, but to decide whether to allow employees to receive worker's compensation. If a worker had the genetic condition, it was preexisting and the company would not have to pay worker's compensation because the carpal tunnel syndrome was not work related.

RESULTS:

The case was eventually settled without a trial. In the settlement, Burlington agreed to do the following:

1. Immediately stop the genetic testing
2. Pay the employees' legal costs and offer them an apology
3. Turn over all test results to the employees
4. Promise to keep all information obtained strictly confidential

QUESTIONS:

1. List three ways in which these two cases are similar.
2. In Burlington, the company seemed to be doing something that might help its employees. Do you think this is the case? Why or why not?
3. In Norman-Bloodsaw, the test results were done at the time of employment and after hiring. Does this make a difference? Why or why not?
4. If the law required everyone to have a genetic test before they could apply for a marriage license, would that be discrimination? If so, which of these two cases could attorneys use to argue their point? Why?
5. Compare the two cases in the following table:

Case	State or Federal	Genetic Condition	Main Issue
Burlington			
Norman-Bloodsaw			

6. Do you agree with both case results? Why or why not?
7. How would you change these cases if you could?

✪ THE ESSENTIAL TEN

1. Each gene encodes a gene product called a protein. [Section 5.1]
   ▮ There are thousands of types of proteins.

2. Proteins give us the phenotype that makes us unique. [Section 5.1]
   ▮ Each protein has a specific function within the body.

3. Proteins are composed of one or more chains of amino acids. [Section 5.1]
   ▮ There can be hundreds of amino acids in a protein.

4. The DNA sequence of a gene is copied into the nucleotide sequence of messenger RNA (mRNA). [Section 5.2]
   ▮ Messenger RNA is similar to DNA.

5. In the cytoplasm, each amino acid specified by an mRNA codon is linked to other amino acids to make a protein. [Section 5.3]
   ▮ The amino acids are assembled in the correct order so the protein can work correctly.

6. Promoter sequences turn genes on or off. [Section 5.4]
   ▮ Proteins can also bind to a cell's membranes and turn genes on and off.

7. Genes can be changed by mutations, which change the function of the proteins they produce. [Section 5.5]
   ▮ The changes in these proteins can cause disease.

8. A gene mutation can cause one amino acid of a protein to be substituted for another amino acid. [Section 5.5]
   ▮ If this occurs, then the protein does not work correctly.

9. The incorrect protein can produce disease symptoms, as in sickle cell anemia. [Section 5.6]
   ▮ Most genetic conditions are caused by something as small as one change in the amino acid chain.

10. In sickle cell anemia and other genetic disorders, a mistake in *one* base pair in the DNA can cause an abnormal protein and an abnormal phenotype. [Section 5.7]
    ▮ This can cause many changes in the body of a person with these conditions.

KEY TERMS

alleles (p. 85)
alpha globin (p. 92)
amino acid (p. 86)
anticodon (p. 88)
beta globin (p. 92)
codon (p. 87)
essential amino acids (p. 86)
enhancer (p. 89)
gene regulation (p. 89)
gene switching (p. 93)
hemoglobin (p. 91)
messenger RNA (mRNA) (p. 87)
mutagens (p. 90)
mutation (p. 90)
peptide bond (p. 88)
phenotype (p. 85)
point mutation (p. 93)
promoter (p. 88)
protein (p. 85)
RNA polymerase (p. 88)
start codon (p. 88)
stop codon (p. 88)
termination sequence (p. 88)
transcription (p. 87)
transfer RNA (tRNA) (p. 88)
translation (p. 88)

REVIEW QUESTIONS

1. Using the codon/amino acid list, make up a protein by listing 10 codons and their respective amino acids.
2. What is the difference between transcription and translation?
3. Why does the major part of protein formation occur in the endoplasmic reticulum?
4. Why are mutations important?
5. In many single-gene defects, a change in *one* base pair can change the outcome. Explain how this happens. List the steps.
6. In sickle cell anemia, where are the mutations in the hemoglobin gene?
7. What are repeats and how do they cause disease?
8. What can cause mutations?
9. If a mutation happens in your grandmother, does it pass down to you? Explain your answer.
10. Why couldn't we live on the same diet as a cow?

What Would You Do If . . . ?

1. You were a member of a state legislature that was voting on testing all married couples for the sickle cell gene?

2. You were a member of a state legislature that was voting on testing all couples applying for a marriage license for the sickle cell gene?

3. You were Marcia and Mark Johnson? (See Case A and Case B.)

4. You were the genetic counselor working with the Johnsons during Marcia's second pregnancy? (See Case B.)

5. You were asked to find a way to prevent the Johnsons from having another child with sickle cell anemia before Marcia became pregnant again? (See Case A.)

APPLICATION QUESTIONS

1. We have discussed sickle cell anemia in this chapter. Do you think that Marcia and Mark Johnson's decision would have been more difficult if their children had a different condition? Which ones?

2. When offered a choice of having a test to find out about their fetuses, do you think that most people would accept? Why or why not?

3. Research a few stories of couples who have raised children with sickle cell anemia. Summarize their stories.

4. Has the treatment for sickle cell anemia changed over the last 10 years?

5. A person who is a carrier for sickle cell anemia is said to have *sickle cell trait*. Research what this is and summarize it. Find three other conditions where the gene and its products have been identified. Fill in the chart below:

Look back at Table 5.2. You can see that some of the mRNA codons are used in many different proteins. Estimate how many possible combinations of codons you would need to make a protein of amino acids that do not duplicate.

6. In Huntington disease, the more repeats a person has, the earlier the onset of symptoms. Based on what you know about the science of repeats, why do you think this happens?

7. If hydroxyurea is used as a treatment for sickle cell anemia, would this stop babies from being born with the condition? Why or why not?

Name of condition	What chromosome is the gene found on?	What is the protein?	When was it discovered?

ONLINE RESOURCES

Preparing for an exam? Log on at academic.cengage.com/login for a pre-test, a personalized learning plan, and a post-test in CengageNOW's Study Tools to help you assess your understanding.

If assigned by your instructor, the Case A and Spotlight on Law activities for this chapter, "Marcia Johnson" and "Discrimination Based on Genetic Information," will be available in CengageNOW's Assignments.

Toronto Star News, May 28, 2007

Toronto Playwright Uses Her Own Experience to Enlighten and Entertain

Nicholas Davis

Christine Nicole Harris was an active child. She did track and field, piano, theatre arts, and ballet. But, as she grew, things changed.

When she got older, she had to stop most of her physical activity. As a teen she cut back a lot. This was because of a disease she was diagnosed with as a baby. Her parents found out she had sickle cell anemia shortly after she was born.

When they came here in the late 1970s from Jamaica, they had never heard of sickle cell anemia. And after they had her, they decided to not have any more children.

To access this article online, go to academic.cengage.com/biology/yashon/hgs1.

QUESTIONS:
1. Why do you think Christine's parents decided not to have any more children?
2. Christine is considered a role model for other people with sickle cell anemia. Do you think she has an obligation to share her story?

San Francisco Examiner, November 5, 2007

Hospitals Will Begin Banking Blood from Newborns

John Upton

California hospitals will start banking blood from umbilical cords and placentas, with permission, to help cure blood diseases and other illnesses and as a source of stem cells.

This will create a California government-managed bank of donated cord blood. This will be used to find donor cord blood to match with patients who have any of roughly 70 diseases including sickle cell anemia and immune deficiencies.

To access this article online, go to academic.cengage.com/biology/yashon/hgs1.

QUESTIONS:
1. Do you think that some parents might not give the permission to take cord blood from their babies? Why?
2. If a law were passed that said that all babies' cord blood had to be taken at birth, do you think people would be for it or against it? Why?

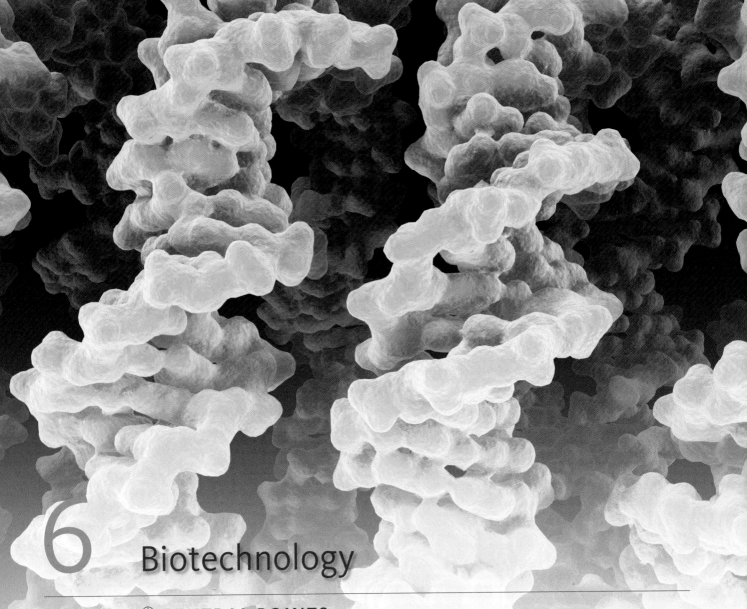

6 Biotechnology

★ CENTRAL POINTS

- Recombinant DNA technology can join DNA from different organisms.

- Biotechnology uses recombinant DNA to make products.

- Bacteria, plants, and animals can be modified using biotechnology.

- The safety of transgenic organisms continues to be debated.

- Human proteins can be produced by biotechnology for disease treatment.

- Many biotechnology inventions have been patented.

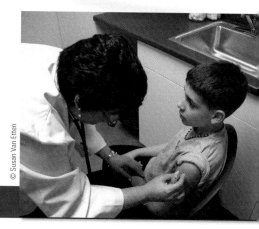
© Susan Van Etten

CASE A A Taller Son for Chris

Chris Crowley is concerned about his 10-year-old son, Mike, who is short—not just short for his age, but very short. Just today Mike came home from school crying. Some kids had made fun of him and pushed him around. This was not the first time.

Chris had gone through this himself. At his final height he was only 5′ 6″ (1.6 m), fairly short for a man. During his teen years, his friends had made fun of him and he didn't want Mike to go through that.

Chris had recently read an article about human growth hormone (hGH), a drug that might help Mike. hGH is safer and more available because now it is made using re-combinant DNA technology. Children who were given injections of the drug before puberty sometimes grew 5–6 in. (12.7–15.2 cm). Athletes also used it to increase their strength for certain sports. Chris thought that if Mike could grow a few inches and be stronger, it would give him confidence and possibly a better life.

When Chris brought this up with Mike's pediatrician, Dr. Sanchez, he was surprised when she said that Mike wasn't a candidate for hGH treatment. The reason for this was that Mike was still within the "normal" range of height for his age, and the hormone is used to treat only the shortest 1–2% of all children. She explained that even though Mike was shorter than his friends, he might still grow to be in the normal range of adult height. Moreover, the side effects of this drug are still unknown. Chris left the doctor's office feeling somewhat confused and let down.

Some questions come to mind when reading this case. Before we can address these questions, let's look at what biotechnology and recombinant DNA are, how they are used to create drugs, and their other uses.

6.1 What Is Biotechnology?

Biotechnology is the coupling of genetic technology to biological systems to create products and services. Biotechnology has many uses, one of which is making human proteins for medical uses, including human growth hormone (hGH).

Before the development of biotechnology, proteins used to treat disease—such as insulin, hGH, and blood-clotting factors—were collected from many sources, including animals in slaughterhouses, human cadavers, and donated human blood. In some cases, proteins from these sources exposed people to serious and potentially fatal risks.

Until 1985, children with growth problems were treated with growth hormone recovered from pituitary glands removed from human cadavers. This limited the available supply, and only a small number of children could be treated. In addition, more than two dozen of the 7,000 children treated contracted a deadly brain condition because the hGH was contaminated with a disease-causing agent similar to the one that causes mad cow disease.

In 1985, hGH was produced successfully using biotechnology. The combination of technology with genetics produces potentially unlimited amounts of growth hormone with no possibility of contamination with disease-causing agents. In fact, because hGH is now available in a safe form, the U.S. Food and Drug Administration (FDA) has expanded the number of conditions that can be treated with this hormone; it can now be used to treat children with short stature (such as Chris Crowley's son) and not just those with serious growth disorders.

How is biotechnology used to make human growth hormone?

Using a method called **recombinant DNA technology,** scientists transferred the gene for hGH from a human cell to a bacterial cell, creating a **transgenic organism** (an organism that carries a gene from another species). The transgenic bacterial cell and its descendants manufacture hGH, which is recovered and purified for medical use.

Recently, a group of researchers successfully transferred the gene for hGH into the DNA of a cow, which then secreted the hGH in its milk. The hormone can then be extracted from the cow's milk. It is estimated that a herd of 15 transgenic cows would be able to meet the worldwide demand for hGH. Genes for other medically important human proteins are being transferred into animals or plants, where they are synthesized in large quantities and extracted for use in treating diseases.

6.1 Essentials

- Recombinant DNA techniques can be used to transfer genes among species.

6.2 What Is Recombinant DNA Technology?

Recombinant DNA can be defined as the combination of DNA from two or more different organisms. Several steps are required to create human proteins such as hGH in trans-

genic organisms; these steps are part of recombinant DNA technology. To see how this process works, follow the steps in the drawing below.

- DNA is extracted from human cells.
- The human DNA is treated with a protein called a **restriction enzyme,** which cuts the DNA into fragments at specific sites (1). Often these cuts leave an overhanging single-stranded region called a "sticky end."
- A circular DNA molecule (called a **plasmid**) is also cut with the same restriction enzyme (2). Many species of bacteria carry these small, circular DNA molecules. Researchers have modified plasmids to create carrier molecules called *vectors* that are used in recombinant DNA technology. This plasmid DNA will carry inserted human DNA fragments into bacterial cells.
- The fragments of human DNA and fragments of plasmid DNA (called **vectors**) are mixed together and join to form a **recombinant DNA molecule** (3–4).
- Plasmids can move across the cell wall and enter the bacterial cells. Once inside, they copy themselves. If a plasmid vector contains an inserted human gene, this recombinant DNA molecule will be carried into the bacterial cell (5).
- Plasmids with the inserted human gene copy themselves inside the host cell; in some cases, the bacterial cells then express the human gene and synthesize the human protein, which can be recovered and used for medical treatments, vaccines, or other purposes.

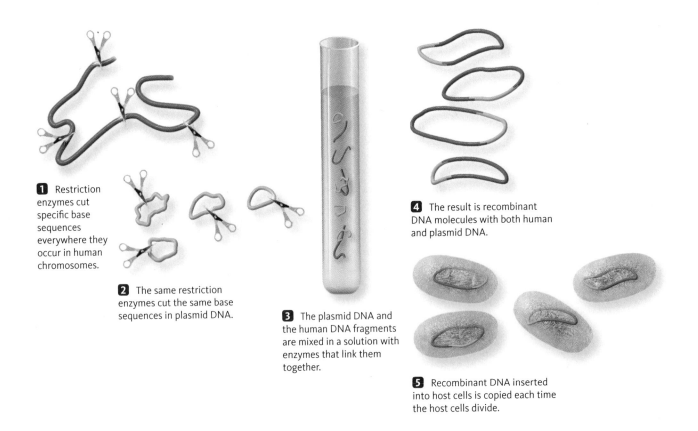

1 Restriction enzymes cut specific base sequences everywhere they occur in human chromosomes.

2 The same restriction enzymes cut the same base sequences in plasmid DNA.

3 The plasmid DNA and the human DNA fragments are mixed in a solution with enzymes that link them together.

4 The result is recombinant DNA molecules with both human and plasmid DNA.

5 Recombinant DNA inserted into host cells is copied each time the host cells divide.

How do restriction enzymes and vectors work?

DNA and the genes it contains can be cut into specific fragments using restriction enzymes. These proteins, isolated from certain strains of bacteria, attach to DNA and move along the molecule until they find a specific base sequence called a *recognition site*. Once at the site, the restriction enzyme cuts through both strands of the DNA. Each restriction enzyme has specific and different recognition sites. More than 1000 different restriction enzymes have been identified, and each has its own recognition and cutting site (Table 6.1).

To produce human growth hormone, DNA from human cells is isolated, cut with a restriction enzyme, combined with a vector, and inserted into a bacterial cell. The bacteria carrying the hGH gene are identified, isolated, and used to synthesize hGH. This process is used to produce many medically important proteins, including insulin, clotting factors, in addition to hGH.

6.2 Essentials

- Recombinant DNA is created when DNA from two or more different organisms are joined together.
- Restriction enzymes bind to DNA and recognize specific DNA sequences.
- Restriction enzymes cut DNA at the recognition sequences into fragments.
- Biotechnology is the commercial use of recombinant DNA techniques to produce drugs and medical treatments.

6.3 How Is Biotechnology Used to Make Other Transgenic Plants and Animals?

In addition to making medically important proteins such as hGH, biotechnology is used in many other ways. One of these is making **transgenic crop** plants with new characteristics (■ Figure 6.1). Transgenic crops have been grown in the United States since 1996. Genes transferred to the crop plant can originate in another plant, an animal, or even fungi or bacteria. One or more new genes are used to give the transgenic plant a unique trait, such as resistance to **herbicides** (chemical weed killers), insects, or viral or fungal diseases. Transgenic plants resistant to herbicides are now available worldwide, and are being used in almost two dozen countries. Transferred genes can also increase the nutritional value of crops, helping eliminate dietary deficiencies that are widespread in many parts of the world.

Table 6.1 Restriction Enzymes

Three restriction enzymes, their cutting sites, and the organisms they come from.

Enzyme	Cutting Sites	Source Organism
EcoRI	GAATTC / CTTAAG	Escherichia coli
HindIII	AAGCTT / TTCGAA	Haemophilus influenzae
BamHI	GGATCC / CCTAGG	Bacillus amyloliquefaciens

CASE A QUESTIONS

Now that we understand how recombinant DNA technology works, let's look at some questions related to Mike's short stature.

1. After the doctor's visit, Chris decided he must deal with Mike's short stature without the use of hGH. Suggest four ways that Chris could deal with Mike's short stature.

2. What do you think Chris should do?

3. Did Dr. Sanchez do the right thing in denying Mike the hGH treatment?

4. Should parents be able to make all medical decisions for their children?

5. If Dr. Sanchez administered hGH to Mike, what might happen?

6. Should Mike be included in the decision whether or not to undergo hGH treatment?

7. Human growth hormone is legal in the United States for treating those with low hGH levels and children who are considered too short. If hGH has other potential uses, should it be used for those purposes? Why or why not?

8. Athletic associations have discussed whether hGH use by athletes might give them an advantage over others, and therefore may outlaw it. What do you think?

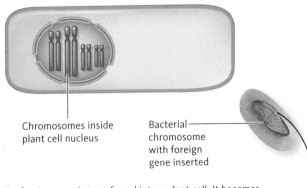

Chromosomes inside
plant cell nucleus

Bacterial
chromosome
with foreign
gene inserted

1 The foreign gene is transferred into a plant cell. It becomes incorporated into one of the plant's chromosomes.

2 The plant cell divides to form an embryo that develops into a mature transgenic plant as shown below.

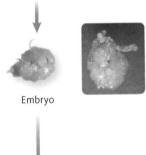

Embryo

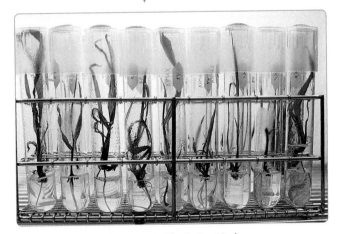

■ **Figure 6.1 How Transgenic Plants Are Made**
A gene is incorporated into a bacterial plasmid and used to transfer a genetic trait into a plant cell. These plants are called genetically modified (GM) organisms. Two traits transferred into crops this way are resistance to insects and herbicides.

The graph at the top of the next column shows the dramatic increase in the use of transgenic crops between 1996 and 2006 in the United States. Products made from corn and soybeans, as well as cottonseed and canola oils, currently account for almost all foods that contain transgenic ingredients.

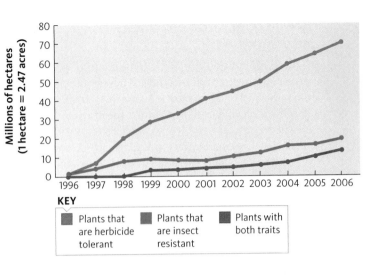

KEY

Plants that are herbicide tolerant	Plants that are insect resistant	Plants with both traits

Some other examples of transgenic plants and animals are shown in the chart on the facing page.

6.3 Essentials

- **Transgenic bacteria, plants, and animals are created with recombinant DNA techniques.**

6.4 Are Transgenic Organisms Safe?

The development and use of transgenic crops has generated controversy in many regions, including the United States, Europe, and developing African nations. Many questions have been raised about the safety and environmental impact of transgenic crops, often called *genetically modified (GM) crops*. Several organizations and movements actively oppose the introduction and use of transgenic crops. Controversies about transgenic crops focus on safety, labeling, and the rights of farmers, among others. As transgenic crops become more common and new ones are developed, we clearly need to consider the health and environmental risks, and resolve the related economic and social issues.

What do we know so far about these risks of transgenic crops?

Most food produced from transgenic crops contains one or more proteins encoded by a transferred gene. Standards for evaluating the safety of foods from transgenic crops are in place in many European countries and are being developed and implemented in the United States. In general, transgenic crops are considered safe if the proteins produced by the transgenes are not poisonous and do not cause allergies any more than plants crossbred in the normal breeding method.

PHOTO GALLERY
The genetically modified Flavr Savr tomato was created so it could stay on the vine longer and ripen. It tasted fresher and could travel long distances without softening. Consumer demand for this tomato was low (some thought it still tasted like the usual store-bought tomato), and the product was withdrawn.
© Tom Myers/Photo Researchers, Inc.

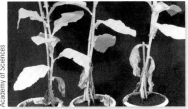

Photo courtesy of Professor Qing-Hu Ma, Institute of Botany, Chinese Academy of Sciences

Transgenic tobacco plants are also being used to produce hGH. Most work with transgenic plants to make human proteins is still in early stages of development, but soon fields of transgenic plants or small herds of transgenic animals may replace laboratories for making human proteins used in medical treatments.

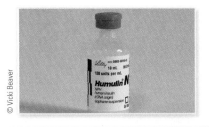

© Vicki Beaver

One of the most successful uses of recombinant DNA technology is the synthesis of human insulin. Many people who have diabetes must take the protein insulin because their bodies do not produce it. Before insulin was made using recombinant DNA, it was extracted from the pancreas of pigs and cows. Some people with diabetes had allergic reactions to pig and cow insulin. Now that recombinant human insulin is available, such reactions are rare.

© Golden Rice Humanitarian Board

Golden rice is a transgenic strain of rice that contains two genes from daffodils and one gene from bacteria. With these added genes, the rice plants synthesize beta-carotene, a compound that turns the rice a golden color. When the rice is eaten, beta-carotene is converted into vitamin A. In many Asian countries there is a dietary deficiency of vitamin A. Golden rice strains are now available for planting in Southeast Asia.

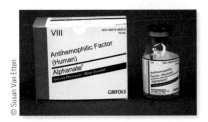

© Susan Van Etten

Factor VIII is a protein that helps form blood clots. People with hemophilia need Factor VIII because it is not present in their blood. Recombinant DNA techniques have been used to create human Factor VIII, so people with hemophilia do not have to use blood donors to get this needed protein. Back in the 1980s, many people with hemophilia contracted HIV from tainted blood products.

© Workbook Stock/ Jupiter Images

The human tissue-typing gene complex called HLA has been transferred to pig embryos. The adult pigs that result have organs compatible with humans. Transplanting these organs (called *xenotransplantation*) to needy recipients can increase the number of organs available for transplantation.

For example, consider herbicide-resistant crops. The bacterial protein encoded by the transgene is broken down in our digestive system when the plant is eaten. In studies with mice, the protein was not poisonous even in very high doses. This protein does not contain any amino acid sequences that are similar to any known toxins or allergens. After 10 years of widespread use, no human health risks have been identified.

However, several environmental risks related to transgenic crops have been identified. These include the transfer of transgenes to wild plants by crossbreeding. This may ultimately reduce biodiversity, with a possible negative impact on the environment.

This problem and others related to transgenic crops are the same as those we face using conventional crops. Even though transgenic crops have a different genetic history from that of conventional crops, their effects on the environment are very similar.

As gene transfer technology becomes more sophisticated, novel combinations of traits may be developed, requiring specific management plans. For example, if production

PHOTO GALLERY
A cotton plant that was eaten by insects and a transgenic plant that is insect resistant.
© Agricultural Research Service/USDA.

of pharmaceutical products in plants becomes common, it will be necessary to develop safeguards to keep such plants out of the food supply and to ensure that transgenes are not transferred to wild plants.

Humans have been genetically modifying agricultural plants and animals for more than 10,000 years using crossbreeding to produce the diversity of domesticated plants and animals we depend on for our food. Biotechnology has only changed the way and the rate at which these changes are made. It has also expanded the range of species that can donate genes. To continue improving plants and animals by both conventional breeding and by biotechnology, scientific questions, economic issues, health safeguards, and public perceptions must be addressed by research, testing, and public education.

6.4 Essentials

- Safeguards are needed in developing and using transgenic organisms to produce food and drugs.

6.5 Can Transgenic Organisms Be Used to Study Human Diseases?

Information gained from the **Human Genome Project** (discussed in Chapter 9), which has sequenced all the human genes, combined with that from plant and other animal genomes shows that many genes found in the human genome are also present in other species, including the mouse and even the fruit fly, *Drosophila.* Because of these similarities, a growing set of animals is being used to explore mechanisms of human disease and to test drugs developed to treat these disorders. These **animal models** of human disease are important in research. Without an animal model, it is often difficult to study new drug treatments and the causes of a disease. Models of specific human diseases can be created by transferring human disease genes into animals. These transgenic animal models can do all of the following:

PHOTO GALLERY
This mouse was genetically engineered to be able to eat and not gain weight.
© Courtesy Salih Wakil, Ph.D., Baylor College of Medicine

- produce an animal with symptoms that mirror those in humans
- be used to study the development and progress of a disease
- be used to develop and test drugs that hopefully will cure or treat the animal model of the human disease

Eventually, this information and the drugs developed using animal models will be used to treat human diseases.

Because of the close genetic similarity between mice and humans, mice have been used extensively as animal models of human diseases. Here are some examples:

GENETICALLY ENGINEERED MICE

The *rhino mouse* is used to study immune deficiency conditions.

The *curly tail mouse* is used to study neural tube defects such as spina bifida.

The *obese mouse* is used to study products that can help with weight loss.

A transgenic mouse model of Huntington disease (HD) is used to study this fatal genetic disorder. (Recall from Chapter 4 that members of Alan's family had this neurodegenerative disorder, which is inherited as an autosomal dominant trait.) HD mice (which have the human allele for HD inserted in their DNA) are being used to study what happens to the brain in the earliest stages of the disease, something that is impossible to do in humans. In addition, HD mice are used to link changes in brain structure with changes in behavior.

Research on these mice has identified several molecular mechanisms that play important roles in the early stages of the disease. These model organisms are also being used to screen drugs to identify those that improve symptoms or reverse brain damage. Several candidate drugs have been identified; these are now being tested in human clinical trials as experimental treatments for HD. Similar methods have been used to construct animal models of other human genetic disorders and infectious diseases.

6.5 Essentials

- **Transgenic animals have been used to create animal models for human diseases.**

In 1987, vandals slipped past guards, climbed over a high fence in California, spread rock salt, and sprayed an herbicide (a chemical that kills plants) on a large strawberry patch. Over 2,000 strawberry plants were destroyed, and millions of dollars' worth of work was lost. This may seem like a strange story except for the fact that these strawberry plants had been treated with genetically engineered bacteria.

These bacteria, a strain of *Pseudomonas syringae* called "ice-minus", had been spread on the leaves of strawberry plants to prevent ice from forming on the leaves when the temperature dropped below freezing. Use of this transgenic bacterial strain extended the harvesting time of the strawberries and reduced plant loss caused by freezing. The strawberry plants and the strawberries themselves were not genetically modified.

The vandals were arrested and said they would do it again if they had a chance.

1. Why would these people want to destroy the strawberry fields?
2. This type of bacteria seems to be helpful to farmers; shouldn't we do all we can to help them improve crops and keep food prices under control? Why or why not?
3. Scientists have also produced transgenic corn plants by implanting a gene from fireflies that makes the plants glow when the firefly gene is expressed. What advantages do you see for the corn farmer from this change?
4. Erwin Chargaff wrote in *Science* on June 4, 1976, "Have we the right to counteract, irreversibly, the evolutionary wisdom of millions of years, in order to satisfy the ambition and the curiosity of a few scientists?" What are your thoughts on his statement? Are transgenic organisms changing the course of evolution?

6.6 What Are the Legal and Ethical Issues Associated with Biotechnology?

In addition to the scientific and societal problems with transgenic animals and plants discussed earlier in this chapter, there are legal and ethical issues as well. One involves patenting these organisms and the genes that are identified.

The U.S. Patent and Trademark Office (USPTO) that decides on the patenting of new inventions has rules about what can be patented. Anything patentable must be novel, not obvious, and useful. These rules also state that anything that is naturally occurring cannot be patented. However, in 1972, after a legal decision in a case called Diamond v. Chakrabarty (447 U.S. 303 (1980)), all that changed. Dr. Chakrabarty developed a bacterial strain that could break down oil. To do this, he crossed bacterial strains carrying different plasmids (gene-carrying DNA molecules found naturally in bacteria), eventually placing four different plasmids into a single strain of *Pseudomonas* bacteria. Each plasmid carried a gene that could degrade a different component of oil. Placed together in one strain, these bacteria could be used to clean up large oil spills.

BY THE NUMBERS

1 in 3500

Number of children who have an hGH deficiency.

60–70%

Percentage of food in stores that contains some transgenic plant material.

90%

Percentage of human disease genes also present in mice.

400

Number of medical products currently being developed using biotechnology.

$16,000 per year

Cost to a patient using Embrel (a biotech drug) for arthritis.

500,000

Number of children who die each year from vitamin A deficiency.

191

Number of amino acids in hGH.

At first, the USPTO refused to patent this bacterial strain on the grounds that the bacteria were naturally occurring, but the argument of Chakrabarty's attorneys won the case. Their argument, still used today, was that the strain containing all four of these plasmids in a single type of bacteria was unique and completely different from naturally occurring *Pseudomonas*. The patent was issued in 1972.

In a later case, Harvard University applied for and received a patent on a transgenic strain of mice called the **OncoMouse**, shown in the photo below. These mice carry a human cancer gene (**oncogene**) in their DNA. (In medical terminology, *onco-* means "cancer.") As a result of carrying this gene, the mice are more susceptible to cancer, allowing them to be used to test chemicals for their ability to cause cancer. This case was the first in which a transgenic animal was patented.

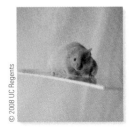

In the United States, the patent for the OncoMouse was issued in 1988. But in Canada, the case <u>Harvard College v. Canada (Commissioner of Patents)</u> was not decided until 2002. The Canadian courts ruled that a patent for a mammal could not be issued. Although recombinant bacterial strains have been patented in Canada, no patents for transgenic vertebrates have been issued.

The following chart addresses some of the ethical and legal questions in the patenting of transgenic organisms and genes.

6.6 Essentials

- **Many ethical questions about the use of recombinant DNA technology are being debated.**

Question	How are these questions decided?	Related case or legal issue
Can we patent genes that are discovered in human DNA?	When researchers apply for patents with the USPTO, judges who specialize in patent law decide each case.	Biotech companies are discovering genes every day and acquiring patents for them, even though the genes are naturally occurring. Their argument, similar to the one used in <u>Chakrabarty</u>, is that when a gene is isolated and removed from an organism, the situation is different than that of a gene that occurs naturally in the DNA of a species.
Can we patent sections of genes that are found in human DNA?	Such patents are also allowed, even when the gene is not identified or its use is unknown.	Using the argument that when removed from the cell and its DNA, a section is different than that found in nature, biotech companies have gained ownership of sections of DNA and collect fees from other companies wanting to use them. Some of these are only parts of genes that have not been identified.
Can the methods used in recombinant DNA techniques be patented?	Methods or processes can be patented if they are useful and unique.	When such methods are patented, no other company or lab can use the process without paying a fee or getting permission from the owner of the patent.
Should we be manipulating the genome of animals, plants, and people?	This ethical question is being debated every day. It is often said that this will create organisms that are very different from those that came about by evolution. Scientists usually have autonomy to work on their own, especially in privately owned biotech companies.	Laws that control what can be done in biotech labs may be written, and some already exist in countries other than the United States. One serious question here might be, who can control science?

The Asilomar Conference,
February 1975
Asilomar Conference Center,
Monterey Peninsula, CA

In the 1970s, recombinant DNA was in its infancy. After the discovery of the structure of DNA and the additional information that all living things had the same DNA, the idea of recombinant DNA seemed to make sense. However, some experiments made scientists think that there might be problems.

The Asilomar Conference was called by Dr. Paul Berg from Stanford and other prominent scientists of the day. Berg was one of the first individuals to develop recombinant DNA technology. In the experiment he designed in 1972, he used a restriction enzyme to cut the DNA of the monkey virus SV40, and then he used methods similar to those discussed in this chapter to join fragments of SV40 DNA to another virus, known as bacteriophage lambda. The final step would have involved placing the recombined genetic material into a laboratory strain of *E. coli* as a host cell to allow the recombinant DNA molecule to be copied. This last step, however, was not completed.

Berg did not complete this step because of pleas from several fellow investigators who feared that biohazards might be associated with the host cell carrying this recombinant molecule. SV40 was known to cause cancerous tumors in mice, and because *E. coli* is found in the human intestinal tract, investigators feared that the final step would create cloned SV40 DNA that might escape into the environment and infect laboratory workers, who could then develop cancer.

A group of leading researchers concerned about Berg's experiment sent a letter to the president of the National Academy of Sciences requesting that he appoint a committee to study the biosafety ramifications of this new technology. This committee met in 1974 and concluded that an international conference was necessary to resolve the issue, and that until then, scientists should voluntarily stop all experiments involving recombinant DNA technology.

The full conference was called in 1975. Its main goal was to address the potential hazards presented by recombinant DNA technology. It was one of the first scientific conferences to invite testimony of the public. A group of about 140 professionals (primarily biologists, but also including lawyers and physicians) participated in the conference to draw up voluntary guidelines to ensure the safety of recombinant DNA technology. Some of these guidelines were as follows:

1. Organisms containing recombinant DNA must be contained.
2. The level of containment should match the risk of the organism to humans, animals, and plants. These risks were ranked from low to high according to their potential for causing human disease.
3. Physical barriers should be used to minimize the escape of transgenic organisms.
4. The following types of experiments were prohibited:
   a. the use of any organisms that could live in the human body
   b. transfer of any gene that produces poisons
   c. synthesis of any products harmful to humans, animals, and plants

QUESTIONS:

1. Many scientists who participated in the conference felt that by writing their own guidelines, they avoided legislation that would control what was done. How could legislation work for the benefit and detriment of science?
2. The conference was held around the time of the Watergate break-in and the ultimate resignation of President Nixon. Do you think this was related to the importance of the conference?
3. Should scientists control their own work?
4. With the stem cell controversy currently being an issue with significant political overtones, do you think that guidelines written by scientists would eliminate or reduce the controversy?
5. Who should control science? Give three reasons for your answer.
6. The book *The Andromeda Strain* by Michael Crichton was published in 1969. Research its plot and then answer this question: Could the publishing of this *fiction* book have affected this conference? Why or why not?

⊛ THE ESSENTIAL TEN

1. Recombinant DNA techniques can be used to transfer genes among species. [Section 6.1]
   ▌ In some cases, the transferred gene can be expressed and produce a protein product.

2. Recombinant DNA is created when two different types of DNA are joined together. [Section 6.2]
   ▌ These fragments can be small sections of DNA that carry one or more genes.

3. Restriction enzymes are used to cut DNA into fragments. [Section 6.2]
   ▌ Restriction enzymes were one of the major scientific breakthroughs in the development of biotechnology.

4. Restriction enzymes recognize specific DNA sequences. [Section 6.2]
   ▌ Being specific, these enzymes can be used with any DNA molecule.

5. Biotechnology is the commercial use of recombinant DNA techniques to produce drugs and medical treatments. [Section 6.2]
   ▌ Biotechnology has become a multibillion dollar industry.

6. The first bacterial strain to be patented was one that degrades oil; this case is a landmark in patenting organisms. [Section 6.2]
   ▌ Even though oil-eating bacteria are present in our natural surroundings, they had been changed to a different strain in the laboratory by the introduction of a unique combination of genes.

7. Transgenic bacteria, plants, and animals can be created with recombinant DNA techniques. [Section 6.3]
   ▌ Transgenic organisms are used in agriculture, industry, and medicine.

8. Safeguards are needed in developing and using transgenic organisms to produce food and drugs. [Section 6.4]
   ▌ Scientists are studying the long-term effects of biotechnology on the environment and species diversity.

9. Transgenic animals have been created for use as animal models to study human diseases. [Section 6.5]
   ▌ Animal models of human disease are used to evaluate the action of drugs to treat disease.

10. Many ethical questions about the use of recombinant DNA technology are being debated. [Section 6.6]
    ▌ When transgenic animals and plants are developed, they have been genetically altered, and, to some extent, might be considered "man-made."

ONLINE RESOURCES

Preparing for an exam? Log on at academic.cengage.com/login for a pre-test, a personalized learning plan, and a post-test in CengageNOW's Study Tools to help you assess your understanding.

If assigned by your instructor, the Case B and Spotlight on Ethics activities for this chapter, "Strawberries on Trial (GMO)" and "The Asilomar Conference," will be available in CengageNOW's Assignments.

KEY TERMS

animal models (p. 108)
biotechnology (p. 103)
Drosophila (p. 108)
herbicides (p. 105)
Human Genome Project (p. 108)
OncoMouse (p. 109)
oncogene (p. 109)
plasmid (p. 104)
recombinant DNA (p. 104)
recombinant DNA molecule (p. 104)
recombinant DNA technology (p. 104)
restriction enzyme (p. 104)
transgenic crops (p. 105)
transgenic organism (p. 104)
vector (p. 104)

REVIEW QUESTIONS

1. Why is the technology discussed in this chapter called *recombinant DNA*?
2. What are the major steps in creating recombinant DNA?
3. Humans have been genetically modifying plants using crossbreeding for a very long time. How is this different from creating plants using recombinant DNA?
4. What is the importance of an animal model for studying human disease?
5. Why is *E. coli* used in recombinant DNA work?
6. It has been said that using recombinant DNA to produce clotting factor saved the lives of thousands of people with hemophilia. How did this happen?
7. The discovery of restriction enzymes was the beginning of the recombinant DNA era. Why was this discovery necessary before DNA from different sources could be mixed to form recombinant DNA molecules?
8. Give an example of a vector. Why is it referred to as a vector?
9. Do some research and find out if any recombinant DNA–derived foods are on the market. List them.
10. Explain why the patenting of recombinant DNA organisms was so difficult in the beginning.

APPLICATION QUESTIONS

1. A significant number of people are against using recombinant DNA technology in food production and other areas. List the pros and cons of this technology.
2. Research some of the groups that have been outspoken against recombinant DNA technology and discuss their reasoning.
3. Before recombinant human insulin was developed, people with diabetes used the insulin extracted from cows and pigs. How was this insulin created?
4. Some athletes have admitted using hGH to enhance their performance. But recently, athletic competitions have banned its use. How can officials determine whether an athlete uses this product?
5. Many people believe that if a product is developed and can help people feel better about themselves, it should be made available. Do you think this applies to the use of Botox for cosmetic reasons? Why or why not? Give scientific information to support your answer.
6. Research the history of the discovery of restriction enzymes.

7. Think of three conditions that might be treated in the same way that hemophilia is treated today. List them and explain how recombinant DNA technology could be used to develop a treatment.

8. If biotechnology is used to treat a condition, does it matter how that condition is inherited?

9. Think of five ways we might use recombinant DNA technology in the future. Be realistic.

What Would You Do If . . . ?

1. You were a legislator being asked to vote on a law to outlaw the patenting of transgenic animals?

2. You were a parent with a child with a serious genetic condition and were asked to have your future embryos examined in the lab to help identify the gene for this condition?

3. You saw a GM tomato (see Photo Gallery) in the market and were curious about its taste?

4. You were a legislator being asked to vote on labeling of transgenic food?

5. Your local grocery store hung up a sign that read, "We do not carry GM food"?

In The Media

Medical News Today, March 31, 2007

Wyeth Receives FDA Approval

Wyeth Pharmaceuticals has announced that the U.S. Food and Drug Administration (FDA) has approved changes for its drug BeneFIX® Coagulation Factor IX (Recombinant). These product changes should simplify patient usage of the drug.

BeneFIX stops or prevents bleeding in people with hemophilia B by supplementing or replacing clotting factor IX when patients do not have enough factor IX of their own.

With this new form of BeneFIX, patients can use less of the drug and therefore do not have to inject the large amounts they previously used. In addition, some patients will not have to use needles when preparing the drug.

To access this article online, go to academic.cengage.com/biology/yashon/hgs1.

QUESTIONS:

1. Do you think that patients will benefit from this new change in the drug? How?
2. By reworking this drug, Wyeth can extend its patent on BeneFIX. Why do you think the extension is allowed by the FDA?

Wired, March 22, 2007

Better Teeth Through Biochemistry

Charles Graeber

A genetically engineered mouthwash may make it possible for you to have no cavities *ever* again.

Acid made by bacteria in your mouth causes cavities. So no acid, no cavities. The genetically engineered bacterial strain, a form of *Streptococcus*, has had the acid-producing gene deleted. This strain has been patented as OraGen. All the bacteria need to activate them is some sugar.

If approved, it would work like this: A dental technician squirts a syringe of genetically modified mouthwash across your teeth. You sit for 5 minutes, chew on some sugary candy to activate the new bacteria, and then leave.

To access this article online, go to academic.cengage.com/biology/yashon/hgs1.

QUESTIONS:

1. How do you think this would affect dentistry?
2. Would you use OraGen?

7 Genetic Testing and Prenatal Diagnosis

✪ CENTRAL POINTS

- ∎ Many types of genetic testing exist.

- ∎ Testing is done to identify genetic disorders in fetuses, newborns, and adults.

- ∎ Cells and DNA are analyzed for heritable disorders.

- ∎ Phenylketonuria (PKU) is diagnosed via blood samples taken from newborns.

- ∎ Adults can be tested for many genetic disorders.

- ∎ Tests are often done on large groups to obtain genetic information.

- ∎ Some genetic conditions can be treated.

- ∎ Test results often create privacy issues.

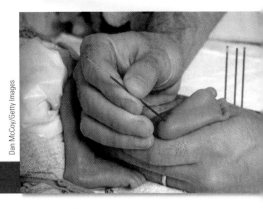

CASE A Hospital Tests Babies

Al and Victoria are in their early 30s and had been married five years when their first child was born. Victoria's pregnancy was perfectly normal, and she had continued working until one week before the baby's birth. She said she had never felt better. At birth, Al and Victoria's son appeared perfectly normal.

A few days after they took the baby home, the doctor called to say that the baby needed more tests. He asked that both parents come to the office visit because he wanted to talk to them. In his office the doc-tor said that a blood test done at the hospi-tal showed that the child had a genetic dis-order called phenylketonuria (PKU). If their child was not treated using a special diet, he would develop mental retardation. Al and Victoria had never heard of PKU, and this news disturbed them. They immediately called around to family members and found that only a few relatives on either side of the family had ever heard of PKU, and none knew of any family member with this disor-der. Al and Victoria wondered how a genetic disease that no one in their family had ever had could suddenly appear in their child.

They made an appointment for another visit with the doctor, and in the meantime, began reading about PKU.

Some questions come to mind when read-ing this case. Before we can address these questions, let's look at what genetic testing is, how it is used to diagnose genetic dis-eases, and the future prospects for this technology.

7.1 What Is Genetic Testing?

This chapter will discuss several methods used to determine whether someone has or is at risk for a genetic disease. Before exploring the rationale, methods, and issues involved in these methods, let's define the differences between the two major types: genetic testing and genetic screening. **Genetic testing** determines whether someone has a certain genotype—in other words, what are his or her genes? Genetic testing identifies people in the following groups:

1. those who may have or may carry a genetic disease
2. those who are at risk of having a child with a genetic disorder
3. those who may have a genetic susceptibility to drugs and environmental agents

Genetic screening is done on large populations rather than on individuals. The goals of genetic screening are more limited; it identifies people in the following groups:

1. those who may have or may carry a genetic disease
2. those who are at risk of having a child with a genetic disorder

Genetic testing is most often a matter of individual choice, whereas genetic screening is often mandated by law. Genetic testing can have a serious impact on the lives of individuals and their families, as in the following situations:

1. Identification of a person who has or is at risk for a genetic disorder often leads to the discovery of other affected or at-risk family members.
2. Testing for some disorders can identify a person who is healthy now but will develop serious or fatal genetic disorders in later life. When the condition is fatal—for example, Huntington disease, as in Alan's family (see Chapter 4)—this information often has serious personal, family, and social effects.
3. The results of genetic testing may have a direct impact on the children or grandchildren of the person being tested.

Genetic counselors work with patients who have had genetic testing to help them make decisions about the results.

What types of genetic testing are available?

Several forms of genetic testing are currently being used:

1. **Prenatal diagnosis**—to determine what genes a fetus carries, such as testing for sickle cell anemia.
2. **Carrier testing**—to test members of a family with a history of a genetic disorder such as cystic fibrosis to determine their chances of having an affected child.
3. **Presymptomatic testing**—to identify individuals who will develop disorders in midlife, such as Huntington disease or polycystic kidney disease.

7.1 Essentials

- Genetic testing determines what genotype a person has.
- Genetic testing can be done on fetuses, newborns, children, and adults.

7.2 Why Is Prenatal Genetic Testing Done?

Prenatal genetic testing is used to detect genetic disorders and birth defects in a developing fetus. More than 200 single-gene disorders can be diagnosed by prenatal testing. Some of these are listed in Table 7.1.

In most cases, testing is done only when there is a family history or some other risk is identified (such as the age of the mother). If there is a family history of an autosomal recessive disorder such as sickle cell anemia, the parents may be tested to determine whether they are heterozygous carriers of the disease before they conceive. If tests reveal that both parents are carriers, the fetus has a 25% chance of being affected with the condition.

One type of testing visualizes the fetus using a technique known as **ultrasound,** a method that can provide some information about the health of the fetus. But for conditions caused by chromosomal aberrations, such as Down syndrome (trisomy 21), chromosome analysis is the most direct way to detect an affected fetus. Because the risk of Down syndrome increases dramatically with the age of the mother (see Chapter 3), chromosomal analysis of fetal cells is recommended for all pregnancies in which the mother is age 35 or older. Prenatal testing for any genetic disease requires a sample of cells from the fetus, which can be obtained by using **amniocentesis** or **chorionic villus sampling (CVS).**

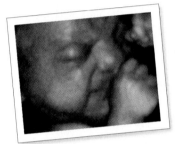

PHOTO GALLERY
The newest ultrasound available to patients is three-dimensional and can show details of the fetus's facial features.
Dr. Benoit/Mona Lisa/PhotoTake

Table 7.1 Genetic Disorders		
Disorder	**Incidence**	**Inheritance Pattern**
Cystic fibrosis	1 in 3300 Caucasians	Autosomal recessive
Congenital adrenal hyperplasia	1 in 10,000	Autosomal recessive
Duchenne muscular dystrophy	1 in 3500 male births	X-linked recessive
Hemophilia A	1 in 8500 male births	X-linked recessive
Alpha and beta thalassemia	Varies	Autosomal recessive
Huntington disease	4–7 in 100,000	Autosomal dominant
Polycystic kidney disease	1 in 3,000	Autosomal dominant
Sickle cell anemia	1 in 400 African Americans	Autosomal recessive
Tay-Sachs disease	1 in 3600 Ashkenazi Jews and French Canadians; 1 in 400,000 in the general population	Autosomal recessive

How is ultrasound done?

The procedure for ultrasound is shown in the photos below. Ultrasound has some use in prenatal diagnosis for genetic disorders. It can be used to identify trisomy 13, trisomy 18, and to a lesser extent, trisomy 21 (Down syndrome), because these disorders each have unique physical traits that can be seen via ultrasound. For example, in Down syndrome, affected children often have a thick fold of skin in the neck region. This can be seen in ultrasound, and amniocentesis can confirm the diagnosis. An ultrasound image of a fetus with a neck fold (outlined in red) is shown below.

ULTRASONOGRAPHY

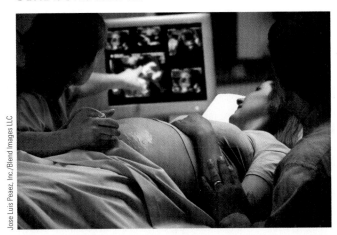

Jose Luis Peaez, Inc./Blend Images LLC

Ultrasonography (ultrasound) is a noninvasive technique based on sonar technology originally developed for the military to find submarines under water. A **transducer** is placed on the abdomen over the uterus. The transducer emits pulses of high- frequency sound; the reflected sound waves are converted into images and displayed on a screen. Many women routinely have ultrasounds during pregnancy to monitor the development of the fetus, see what the fetus looks like, and check for problems.

Ultrasound can also be used to determine the sex of the fetus and to diagnose multiple pregnancies. In the ultrasound below, the extra skin of the neck fold is outlined in red.

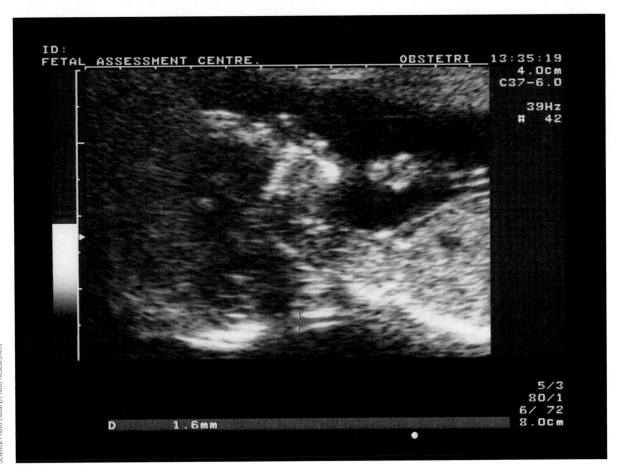

Science Photo Library/Photo Researchers

How is amniocentesis done?

The procedure for amniocentesis is shown in the photos below and was briefly described in Chapter 3. More than 100 disorders can now be diagnosed by amniocentesis. The cells retrieved by this method can be analyzed to detect biochemical disorders, chromosomal abnormalities, and the sex of the fetus.

When is amniocentesis used?

Amniocentesis carries a small risk of maternal infection and a slight increase in the probability of a spontaneous abortion. To offset these risks, amniocentesis is normally used only under certain conditions:

1. *When the mother is age 35 or older.* Because the risk of having a child with chromosomal abnormalities increases dramatically after age 35, amniocentesis is recommended for pregnant women who are age 35 or older. Most amniocentesis procedures are performed because of advanced maternal age.
2. *When the family has a previous child with a chromosomal abnormality.* The recurrence risk in such cases is 1–2%.
3. *When a parent has a chromosomal abnormality.* This can cause an abnormal karyotype in the child, and amniocentesis should be considered.
4. *When the mother is a carrier of an X-linked disorder.*

AMNIOCENTESIS

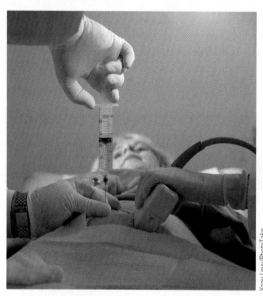

Yoav Levy/PhotoTake

As discussed in Chapter 3, amniocentesis is one way fetal cells can be collected for analysis. First, the fetus and placenta are located by ultrasound, and a needle is inserted through the abdominal and uterine walls (avoiding the placenta and fetus) into the amniotic sac surrounding the fetus. Approximately 10–30 ml of fluid are removed. Amniotic fluid is mostly fetal urine containing cells shed from the fetus's skin. These cells are isolated from the fluid by centrifugation. The cells are grown and used to create a karyotype (see Chapter 3).

Amniocentesis is usually not performed until the 16th week of pregnancy. Before this time, there is not enough fluid to sample.

How is chorionic villus sampling (CVS) done?

The procedure for chorionic villus sampling is shown on the next page and was briefly described in Chapter 3. Chorionic villus sampling (CVS) has several advantages over amniocentesis:

1. CVS can be performed earlier in the pregnancy (8 to 10 weeks) than amniocentesis (16 weeks).
2. Because placental cells are already dividing, karyotypes are available within a few hours or a few days. With amniocentesis, fetal cells must be stimulated to divide and grow in the laboratory for several days before a karyotype can be prepared.

When is CVS used?

CVS is used less frequently than amniocentesis because of its increased risk of spontaneous abortion. CVS offers early diagnosis of genetic diseases, and if termination of pregnancy is elected, maternal risks are lower at an earlier stage of pregnancy. The conditions under which CVS is used are similar to those for amniocentesis, but because it can be used earlier in pregnancy, some couples choose CVS if they know they are at risk of having a child with a genetic condition.

Are there less invasive methods for prenatal genetic testing?

Realizing that amniocentesis and CVS carry a 0.5–2% risk of miscarriage, researchers are working to develop newer and less invasive ways of prenatal testing. One of these less invasive methods involves recovering fetal cells from the mother's blood, which minimizes risk to both the mother and the fetus.

CHORIONIC VILLUS SAMPLING

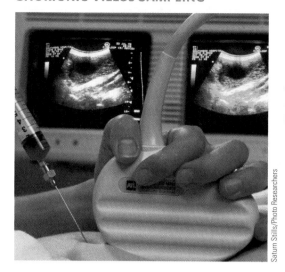

Saturn Stills/Photo Researchers

As discussed in Chapter 3, during the CVS procedure a flexible catheter is inserted through the vagina or abdomen into the uterus, guided by ultrasound images.

A sample of the **chorionic villi** (fetal tissue that forms part of the placenta) is removed by suction. Enough material is usually obtained to allow karyotype preparation, biochemical testing, or extraction of DNA for molecular analysis.

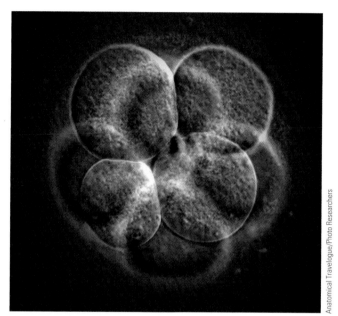

Anatomical Travelogue/Photo Researchers

■ **Figure 7.1 A Developing Embryo**
An embryo in the six- to eight-cell stage.

Several types of fetal cells enter the maternal circulation, including the following:

1. placental cells
2. white blood cells
3. immature red blood cells with nuclei

These cells probably enter the bloodstream in detectable amounts between the 6th and 12th weeks of pregnancy.

A problem with this method is that only 1 in every 100,000 cells in the mother's blood is a fetal cell. Collecting enough fetal cells from a blood sample is one of the challenges facing those working to develop this technique. Fetal cells collected this way have been used to diagnose some genetic disorders, including sickle cell anemia, and chromosomal abnormalities, but the method needs further development before being widely adopted.

Can embryos be tested before they are implanted in the mother?

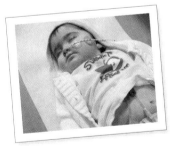

PHOTO GALLERY
A child with Tay-Sachs disease, in which brain and nerve cells slowly deteriorate and the person dies at approximately age 2.
Seungjae Seo/The Minnesota Daily

In Chapter 1, we discussed the use of preimplantation genetic diagnosis (PGD) for sex selection. This test can also be used to diagnose genetic disorders in the earliest stages of embryonic development. Here we will focus on the use of PGD to detect genetic disorders.

Before using PGD, IVF must be performed (see Chapter 2). During this procedure, eggs must be collected, fertilized, and allowed to develop in a culture dish for several days. On about the third day after fertilization, the embryo consists of six to eight cells (■ Figure 7.1). For PGD, one of these cells, called a **blastomere,** is removed (■ Figure 7.2).

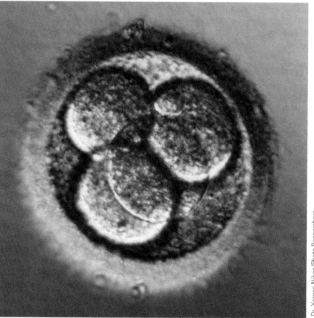

Dr. Yorgos Nikas/Photo Researchers

■ **Figure 7.2 Blastomeres**
Each cell in the early embryo is called a blastomere; a number of blastomeres are shown here.

DNA is extracted from this cell and tested, using techniques that increase the amount of DNA available, to determine whether the embryo has a genetic disorder. Blastomere testing can be used for many common autosomal recessive and dominant disorders, such as Tay-Sachs disease, and most X-linked disorders, including muscular dystrophy and hemophilia. Embryos that do not have a genetic disorder are implanted in the mother's uterus for development.

7.2 Essentials

- Prenatal testing is usually done to detect genetic disorders in the fetus when there is a risk of such conditions.
- Preimplantation genetic diagnosis (PGD) can detect genetic abnormalities *before* the embryo is implanted into the mother.

7.3 How Are Fetal Cells Analyzed?

PHOTO GALLERY
Karyotypes were once constructed by lab technicians who cut and pasted the chromosomes on paper; now it is all done by computer.
Hop Americain/Photo Researchers

The fluids and cells obtained by amniocentesis, CVS, and PGD can be analyzed using several different methods including karyotyping (see Chapter 3), biochemical analysis, and recombinant DNA techniques (see Chapter 6).

Direct DNA analysis is the most specific and sensitive method currently available. The accuracy, sensitivity, and ease with which recombinant DNA technology can be used to identify a genetic disease and susceptibilities carried by an individual have raised a number of legal and ethical issues that have yet to be resolved; these will be discussed later in this chapter.

7.3 Essentials

- Prenatal testing can determine the genotype of a fetus by looking at the cells retrieved using several different procedures.

7.4 How Can Prenatal Tests Diagnose Phenylketonuria (PKU)?

Al and Victoria's baby was tested for PKU at birth; all newborns in the U.S. are tested for this disorder. However, if there is a family history of this disorder, the test is often done prenatally. PKU is a metabolic disorder that is present in people across all ethnic and racial groups. It is inherited as an autosomal recessive trait; therefore, both parents must be heterozygous carriers for their child to inherit the disorder.

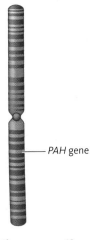

■ **Figure 7.3** *PAH* on a Chromosome Map
This chromosome map shows the location of the PKU gene (address 12q24.1) on chromosome 12.

— *PAH* gene

Chromosome 12

The gene that causes PKU is called *PAH* and is located on chromosome 12 (■ Figure 7.3). People with this condition cannot convert an important amino acid, **phenylalanine,** into another amino acid, **tyrosine.** The en-zyme that converts phenylalanine into tyrosine is called phenylalanine hydroxylase (PAH). In PKU, a mutation in the *PAH* gene inactivates the enzyme. Therefore, phenylalanine cannot be broken down and builds up in the body's tissues, causing damage to the brain and other organs.

PKU is a genetic disorder, but it is also an environmental disease. If phenylalanine is removed from the diet, the phenotype is normal, and there is no mental retardation. This is a successful treatment for the condition. Because the brain is fully mature by the early teen years, PKU homozygotes can usually switch to a normal diet.

However, if a woman who has PKU gets pregnant, she must follow the PKU diet very carefully; otherwise, the child may be seriously affected because the excess amount of phenylalanine in the mother's blood would pass into the fetus, causing serious mental retardation.

Can we test everyone for conditions such as PKU?

Genetic testing on a large scale is not always possible. For some disorders, such as the sickle cell anemia that Martha Johnson's twins had (see Chapter 5), a single mutation is always responsible, so testing is efficient and uncovers all cases. Many different mutations have been identified in the PKU gene, and testing for all of these mutations is impractical. However, most states test newborns for PKU using a simple blood test that is not based on DNA analysis. The test looks for an increase in phenylalanine in the baby's blood and does not analyze the *PAH* gene itself. Above is a photo of a *Guthrie card* with blood samples taken from a child soon after birth. The blood is transferred to the card and tested for PKU. The card is kept as part of the child's medical record.

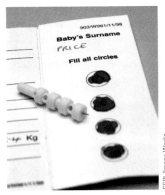

SSPL/Image Works

7.4 Essentials

- PKU is one of the few genetic conditions that has a treatment.

7.5 Can Adults Be Tested for Genetic Conditions?

Some conditions—such as Huntington disease, which runs in Alan's family (see Chapter 4)—show up much later in life but can be tested for earlier, before symptoms appear. A genetic predisposition to breast cancer (discussed in Chapter 11), also falls into this category. These presymptomatic tests can detect genetic disorders before the disease develops. Alan's family and others like them are candidates for this type of testing. Genetic tests can identify a number of these disorders, called adult-onset disorders. They include Huntington disease (HD), amyotrophic lateral sclerosis (ALS), polycystic kidney disease (PCKD), and others. As an example, the symptoms of PCKD, which affects about 1 in 1,000 individuals, usually appear between ages 35 and 50. This disease is characterized by the formation of cysts in one or both kidneys;

these cysts grow and gradually destroy the kidney. Below is a photograph of a normal kidney (left) and one with PCKD (right).

Treatment options for PCKD include kidney dialysis or transplantation of a normal kidney, but many affected individuals die prematurely before a transplant becomes available. Because PCKD is a dominant trait, anyone who is heterozygous for the mutant gene will be affected. Genetic testing can determine which family members carry a mutant allele and will develop this disorder. Testing for these and other disorders that appear in adults can be done prenatally or at any age before (or after) the condition appears. The decision on whether to be tested for these traits is one's own decision.

Arthur Glauberman/Science Photo Library —(represented by Photo Researchers)

7.6 What Is a Genetic Screening Program?

Some genetic screening programs, such as those for newborns, are mandated by law in the United States. All states and the District of Columbia require newborns to be tested for a range of genetic disorders, although the number and types of disorders screened for vary significantly from state to state. These programs began in the 1960s with screening for PKU and gradually expanded to all states and a wide range of genetic disorders. Al and Victoria's baby was tested under the screening program in their state.

Many states screen for only 3–8 disorders. However, using new technology, labs can now screen for 30–50 metabolic disorders from a single blood sample. The number of tests mandated in each state is shown in the map below. Parents should be aware of which disorders are included in their state's screening program.

Are there screening programs to detect adult carriers of genetic disorders?

As discussed in Chapter 4, a person can carry a gene for a recessive genetic condition but remain disease free. These individuals are heterozygous for the disorder and are called *carriers.* Carriers of certain recessive disorders, such as PKU and sickle cell anemia, are found in higher frequencies in certain ethnic groups.

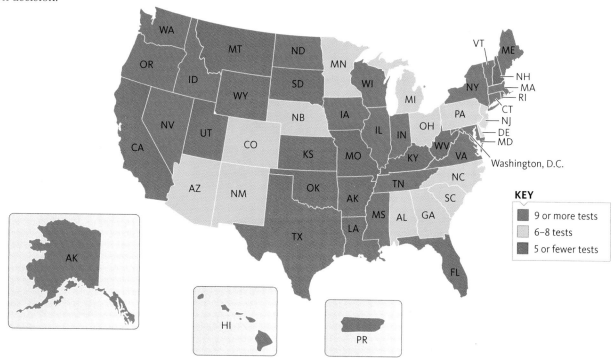

KEY

- 9 or more tests
- 6–8 tests
- 5 or fewer tests

Population screening for carriers of genetic disorders is possible only under certain circumstances, including the following:

1. The disease must occur mainly in defined populations. For example, sickle-cell carriers are most often found among African Americans. However, PKU, the condition present in Al and Victoria's baby, occurs across all racial and ethnic groups and would not qualify for carrier testing.
2. Tests for carriers must be available, fast, and fairly inexpensive.
3. Screening for these disorders must give at-risk couples several options for having only unaffected children.

At this point, there are no mandated programs for carrier screening; however, one disorder that meets these three conditions is **Tay-Sachs disease,** a fatal autosomal recessive trait that affects 1 in 360,000 individuals. This disease is associated with a disorder in the lysosomes in the body's cells (see "Biology Review: Cells and Cell Structure"). It can lead to mental retardation, blindness, and death by age 3 or 4. In Jews of Eastern European ancestry (Ashkenazi Jews), the frequency of this disorder is almost 100 times higher than that for the general population (■ Figure 7.4). In the 1970s, voluntary carrier screening programs were initiated in Ashkenazi

■ **Figure 7.4 A Jewish Couple**
Members of Orthodox Jewish communities have embraced genetic testing for Tay-Sachs disease and its carrier status.

BY THE NUMBERS

1 in 2600
Number of births of children with PKU among people of Irish/Scotch descent.

1 in 30,000
Number of births of children with PKU in Sweden.

1 in 10,000
Number of births of children with PKU in the United States.

1 in 2600
Number of births of children with PKU in Turkey.

1 in 150
Number of miscarriages after amniocentesis.

1 in 100
Number of miscarriages after CVS.

3 out of 100
Number of genetic disorders found by prenatal diagnosis.

populations in the United States. More than 300,000 individuals were tested in the first 10 years of screening, and 268 couples were identified in which both members were carriers and had not yet had a child. This test is not DNA based; rather, it tests for the level of an enzyme in the blood. These simple blood tests have been effectively used for years.

In the Ashkenazi community, Tay-Sachs screening programs are coupled with counseling sessions that provide information about the risks of having an affected child, the availability of prenatal testing, and reproductive options. These programs are voluntary, although some states require that couples be informed about Tay-Sachs testing. Before the screening programs began, each year 50–100 children were born with Tay-Sachs disease in the United States. This number decreased to fewer than 10 such births each year after screening programs were implemented.

Another population screening program began when the U.S. Congress passed the National Sickle Cell Anemia Control Act in 1972. This law was designed to establish a screening program to identify carriers of sickle cell anemia (■ Figure 7.5). Each state received funds to set up a screening program. Some states established compulsory programs requiring that all African-American children be tested before attending school; others required testing before obtaining a marriage license. Testing was also done on professional football players and applicants to the U.S. Air Force Academy, because carriers were not allowed to enter the academy.

■ **Figure 7.5 Sickle-Cell Screening**
Families can be tested to find out whether they are carriers of the sickle-cell gene.

CASE A QUESTIONS

Now that we understand how prenatal testing is done and why, let's look at some of the issues raised in Al and Victoria's case.

1. Why was Al and Victoria's baby tested for PKU?

2. Neither Al's family nor Victoria's family has a history of PKU. Should Al and Victoria's baby have been tested at all?

3. Should Al and Victoria have the right to refuse this testing for their child? Why or why not?

4. Based on what you know about the law, do Al and Victoria have a right to sue the hospital?

5. PKU is a treatable genetic disorder. Affected children must follow a strict diet that prevents many of the condition's symptoms. Does this change your opinion about testing the baby?

6. Testing of newborns for PKU is done in every state; some states offer more extensive testing for newborns. Al and Victoria live in one of these states. Should they have been notified of the testing? Why or why not?

This screening program generated unanticipated problems. In 1981, the Air Force policy that denied sickle-cell carriers admission to the Academy was reversed under threat of a lawsuit. In other cases, carriers were reportedly turned down for insurance and employment, even though carriers have no inherent health problems.

Some sickle-cell screening programs were criticized for not maintaining confidentiality of records and not providing counseling to those identified as heterozygotes. In the late 1970s and early 1980s, many of these screening programs were cut back or reorganized, and only a few states currently offer, but don't mandate, testing for adults, although all 50 states and the District of Columbia screen newborns for sickle cell anemia.

7.6 Essentials

- Genetic screening tests many individuals in a population.

7.7 Can Any Genetic Conditions Be Treated?

Many of the conditions we have discussed can be treated to alleviate symptoms, but none of them have a cure. In the next column, Table 7.2 lists some conditions and their treatments, if available.

Table 7.2 Treatment for Genetic Conditions

Name of Condition	Is Treatment Available?	Treatment
PKU	Yes	Diet that is low in phenylalanine
Sickle cell anemia	Yes	Bone marrow transplants, blood transfusions
Hemophilia	Yes	Recombinant DNA factor VIII
Cystic fibrosis	Yes	Inhalers, antibiotics
Polycystic kidney disease	Yes	Kidney transplant, dialysis
Tay-Sachs disease	No	
Huntington disease	No	
Muscular dystrophy	No	
Trisomies (Down syndrome, trisomy 18, etc.)	No	

7.7 Essentials

- A few genetic conditions can be treated but not cured.

CASE B Company Gives Big Party

A wonderful party was in full swing. Representative Roland Smith had been to many parties since being elected to Congress from his state, but, this one was the best. The party was hosted by MegaGene, a large pharmaceutical company that had just developed a test to detect the mutant gene for a disease that was rare and serious and had no treatment or cure.

Finally, by the end of the evening, MegaGene came to the point. Company representatives suggested that Mr. Smith sponsor a bill the company had drafted to require that every baby born in the United States be tested for this gene.

1. Why was MegaGene throwing this party? Give three reasons.

2. Why did the company draft the law? Give two reasons.

3. What should Congressman Smith do if asked to sponsor such a bill?

4. Today all babies born in hospitals are tested for several to dozens of genetic

istockphoto.com

diseases by law (which condition is dependent on the state). Should parents be allowed to refuse this testing? Why or why not?

5. What do you think will happen, years from now, when we have tests for all the human genes?

7.8 What Are the Legal and Ethical Issues Associated with Genetic Testing?

The use of genetic testing and screening has a number of accompanying ethical and legal problems. Because these methods examine and record a person's genetic makeup, privacy is extremely important. Both federal and state laws require that medical records be kept private and their contents not be revealed to anyone without the patient's explicit consent.

Information from these tests can affect a patient's health insurance, job security, and even personal life. For example, if an insurance company discovers that a family carries the gene for Huntington disease, it may not insure them. In another example, would a person be willing to marry someone who carries the gene for Huntington disease or the gene for polycystic kidney disease?

These questions are dealt with either personally, on a one-to-one basis, in the courts, or in the legislature. In some cases, laws have been passed prohibiting discrimination against someone because of his or her genetic makeup. During the 1970s, many people were discriminated against based on their carrier status for sickle cell anemia, and many screening laws were repealed.

On another level, companies that work to create a test for a specific disease gene spend millions of dollars on research and development. For this reason, they are looking to sell this test to as many potential customers as possible. If the company's test were used in a screening program of newborns, this would be one way to recover the money spent on research and development.

The following chart addresses some of the ethical and legal questions in testing and screening.

Question	How are these questions decided?	Related case or legal issue
How should test results be used?	Laws control the use of information, but insurance companies ask potential insurees to sign a release for their medical records.	The Kennedy-Kassebaum law stops insurance companies from discriminating against patients with genetic conditions.
Should we test for conditions that have no cure or treatment?	The decision for this testing (except as mandated by screening laws) is made by the individual after meeting with a genetic counselor or a physician.	Newborns and children have no right to refuse testing; however, adults can refuse, and the law has upheld this right.
Should the government decide who should be tested?	Since the 1970s, the U.S. government has mandated testing for all newborns born in a hospital. Babies born at home may be missed. The numbers of tests is increasing.	Parents have sued to be excluded from the testing and have won on religious and moral grounds. But most parents don't even know the testing occurs.
Should insurance companies or employers be allowed to require genetic testing of their potential customers or employees?	This has happened only in a few cases. The argument is usually that it is for the customer's or employee's own good.	In EEOC v. Burlington N. Santa Fe Railway Co. (see Spotlight on Law, Chapter 5), the company wanted to test all employees for the gene that causes carpal tunnel syndrome. The court decided that the testing violated the Americans with Disabilities Act, and the company immediately stopped the testing.

7.8 Essentials

- Genetic testing must be done under circumstances that ensure that the results will remain private.

This Spotlight will examine the nationwide genetic testing being done in Iceland. Back in 1996 an Icelandic scientist, Kari Stefansson, started a company called deCODE genetics to find disease genes in the human genome. It seemed logical that if he could identify a large population of closely related people, any genetic differences related to disease would stand out.

Iceland has a relatively small, homogenous population of approximately 290,000. Few people have immigrated there, and even fewer have left. Iceland keeps meticulous records of families as well as comprehensive health records that date back to the early 1900s. Below is a photo of some people from Iceland.

Sigurdur J. Olafsson/Icelandic Photo Agency (IPA)

Stefansson believed that if he could get DNA samples from everyone in Iceland, as well as their health and genealogy records, he could identify the disease genes more easily. In the beginning, he asked for volunteers to supply samples for his DNA database. However, to speed up the process of gene hunting, he approached the Icelandic government with an idea. He proposed that the government ask each citizen of Iceland to come forward and give a sample of their DNA for this database. After a great deal of debate, a law was written, and on December 17, 1998, the Health Sector Database Act was voted into law by the Icelandic Parliament.

For a one-time fee of $950,000 paid to the Icelandic government, deCODE was given complete access to family histories, medical records, and genetic information for the entire population of Iceland. According to the law, deCODE would construct a database (at a cost of $10 million–$20 million) of DNA samples from members of the population. Differences in DNA sequence would be used to search for disease-associated genes. In exchange, deCODE will receive and retain all profits derived from its operating license.

To ensure privacy of medical records, the law forbids deCODE from sharing medical information with anyone. In exchange, Icelanders have been promised free access to genetic tests and medications developed from this project. deCODE has signed drug development contracts with several pharmaceutical companies, including Merck, Roche Diagnostics, and Hoffmann–La Roche.

PHOTO GALLERY
Here is where deCODE genetics
does much of its work in Iceland.
Icelandic Photo Agency/Alamy Limited

These are some of deCODE's accomplishments:

1. 1997: localized genes associated with familial essential tremor, osteoarthritis, preeclampsia, Alzheimer disease, Parkinson disease, rheumatoid arthritis, obesity, and anxiety. Other genes for these conditions remain to be identified.

2. 2002: discovered the first genes linked to schizophrenia, common stroke, and peripheral arterial occlusive disease.

3. 2003: tested drugs to treat peripheral arterial occlusive disease, stroke, and schizophrenia; isolated three genes common to predisposed obesity and launched its first clinical trial of a drug called DG031 for the prevention of heart attack.

4. 2004: completed a second round of clinical trials of DG031.

5. 2008: started a third round of clinical trials for DG031 and a second round of trials for a drug that prevents asthma.

deCODE has recently started a program called deCODEme. Using a sample provided by swabbing the inside of the cheek, this online service offers a genome scan to identify genomic variations associated with risks for a number of genetic disorders. Results from the genome scan can also be used to trace an individual's genetic ancestry, compare genome data with other individuals, and make predictions about someone's physical attributes, based on the analysis of the genome scan. As more information from the human genome becomes available, deCODEme plans to expand the range of genetic disorders screened for, and the number of services offered.

QUESTIONS:

1. One of the interesting parts of Iceland's Health Sector Database Act is that it presumes consent of all the members of the population. What problems do you see with this?

2. Should Icelanders have the right to opt out of the testing? Explain.

3. Why should deCODE have a monopoly on the data and the ability to profit from such information? Give both pro and con arguments.

4. Will deCODE or other companies own part or all of an individual's genome? Explain.

5. deCODE has promised Icelanders free medications. Will it be liable for any unforeseen side effects? Explain.

6. At left is a page from an Icelandic phone book. Notice how many of the first and last names are the same. What does this tell you about the genetics of the people of Iceland?

✪ THE ESSENTIAL TEN

1. Genetic testing determines what genotype a person has. [Section 7.1]
   ▌ This information is used in various ways.

2. Genetic testing can be done on fetuses, newborns, children, and adults. [Section 7.1]
   ▌ As more tests become available more people can be tested for conditions.

3. Prenatal testing is usually done to detect genetic disorders in the fetus when there is a risk of such conditions. [Section 7.2]
   ▌ Usually this testing is done early in a woman's pregnancy.

4. Preimplantation genetic diagnosis (PGD) can detect genetic abnormalities *before* the embryo is implanted into the mother. [Section 7.2]
   ▌ Children born after this procedure have no side affects.

5. Prenatal testing can determine the genotype of a fetus by examining the cells retrieved using a number of different procedures. [Section 7.3]
   ▌ These procedures are usually done in accredited laboratories.

6. PKU is one of the few genetic conditions that has a treatment. [Section 7.4]
   ▌ The treatment is a diet low in phenylalanine.

7. Adults can be tested for many genetic conditions. A person may want to find out if he or she is a carrier of a condition before marriage. [Section 7.5]
   ▌ Some genetic disorders are much more prevalent in certain ethnic groups.

8. Genetic screening tests many individuals in a population. [Section 7.6]
   ▌ Screening is done most frequently to newborns to determine certain conditions.

9. A few genetic conditions can be treated, but not cured. [Section 7.7]
   ▌ A question exists as to whether we should test for conditions that have no treatment.

10. The information gleaned from genetic testing must remain private. [Section 7.8]
    ▌ Discrimination might occur against individuals with genetic conditions.

KEY TERMS

amniocentesis (p. 116)
blastomere (p. 119)
carrier testing (p. 116)
chorionic villi (p. 119)
chorionic villus sampling (CVS) (p. 116)
genetic screening (p. 115)
genetic testing (p. 115)
PAH (p. 120)
phenylalanine (p. 120)
prenatal diagnosis (p. 116)
presymptomatic testing (p. 116)
Tay-Sachs disease (p. 122)
transducer (p. 117)
tyrosine (p. 120)
ultrasonography, or ultrasound (p. 116)

ONLINE RESOURCES

Preparing for an exam? Log on at academic.cengage.com/login for a pre-test, a personalized learning plan, and a post-test in CengageNOW's Study Tools to help you assess your understanding.

If assigned by your instructor, the Case A and Spotlight on Society activities for this chapter, "Hospital Tests Babies" and "Iceland and Its Genetic Code," will be available in CengageNOW's Assignments.

REVIEW QUESTIONS

1. Who might be screened for a genetic disease?
2. Who might be tested for a genetic disease?
3. Draw a simple pedigree for a family that might have *one* of the tests listed in Section 7.7. Indicate the condition and those who would have it.
4. As we find more and more genes associated with genetic disorders, what might happen to genetic testing?
5. If parents do not want to have their child tested at birth, what legal action could they bring?
6. What is the difference between testing and screening?
7. List the three types of genetic testing.
8. What condition is prenatally tested for the most?
9. What are the main types of prenatal testing?
10. Explain the process of PGD.

APPLICATION QUESTIONS

1. In Chapter 3, we discussed some prenatal testing and looked at Martha Lawrence's case. After reading this chapter, have you changed your mind about any of your answers in her case?
2. Research the screening laws in your state and list the conditions for which screening is conducted.
3. When a company spends a great deal of money developing a genetic test, what would be the best way for it to earn the money back?
4. Research a new test that is being developed for pregnant women to find out genetic information about their fetuses.
5. Very recently, small fragments of charred bone were found in a grave in Yekaterinburg, Russia. They were identified as two children of Czar Nicolas of Russia. Do some research and find out which children they were.
6. What would be the pros and cons of each of the tests listed in this chapter?

What Would You Do If . . . ?

1. You knew there was a genetic test for breast cancer, which runs in your family?
2. You could have your newborn genetically tested for breast cancer?
3. The state you live in were considering a law requiring officials to collect DNA samples from every newborn, test those samples for more than 100 diseases, and then save them in a DNA data bank?
4. In the future, a small company, ezPGD, has developed an easy test for PGD and wants to market it to doctors, and you were an advertising executive given the project?
5. You and your partner were going to get married next year, and a new state law required all couples getting marriage licenses to be tested to see whether they were carriers of cystic fibrosis or sickle cell anemia?

In The Media

New York Times, April 27, 2007

Scientists Identify 7 New Diabetes Genes

Nicholas Wade

Work on diabetes has been ramped up as the U.S. population gets older. Both university scientists and private companies are looking for a way to control this condition. One approach is to find a link between specific genes and diabetes.

An important diabetes web site and the *New York Times* reported that seven gene variants were found to be more common in patients with type 2 diabetes. This may result in testing for type 2 diabetes risks.

One of the companies that reported finding a gene was deCODE genetics, a company that is researching the genetic code of the population of Iceland.

Source: http://www.nytimes.com/2007/04/27/us/27diabetes.html

QUESTIONS:

1. If many more genes are found to be associated with type 2 diabetes, will finding a cure be more difficult? Why or why not?
2. Now that we know that diabetes has a genetic component, will this allow us to eat everything we want?

Political Gateway, April 16, 2007

Genetic Link to Crohn's Disease Found

Canadian scientists have identified seven genes associated with Crohn's disease. Crohn's disease is one of the most common forms of inflammatory bowel disease, affecting 100–150 people per 100,000 individuals of European ancestry.

The researchers conducting a search of the entire human genome found that three of the seven genes were previously known, but four are newly associated with the disease.

The researchers found that one gene, ATG16L1, may have a role in the degradation and processing of bacteria by the body.

To access this article online, go to academic.cengage.com/biology/yashon/hgs1.

QUESTION:

Do some research on Crohn's disease and discuss how the genetic component of this disease makes a difference in its treatment.

National Human Genome Research Institute, April 25, 2007

U.S. Senate Passes Genetic Information Nondiscrimination Act

The Genetic Information Nondiscrimination Act makes it illegal to deny someone health insurance or job opportunities based on genetic information taken from his or her DNA. The chief sponsor of the bill, Representative Louise M. Slaughter (D-NY) introduced a similar bill 12 years ago. There have been repeated instances in which employers have discriminated against people with family histories of sickle cell anemia, Huntington disease, and various forms of cancer.

President George W. Bush urged Congress to pass this legislation that will protect Americans from having their genetic information about cancer and other diseases used against them by health insurance companies or employers. Congress passed the bill and President George W. Bush signed it into law in May 2008.

To access this article online, go to academic.cengage.com/biology/yashon/hgs1.

QUESTIONS:

1. One interesting thing about genetic information is that the only way to know it is to get a genetic test. If insurance companies want to have this information, they must ask you to be tested or be dropped from coverage. Would you be tested under these conditions?
2. Discrimination is wrong. Do you think we need laws to *force* companies not to do it? Why or why not?

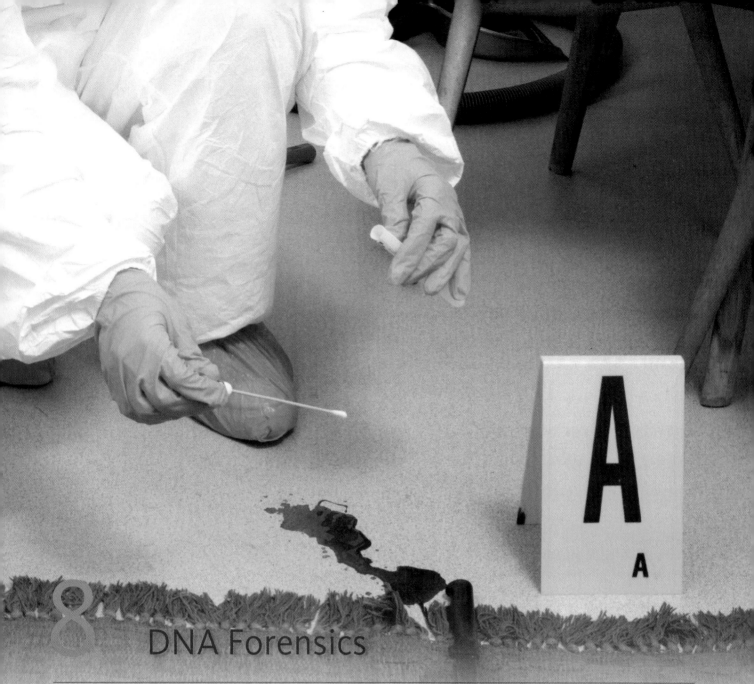

DNA Forensics

⊛ CENTRAL POINTS

- Testing DNA can determine one's identity.

- DNA profiles are constructed in specialized laboratories.

- DNA profiles are used in courts of law.

- Ancestry can be determined by testing DNA.

- A person may refuse to give a DNA sample to the police.

CASE A DNA Frightens Victim

Margaret Sackler was really frightened. It had been five years since William Bern attacked her, and now he was asking to be released from jail on new evidence his attorney had presented. She didn't really understand the evidence, but she had read about it in the paper. A number of prisoners had been released using DNA evidence.

Margaret had been walking in a parking lot outside a mall near her home. It was dark and near closing time, so there weren't too many people around. A man with a gun came up to her and ordered her to hand over her money and jewelry. She had been told never to fool around by trying to protect your valuables, so she gave him everything. But then he started to beat her. She fought back,

scratching and hitting him. His blood was all over her. Then a car passed by, and her assailant got scared and ran. She always wondered what he might have done to her if the car hadn't come by.

She identified him in a lineup and testified at the trial. She had seen his eyes clearly, but he had covered the rest of his face with a mask. She would never forget his eyes. He was found guilty and sentenced to 25 years to life for assault. She finally felt safe.

But, William Bern had always claimed he was innocent, so when he heard about all of these DNA cases, he asked his attorney to test the blood on the blouse Margaret was wearing.

It was old blood, five years old, but with a type of DNA testing called PCR, old DNA present in very small amounts can still be tested.

Margaret was asked to give a blood sample, and all the samples were tested. Bern's DNA did not match the blood on Margaret's shirt. Now he was asking to be released from jail.

Margaret didn't care—she was sure, he was the one. She'd never forget his eyes.

Some questions come to mind when reading about this trial. Before we can address those questions, let's look at the biology behind DNA forensics.

PHOTO GALLERY
Alec Jeffreys, the man who first developed the RFLP test in the United Kingdom.
John Smock/AP Photos

8.1 How Is DNA Tested?

In 1975, in England a scientist named Dr. Alec Jeffreys developed the first method of preparing what we now call DNA profiles and used these to compare profiles from different people. He called this method **DNA fingerprinting.** Soon after the method was developed, he used it to solve two murders in northern England by matching DNA samples from a number of men to a sample found at the crime scenes (see "Spotlight on Law: Narborough Village Murders"). This spectacular discovery revolutionized criminal investigations all over the world.

8.1 Essentials

- DNA forensic testing began in England in 1975.

8.2 What Is a DNA Profile?

In the 1980s, researchers discovered variations in the length of certain DNA sequences called **minisatellites.** Minisatellites are repeated sequences of 10–100 base pairs (see "Biology Basics: DNA"). They are located at many different places on all chromosomes and are used in DNA fingerprinting. Here is what one repeat of a minisatellite looks like:

```
...CCTGACTTAGGATTGCCA...
```

Other, shorter sequences called **short tandem repeats (STRs)** were then discovered. They are even shorter, and are only two to nine base pairs long. Clusters of STRs are found widely distributed on all human chromosomes. Variations in the number of repeats is used in preparing DNA profiles. Here is what an STR of the repeat TTCCC looks like:

...TTCCCTTCCCTTCCCTTCCCTTC...

In addition, **single nucleotide polymorphisms (SNPs)**, pronounced "snips," have been found. They are differences in a DNA sequence that are only one base pair long. Because they consist of only *one* base pair, they can be specific to an individual. Here is what a SNP looks like compared with DNA sequences of others within a population:

...ATCGTCGAGCCTAAATA...
...ATCGTCGAGCCTAAATA...
...ATCGTCGAGCCTAAATA...
...ATCGTCG**T**GCCTAAATA...
...ATCGTCGAGCCTAAATA...
...ATCGTCGAGCCTAAATA...

Each of these DNA variations can be used to create a DNA profile and identify individuals. These profiles are applied in numerous ways, including criminal cases, paternity lawsuits, studies of human evolution, and identification of bodies. Although the technology began with the use of minisatellites, STRs are now routinely used, and the term *DNA fingerprint* has been replaced by the term **DNA profile.**

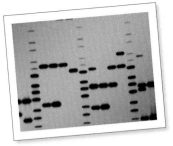

How are DNA profiles constructed?

DNA profiles can be constructed using a number of methods. Two methods commonly used in DNA forensics are **restriction fragment length polymorphism (RFLP) analysis** and the **polymerase chain reaction (PCR).**

How are RFLPs done?

RFLP analysis is done using relatively large amounts of blood or other biological samples. The following steps are involved in this process (follow along using the figure below):

1. DNA is extracted from the three samples.
2. The DNA is cut into fragments using restriction enzymes (see Chapter 6).
3. The DNA fragments are loaded into small slots (called wells) on a gel-like substance called agarose.
4. Electrical current is passed through the gel (which is immersed in a liquid). The current causes the DNA fragments to migrate through the pores of the gel, separating the fragments by size.
5. As they migrate through the gel, the mixture of DNA fragments separate by size. Smaller fragments migrate faster and further than the larger fragments.
6. The resulting pattern of DNA fragments is visualized and photographed. Patterns obtained from one individual can be scanned into a database and compared to patterns obtained from other individuals.

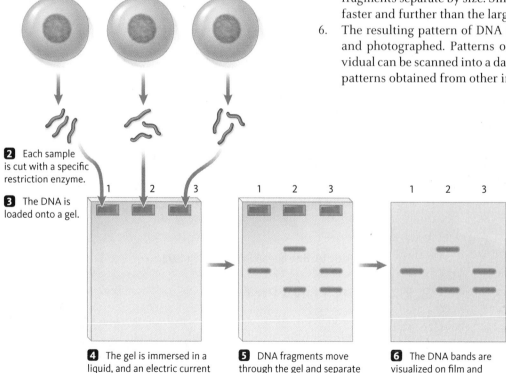

1 DNA is extracted from cells belonging to three different people.

2 Each sample is cut with a specific restriction enzyme.

3 The DNA is loaded onto a gel.

4 The gel is immersed in a liquid, and an electric current is applied across the gel from top to bottom.

5 DNA fragments move through the gel and separate by size. Smaller fragments migrate furthest.

6 The DNA bands are visualized on film and analyzed.

How does PCR work?

PCR uses very small amounts of DNA to produce profiles, working in a way that is similar to DNA replication in a cell (see "Biology Basics: DNA"). During the process of PCR, DNA is heated and cooled. After every round of replication, the amount of DNA is doubled. This can be repeated many times. In this way, a single DNA fragment can be amplified to make millions of copies. The following steps are involved in this process (follow along using the figure below):

1. The DNA is extracted from the sample and placed in a solution.
2. The DNA is heated, and separates into two single strands.
3. The temperature of the solution is lowered, and short nucleotide sequences called **primers** are mixed with the DNA.
4. The primers find and pair with complementary regions on the single stranded DNA. These primers act as the beginning of a new DNA strand.

PHOTO GALLERY
The PCR test made it possible to amplify small amounts of DNA and to look at old evidence as well.
Phanie/Photo Researchers

5. DNA polymerase (the enzyme involved in DNA replication) is added to the solution along with nucleotides (C, A, T, and G). The DNA polymerase uses these nucleotides to synthesize a double-stranded DNA molecule from each of the single-stranded DNA templates.

These steps make up one PCR cycle. The cycle can be repeated over and over. The amount of DNA present doubles with each PCR cycle. The power of PCR allows millions of copies to be made in hours from tiny amounts of DNA.

This table shows how the amount of DNA increases with each PCR cycle.

Table 8.1 Using PCR to Increase DNA	
Cycle	Number of Copies of DNA
0	1
1	2
5	32
10	1,024
15	32,768
20	1,048,576
25	33,544,432
30	1,073,741,824

1 DNA is recovered for PCR testing.

2 DNA is heated and separates into single strands.

3 Primers are added and bind with complementary regions in the DNA.

Primers

4 Primers serve as starting points for new DNA.

5 New double stranded DNA is formed.

a. One PCR cycle

1 Original sample

2 One cycle

3 Two cycles

4 Three cycles

b. Multiple PCR cycles

How are STRs used to make a DNA profile?

Each STR is composed of two to nine nucleotides. Each allele, of an STR contains a unique number of repeated copies of these nucleotides. For example, an STR might contain copies of three nucleotides AGA. An allele of this STR (allele 1) might contain one copy of AGA, another might have three (allele 2), and yet another (allele 3) might have five, as shown here:

Allele 1: ...AGA...
Allele 2: ...AGAAGAAGA...
Allele 3: ...AGAAGAAGAAGAAGA...

Analysis of several STR alleles can show how frequently a specific combination of these alleles is found in a population. Different populations may have different frequencies of these alleles. This analysis is done in a series of steps:

1. DNA samples from a population are analyzed to establish which STR alleles are present in members of the population.
2. The results can be analyzed to see how often specific combinations of alleles are present in a population; this result is called the **population frequency** (discussed in more detail in Chapter 14).
3. The population frequencies for each STR allele are multiplied together to estimate the probability that anyone who carries this specific combination of alleles is a match to the sample being tested.

STR	Allele	Frequency in population	Combined frequency
A	1	1 in 25	
B	2	1 in 100	A1 × B2 (1 in 25 × 1 in 100) = 1 in 2500
C	3	1 in 320	A1 × B2 × C3 (1 in 25 × 1 in 100 × 1 in 320) = 1 in 800,000
D	4	1 in 75	A1 × B2 × C3 × D4 (1 in 25 × 1 in 100 × 1 in 320 × 1 in 75) = 1 in 60 million

The combined frequency shows that the more alleles that are analyzed, the rarer the combination becomes in the population. Later discussion will show how this applies to criminal cases.

Where do DNA samples come from?

DNA profiles can be prepared from very small DNA samples (thanks to PCR) obtained from many different sources, including single hairs, licked envelope flaps, toothbrushes, cigarette butts, and dried saliva on the back of a postage stamp.

Evidence from crimes including murder, rape, break-ins, and hit-and-run accidents can be used. Profiles can also be obtained from very old samples of DNA, increasing its usefulness in legal cases. In fact, DNA profiles have been prepared from mummies more than 2400 years old.

8.2 Essentials

- Three types of DNA variations are used in DNA profiling: minisatellites, short tandem repeats (STRs), and single nucleotide polymorphisms (SNPs).
- Two common methods used in DNA forensics are restriction fragment length polymorphism (RFLP) analysis and the polymerase chain reaction (PCR).
- DNA samples for these tests can come from any number of sources, including cheek swabs, semen, saliva, and blood.
- Each of these tests can be used to analyze DNA and then make a comparison with other samples.

CASE A QUESTIONS

Now that we understand how a DNA profile is done and why, let's look at some of the issues raised in Margaret Sackler's case.

1. If you were the judge, would you give William Bern a new trial? Why or why not?

2. If you were the prosecutor, what arguments would you give for not using the DNA evidence?

3. If you were William's attorney, what arguments would you give to convince the judge that the DNA evidence should be submitted and a new trial granted?

4. Margaret is an eyewitness. Which is more reliable, an eyewitness or DNA testing? Why?

5. Should DNA testing be used by people who have already been convicted to get new trials? Why or why not?

6. In some cases, the DNA match is not perfect but is a 60% match. What could that mean?

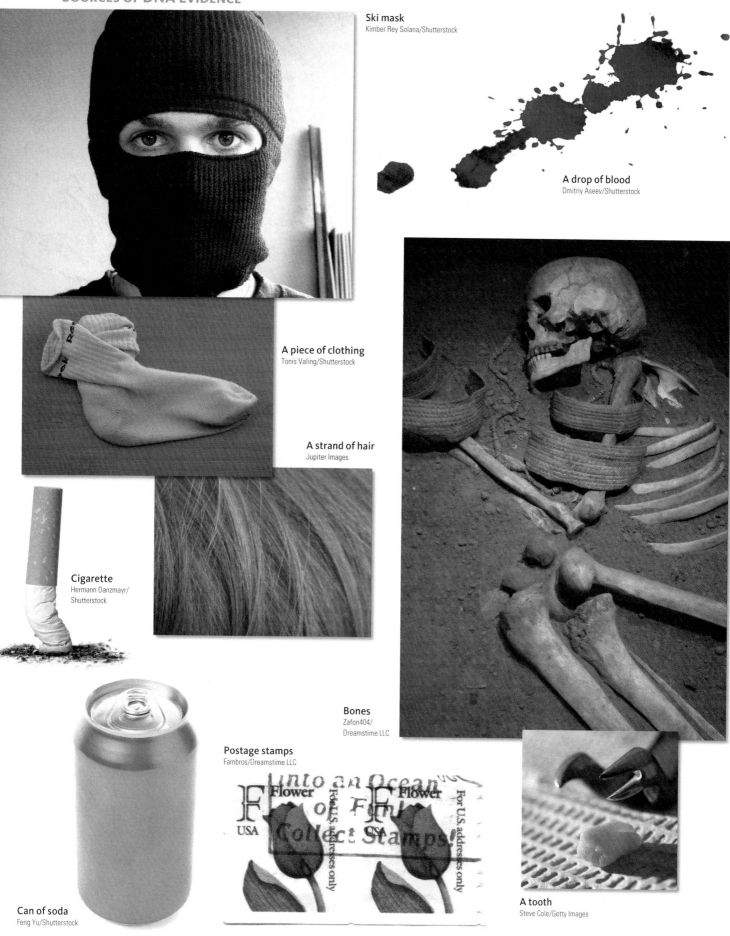

Ski mask
Kimber Rey Solana/Shutterstock

A drop of blood
Dmitriy Aseev/Shutterstock

A piece of clothing
Tonis Valing/Shutterstock

A strand of hair
Jupiter Images

Cigarette
Hermann Danzmayr/
Shutterstock

Bones
Zafon404/
Dreamstime LLC

Postage stamps
Fambros/Dreamstime LLC

A tooth
Steve Cole/Getty Images

Can of soda
Feng Yu/Shutterstock

8.3 When Did DNA Forensics Enter the U.S. Courts?

As discussed in Chapter 1, scientific evidence does not automatically get admitted into a court of law. In the landmark case Frye v. U.S. (Spotlight on Law: Chapter 1), four questions were posed that would decide whether science was "good science" and not "junk science." This test has been applied to DNA forensics, and it passed the Frye test and has been accepted in court.

The first criminal case to ever use DNA forensics began in England in 1983 (see "Spotlight on Law: Narborough Village Murders"). It took a few years before this type of evidence came to the United States. When it did, the use of DNA evidence in the legal system was scrutinized as thoroughly as any other scientific method. Each state had to decide on its own standards for use of the procedure, and the public was generally not aware of DNA forensics for several years.

The first U.S. case to challenge a DNA profile's admissibility was People v. Castro, 545 N.Y.S.2d 985 (S.Ct. 1989). Castro murdered a 20-year-old woman and her 2-year-old daughter. Blood was found on Castro's watchband.

The court determined that DNA identification procedures were generally accepted among the scientific community. But because investigators could show only that the blood was not the defendant's, and not that the blood belonged to one of the victims, it was not admitted into evidence. This, however, was only the beginning of DNA evidence in the U.S. courts.

How are DNA profiles used in the courtroom?

The use of scientific knowledge in civil and criminal law is called forensics. Forensic DNA analysis is usually performed in state and local police crime labs, private labs, and the Federal Bureau of Investigation (FBI) lab in Washington, D.C.

In criminal cases, DNA is often extracted from biological material left at a crime scene. This can include blood, tissue, hair, skin fragments, and semen. DNA profiles are prepared from evidence and compared with those of the victim and any suspects in the case. Shown at right is one of these comparisons.

BY THE NUMBERS

10,000
Number of criminal cases that use DNA forensics in the United States each year.

artpartner-images.com/ Alamy Limited

75%
Percentage of those cases that involve sexual assault.

20,000
Number of civil cases that involve DNA forensics in the United States each year.

1 in 100 trillion
Chance that two people will have the same 13 alleles in CODIS.

6.6 billion
Number of people on Earth.

Most testing in the United States uses a panel of 13 STRs for preparing DNA profiles. This **Combined DNA Index System (CODIS) panel** is used by law enforcement and other government agencies in comparing DNA profiles.

A DNA sample is tested using STRs from the CODIS panel to develop a profile of DNA. In a criminal case, if a suspect's profile does not match that of the evidence, he or she can be cleared, or excluded, as the criminal. About 30% of DNA profile results clear innocent people by exclusion.

What is a DNA database?

The FBI began cataloging DNA profiles and DNA samples obtained from convicted felons in 1998, and that database now contains more than 1,700,000 profiles. Many states are now collecting DNA samples from anyone arrested for a felony and entering their DNA profiles in a database. As they accumulate large numbers of profiles, DNA databases are becoming more important tools in solving crimes. Over a three-year period in Virginia, for example, matching DNA profiles from evidence at crime scenes with those in the state database solved more than 1,600 crimes in which there were no suspects.

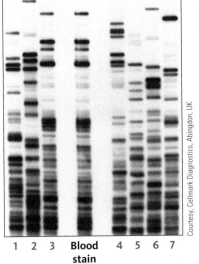

1 2 3   **Blood stain from crime**   4 5 6 7

Courtesy, Cellmark Diagnostics, Abingdon, UK

8.3 Essentials
- DNA forensics entered the U.S. court system in 1989 and is now used in courts all over the country.

8.4 What Are Some Other Uses for DNA Profiles?

DNA profiles have many uses outside the courtroom. **Biohistorians** use DNA analysis to correctly identify bodies and body parts of famous and infamous people whose graves have been moved several times or whose graves have been newly discovered. For example, Czar Nicholas II of Russia (■ Figure 8.1) and his family members were killed in 1917, and buried in unmarked graves. In 1991, remains found in Yekaterinburg, Russia, were identified as the those of the Czar and some of his family. In 2008, bones found at a nearby site were identified as those of two more of his children, accounting for all family members.

DNA profiles are used to identify the remains of military personnel killed in action, as well as people killed in the September 11 attacks in the United States. In the wake of the devastation caused by Hurricane Katrina in the southern United States in 2005, many of the dead were identified by DNA profiles.

Another important use of DNA profiling is paternity identification. Matching samples from the mother, father, and child allows 100% accuracy in identifying fathers. Use of DNA profiles in paternity testing ensures that many children will get the child support that they need. Below is a photo of a paternity test that compares the mother's, father's, and child's DNA:

■ **Figure 8.1 Czar Nicholas II of Russia** Czar Nicholas II of Russia and his family were killed during the Russian Revolution. Years later their remains were identified using DNA forensics.

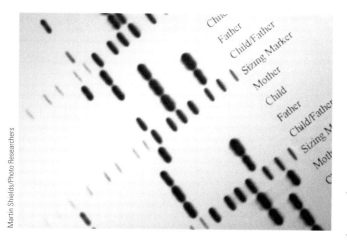

Mitochondria are found only in the cytoplasm of a cell (see "Biology Review: Cells and Cell Structure"). Because the egg is the only gamete with cytoplasm, mitochondria are always passed from a mother to her children (■ Figure 8.2). The mitochondria contain small amounts of DNA. Sperm have no cytoplasm, and therefore men cannot pass **mitochondrial DNA (mtDNA)** to their children. Differences in the sequence found in mtDNA provide a series of markers called **haplotypes.** Each person has the same haplotype as his or her mother, so comparison of haplotypes can help trace maternal ancestry.

Similar haplotypes can be grouped together to form **haplogroups;** these can be used to provide information about the genetic origins of large population groups (European, Asian, African).

Testing of Y chromosomal haplotypes is used to trace paternal ancestry. Because the Y chromosome is passed from father to son virtually unchanged (■ Figure 8.3), it can be used to establish male lineages.

Can DNA profiles trace our ancestry?

The use of DNA in forensic identification in high-profile paternity cases and on television crime shows has raised awareness that DNA can be used to trace an individual's ancestry. Many companies now offer ancestry genetic testing on the Internet. Two types of testing are usually offered: mitochondrial DNA testing and **Y chromosome testing.**

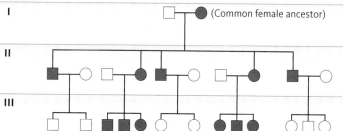

■ **Figure 8.2 Inheritance of Mitochondrial DNA**
This pedigree shows a common female ancestor and the inheritance of mitochondrial DNA through a family. The red circles and squares show how the mitochondrial DNA from the woman in generation 1 is passed from one generation to the next.

DNA testing using mtDNA or Y chromosomes can be used to search databases to provide information about your ancestry.

One of the largest projects to trace ancestry and the patterns of human migration out of Africa and across the globe is being sponsored by the National Geographic Society. This effort is called the Genographic Project. Interested individuals must pay to participate in this project. Those enrolled receive a vial and a swab to collect cheek cells. The Genographic Project uses the results anonymously to construct a detailed chart of human migrations over the last 50,000–100,000 years.

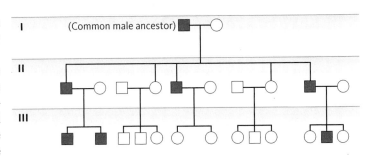

■ **Figure 8.3 Inheritance of the Y Chromosome**
This pedigree shows a common male ancestor and how the Y chromosome was passed on through a family. The red squares show how the Y chromosome is passed from father to son.

8.4 Essentials

- DNA profiles can be used to trace ancestry through the maternal and paternal lines.
- DNA profiles have many other uses in many areas of our society.

CASE B Samples Asked of All

There were very few clues to the murders of two women in a small rural community. The police had little evidence, but they did have skin samples from the killer, found under the fingernails of the victims.

From these samples a DNA profile was created, but there were no suspects. Because the crimes had occurred out in the forest, presumably at night, there were no eyewitnesses. No fingerprints, footprints, or pieces of clothing were recovered from the crime scene. If investigators could find someone whose DNA profile matched the sample, they might solve the case.

The investigator in charge of the case, Lt. DeMato, decided to try a new technique being used in the United Kingdom called a DNA dragnet. He would ask all the male members of the rural community over the age of 17 to give a cell sample (obtained with a cotton swab scraped along the inside of the cheek) and do DNA testing on all of them.

1. What might happen if a man didn't want to come in and give a sample?
2. Suppose that after samples were taken from all area men over age 17, no match was found. What would you do next if you were Lt. DeMato?
3. Amazingly, a match was found in the samples taken in the dragnet. The district attorney questioned whether the dragnet would be accepted as evidence by the court during the trial. Give three arguments the prosecution could use to

have the evidence from the dragnet admitted.

4. Give three arguments that the defense could use to keep the evidence from being admitted.
5. After Lt. DeMato has sent out a letter to all men in the area, he began testing. The 23rd man on the list, Irving Tomston, doesn't respond. A second letter was sent, and there was still no response. What should Lt. DeMato do?

8.5 What Are the Legal and Ethical Issues Associated with DNA Forensics?

The following chart addresses some of the ethical and legal questions in DNA profiling.

8.5 Essentials

- DNA profiling has been used to secure the release of hundreds of wrongfully convicted prisoners.
- Most states and the federal government have constructed DNA databases for use in solving crimes.

Question	How are these questions decided?	Related case or legal issue
If asked to give a sample in a DNA dragnet (in which many people are asked), can you refuse?	In most cases, one has the right to refuse, but police can look into your background if you do not comply.	In some cases, subjects in a DNA dragnet have been harassed and often lost their jobs if they didn't comply. No legal case has yet resulted.
Who must give DNA in a DNA database?	Most states have passed legislation requiring convicted criminals to give samples. The armed services require members to give samples for identification in case of war.	New York recently passed legislation that allows the collection of DNA from people suspected in a crime. The Pentagon has one of the largest DNA databases in the world. In 2000, two marines sued for the right not to contribute DNA to this database. They were compelled to give it or be discharged.
Can DNA free prisoners who were wrongly convicted?	If a prisoner's case was finalized before 1989, he or she may be allowed to have DNA testing. It would depend on the state, the court's decision, and whether the evidence was still available.	More than 300 convicted prisoners have been released by innocence projects around the country. These programs investigate the case and make DNA forensics available for an appeal. Some prisoners have been on death row when released.
Can old evidence be investigated and tested years after a crime has been committed?	This has definitely happened. If the crime is murder, no statute of limitations exists and PCR can be used to increase the amount of DNA available for testing.	In one case, old evidence was used to convict a man of a 40-year-old murder.

PHOTO GALLERY
Barry Scheck began the Innocence Project at Yeshiva University in New York, which has reviewed evidence in hundreds of DNA exoneration cases.
John Bazemore/AP Photos

DNA fingerprints were first used as scientific evidence in a criminal case that began in 1983. In the small English village of Narborough, Leicestershire, England, two girls were raped and murdered within a three-year period. Both girls, Dawn Ashworth and Lynda Mann, were 15 years old and their bodies were found in a field.

The prime suspect was a local boy, Richard Buckland, who after questioning, revealed previously unreleased details about Dawn Ashworth's body. He later confessed to the murder of Lynda Mann but denied any involvement in the first murder.

Convinced that Buckland had committed both crimes, officers from Leicestershire Constabulary contacted Dr. Alec Jeffreys at Leicester University, who had recently developed DNA fingerprinting as a way to match DNA samples. They asked him to match samples recovered from both crime scenes with Buckland. Dr. Jeffreys compared semen samples from both murders against a blood sample from the suspect, which conclusively proved that the same man killed both girls, but it was not Buckland.

Richard Buckland became the first person in the world to be exonerated of murder through the use of DNA profiling. Dr. Jeffreys said, "I have no doubt whatsoever that he would have been found guilty had it not been for DNA evidence. That was a remarkable occurrence."

At this point, the police were at a loss. To find the man who matched the killer's DNA, police conducted a DNA dragnet. They asked more than 4,000 men from the surrounding area to provide DNA samples for analysis. Unbeknownst to the police, the killer paid another man to give a DNA sample in his place. Colin Pitchfork told Ian Kelly, a baker who worked with him, that he could not give a blood sample. Kelly used a doctored passport and successfully gave blood in Pitchfork's name.

Oddly enough, Pitchfork would have escaped detection, but he was caught when Kelly bragged about the deception in a pub. A woman overheard the conversation and went to the police.

This historic case was described in the book, *The Blooding*, by Joseph Wambaugh.

QUESTIONS:

1. This case is amazing for a number of reasons. It was the first to use DNA profiling in a criminal case, and it broke new ground in forensics and in the use of DNA dragnets. What do you think might have happened if investigators had not used Dr. Jeffreys's technique?

2. Today DNA dragnets are used all over the United Kingdom. Why do you think they are not used very much here in the United States?

3. Do some research on the laws concerning DNA dragnets in the United Kingdom. Do you agree with them?

4. In Truro, Massachusetts, in the summer of 2005, Christa Worthington was murdered in her home, and her 2-year-old daughter was found sitting by the body. The entire state was in an uproar because she was a popular newspaper columnist. Police asked all men who lived around Truro in Cape Cod to give DNA samples. The dragnet got a lot of press. Oddly, the killer (who was her garbage collector) gave blood and waited in the area for two years for the samples to be run. What do you think he was thinking?

PHOTO GALLERY
The Blooding is a nonfiction book about the Colin Pitchfork case.
Jacket Cover from THE BLOODING by Joseph Wambaugh. Used by permission of Bantam Books, a division of Random House, Inc.

PHOTO GALLERY
The murder of Christa Worthington in Truro, MA, brought about one of the first DNA dragnets in the United States.
Massachusetts State Police/AP Photos

⊛ THE ESSENTIAL TEN

1. DNA forensic testing began in England in 1975. [Section 8.1]
   ▮ Alec Jeffreys first developed a way to do this testing on blood samples.

2. DNA samples for these tests can come from any number of sources, including cheek swabs, semen, saliva, and blood. [Section 8.2]
   ▮ Any tissue from a person's body can be used to extract cells for testing.

3. Three types of DNA variations are used in DNA profiling: minisatellites, short tandem repeats (STRs), and single nucleotide polymorphisms (SNPs). [Section 8.2]
   ▮ These variations occur in all individuals.

4. Two common methods in DNA forensics are restriction fragment length polymorphism (RFLP) analysis and the polymerase chain reaction (PCR). [Section 8.2]
   ▮ Both of these methods are used in criminal trials.

5. Each of these tests can be used to analyze DNA patterns and then make a comparison with other samples. [Section 8.2]
   ▮ The comparisons are often done with samples found in crime scenes.

6. DNA forensics entered the U.S. courts in 1989 and is now used in courts all over the country. [Section 8.3]
   ▮ This type of testing has revolutionized personal identification.

7. DNA profiles can be used to trace ancestry through both the maternal and paternal lines. [Section 8.4]
   ▮ Ancestry testing has become popular but is still in its infancy.

8. DNA profiles have many other uses in many areas of our society. [Section 8.4]
   ▮ Many think that profiling will be the way to identify people in the future.

9. DNA profiling has been used to secure the release of hundreds of wrongfully convicted prisoners. [Section 8.5]
   ▮ More prisoners are being released every day.

10. Most states and the federal government have constructed DNA databases for use in solving crimes. [Section 8.5]
    ▮ These databases contain valuable information that should be kept private.

KEY TERMS

biohistorian (p. 137)
Combined DNA Index System (CODIS) panel (p. 136)
DNA fingerprinting (p. 131)
DNA profile (p. 132)
haplogroups (p. 137)
haplotypes (p. 137)
minisatellites (p. 131)
mitochondrial DNA (mDNA) (p. 137)
polymerase chain reaction (PCR) (p. 132)

population frequency (p. 134)
primers (p. 133)
restriction fragment length polymorphism (RFLP) analysis (p. 132)
short tandem repeats (STRs) (p. 132)
single nucleotide polymorphisms (SNPs) (p. 132)
Y chromosome testing (p. 137)

ONLINE RESOURCES

Preparing for an exam? Log on at academic.cengage.com/login for a pre-test, a personalized learning plan, and a post-test in CengageNOW's Study Tools to help you assess your understanding.

If assigned by your instructor, the Case B and Spotlight on Law activities for this chapter, "Samples Asked of All" and "Narborough Village Murders," will be available in CengageNOW's Assignments.

REVIEW QUESTIONS

1. The term *DNA fingerprinting* was developed early on in the technology. Make a chart comparing and contrasting DNA profiling and, the older technology, fingerprinting.
2. What are the differences between minisatellites, STRs, and SNPs?
3. Why is PCR more desirable in testing than RFLP?
4. What is the meaning of the name *restriction length polymorphism* and where does it come from?
5. Who invented the PCR technique? Research his background.
6. What moves the DNA through the gel in RFLP testing?
7. What two benefits do DNA profiles have in the courts?
8. Do some research and find out which state has the largest DNA database.
9. At some point, RFLP testing and PCR testing merged. How did this happen?
10. Why is it useful to be able to test smaller and smaller amounts of DNA?

APPLICATION QUESTIONS

1. Research the laws in the United Kingdom that control DNA dragnets and write a short report.
2. Read up on the DNA chip and discuss how it will change criminal detection.
3. If an allele is prevalent in a certain group of people, what does this probably mean?
4. DNA forensics took a number of years to enter the U.S. court system. Why do you think this is?
5. If a person is picked up and questioned for a crime, should his or her DNA be taken? Should it be kept? Give your reasons.
6. Go back to Chapter 1 and re-read Frye v. U.S. How does this case apply to the use of DNA forensics in the US courts?
7. As PCR allows us to test smaller and smaller samples of DNA, how do you think this will change the way police and forensics experts treat a crime scene? List three ways.
8. DNA testing is also done to identify purebred dogs for dog shows. If this is done all the time, what effects might it have on these shows?
9. Some think that a huge DNA database should be formed and include the DNA from all newborns for the future. What is your opinion of this? Why?

Learn by Writing

In Chapter 1 we suggested that you start a blog with members of your class or others who are interested in our topics. Now is the time to revisit your blog and consider some of the questions in Chapters 5–8. Email others you think might be interested and invite them to contribute to the blog.

Here are some ideas to address in your blog:

- Is a single gene defect such as sickle cell anemia more serious than an extra or missing chromosome?
- What compensation should be given to people who are released after years in jail?
- It took many years to identify all the body parts taken from the 9/11 destruction of the Twin Towers in New York City, but DNA matching hasn't been as successful in identifying bodies found in Katrina. Why do you think that is?
- Some have suggested making a DNA database by taking DNA samples from all newborns. Is this a good idea?
- Discuss any other questions or comments you want to make or address with your fellow students.

What Would You Do If . . . ?

1. You were asked to participate in a DNA dragnet?
2. You were asked to vote on a referendum that would create a DNA database for all newborns?
3. You were on a jury that was asked to release a felon from prison after 20 years because of DNA evidence?
4. You were on the Sackler jury?
5. You were given the opportunity to participate in an ancestry test?

In The Media

60 Minutes (CBS), July 15, 2007

DNA: Going Too Far?

An episode of *60 Minutes* discussed the use of DNA profiles to search databases to find relatives of suspects. In the U.K. "shoe rapist" case, it was very effective. Using a DNA match that was not 100%, police tracked down a woman whose DNA profile was in the database. They asked her if she had a brother, and when they located him they found hundreds of shoes under the floor of his home. Subsequent DNA testing showed that he had raped a large number of women.

To access this episode content online, go to academic.cengage.com/biology/yashon/hgs1.

QUESTIONS:

1. Can you see why some people might be against this use of DNA? Why?
2. In the United Kingdom, as discussed in this chapter, many laws control one's ability to refuse DNA testing. Do you think these laws could be passed in the United States? Why or why not?

April 11, 2007

Paternity Test Results Are In, and Larry Birkhead Is the Father of Anna Nicole Smith's Daughter

One of the more sensational paternity test cases of 2007 involved the paternity of a baby born to Anna Nicole Smith, a television reality star. After Smith died, two men were claiming to be the father. Smith was the widow of billionaire J. Howard Marshall. Although her claim to half of Marshall's estate is still in the courts, her baby could inherit millions.

Larry Birkhead, Smith's photographer and longtime friend, was shown to be the father. He said to the press, "I told you."

QUESTIONS:
1. Some people think that heavy coverage of DNA in the press increases the public's knowledge of science. Do you agree? Why or why not?
2. Anna Nicole Smith's daughter was born in the Bahamas, which has different laws concerning paternity. In the United States, the father with matching DNA is not necessarily considered the legal father; a man who has lived with the child also has a claim. Do you agree with this? Why or why not?

Identigene Product Site, launched June 2007

DNA Paternity Test: At Home

With Identigene's home paternity test, you collect DNA by a cheek swab and send the DNA collection kit back to Identigene's DNA testing laboratory, and in a matter of days you receive the DNA test results. Results that are 100% accurate are necessary to exclude someone; those that are 99% accurate or better are sufficient for identifying the father. Identigene claims that these are the most accurate DNA test results available today. It also claims that its results are admissible in most courts of law. Although Identigene adheres to the most stringent standards for testing, some states may differ in their requirements.

One man who was interviewed said, "A girl I dated off and on informed me that I was a dad. What to do?!! I had to know if the child was actually mine. I was sent a kit, I followed the simple instructions, sent the lab samples back and days later, received the DNA results telling me that the child was mine. I can finally look into her eyes and feel at ease knowing I did the right thing."

To access this product site online, go to academic.cengage.com/biology/yashon/hgs1.

QUESTIONS:
1. Making paternity testing easier and more accessible can cause some problems. What problems might arise if it were cheap and readily available?
2. Suppose a woman thinks her old boyfriend is the father of her child. She meets him for coffee, takes his DNA from his coffee cup, and sends it to GeneTree. Do you think this could be used in court? Why or why not?

9

The Human Genome Project

✪ CENTRAL POINTS

- ▮ A large, international project is analyzing the human genome.

- ▮ Researchers are obtaining information by sequencing and mapping all the human genes.

- ▮ Gene mapping will show us where human genes are located on the chromosomes.

- ▮ There have been a number of surprises as the sequence of the human genome was analyzed.

- ▮ Scientists are applying information gleaned from the Human Genome Project (HGP) to medical diagnosis and treatment.

- ▮ Physicians will use your genome to give you better medical care.

- ▮ Gene therapy is one of the applications of the HGP.

- ▮ The ethical and legal aspects of the HGP are still being discussed.

Libby Welch/Alamy Limited

CASE A The Future Tells All

Natalie and Ben Coleman had already been to the obstetrician a number of times. They had been tested to find out their carrier status for many conditions. Natalie gave a blood sample so that the fetus's cells present in her blood could be isolated and the DNA sequenced to analyze all the genes in its genome. This analysis would provide detailed information about their child, including risks for genetic diseases, and other characteristics such as intelligence, looks, and height, just to name a few.

All of their friends had decided to have this fetal genome sequencing when they had children and had told them how exciting it was to find out about their baby before it was born. Many said it helped them "know" their baby way before they could actually interact with it.

Natalie's baby was due to be born in three months, and the doctor had already set a date for its delivery: May 3, 2025. The doctor said it was simple to time births down to the minute.

Sitting in the waiting room, Natalie and Ben were nervous. What would they find out? Who would their child be?

Soon they were called in to Dr. Rudin's office. He was sitting at his desk with a thick ream of paper in front of him. He put his hand on the papers and said, "Well, let's see what we have here."

Some questions may come to mind when reading about this futuristic case. Before we can address those questions, let's look at how the Human Genome Project may make this story a reality.

PHOTO GALLERY
Craig Venter was one of the pioneers in mapping and sequencing the HGP. Now, his goal is to have a $1000 genome sequencer so everyone can have his or her genome sequenced affordably.
AFP/Getty Images

9.1 What Are the Goals of the Human Genome Project?

The **genome** of an organism is the set of hereditary information carried in its DNA. The long-term goal of the **Human Genome Project (HGP)**, launched in 1990, was to discover the order of all the nucleotides in the human genome. This information, known as a **sequence**, looks like a long string of As, Cs, Gs, and Ts. Here is an example:

```
GCAAAAATACAAAAAATCTTGGATTCTATCGATAACAGCCGAGGTGCCAATCCATATGC
TACAAATAAAAAGCTTACTTTGGATACTTTGACAGGTGGACACTCAAAAGAATCTTATT
TGCGAAGTTATATTAATGGCAAACGTATTCCTGAGACTGCCAGAGCTGTAATCGAACCC
TCTATGAATAAAACTGGCTTTATTGAAGTACCATCTTACATTTTAAACAAGTTAAGAGA
TGTTGTCTTTTATAATCACGTTACGAAAGATAACATACTCAAAAGTCTTCAAAACGAAC
AAGCTTTTCTAACATATATCAAAAGTGATCATAATTCTGAAAATCCTTATATGGTTTAT
```

Once the sequence is obtained, it is analyzed to identify the genes. In addition, the project seeks to classify all the proteins made by these genes and understand their functions. The overall and long term goals of the HGP are as follows:

1. to create *maps* of the human genome and the genomes of other organisms
2. to find the location of all genes in these genomes and locate each gene on a map

3. to compile lists of **expressed genes** and **nonexpressed sequences**
4. to discover the function of all genes
5. to identify all proteins encoded by genes and their functions
6. to compare genes and their proteins among species
7. to analyze DNA differences between genomes
8. to set up and manage databases based on the genomes discovered

The history and timeline of the Human Genome Project is shown in ■ Figure 9.1.

Genome Project Timeline

1984 — Discussion, debate in scientific community

1990 — Human Genome Project (HGP) begins on October 1
1991 —
1992 — First genetic map of human genome
1993 — Revised goals call for sequencing genome by 2005
1994 — High-resolution genetic map
1995 — First physical map of genome
1996 — 16,000 human genes cataloged
1997 — National Human Genome Research Institute (NHGRI) created
1998 — Celera Corporation announces plans to sequence the human genome
1999 — Full-scale sequencing begins in HGP
2000 — HGP and Celera jointly announce draft sequence of genome
2001 — Working draft of genome published
2002 — Mouse genome sequenced
2003 — Sequence of gene-coding portion of human genome finished
2004 — Rat and chicken genomes sequenced
2005 — Chimpanzee genome sequenced
2006 — Rhesus monkey genome sequenced
2007 — J. Craig Venter sequences his own genome
2008 — Platypus genome sequenced

■ **Figure 9.1 Human Genome Project Timeline**
This timeline shows the major discoveries of the Human Genome Project. New things are happening every month.

What is the HGP creating?

Not only are scientists sequencing the human genome, but they are also creating **maps** of all human chromosomes. A map shows where all the genes are located on each chromosome. An example of a map is shown in Figure 9.2.

After a sequence is obtained, it must be analyzed to identify genes. Once the genes have been identified, they are placed on the map showing the order and distance between them (■ Figure 9.2).

■ **Figure 9.2 A Map of Chromosome 11**
This chromosome map shows the location of four genes.

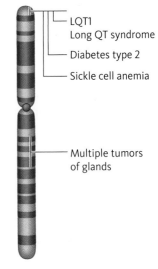

LQT1
Long QT syndrome
Diabetes type 2
Sickle cell anemia

Multiple tumors of glands

Chromosome 11

The human genome contains about 3.2 billion nucleotides (See "Biology Basics: DNA"). The sheer size of our genome required the development of computerized methods of DNA sequencing. The field of **bioinformatics** has created new software to collect, analyze, and store this information; web-based databases and research tools were also created to access genome sequences. The study of genomes by these methods is called **genomics.**

Where is the HGP now?

The portion of the human genome that carries genes was completely sequenced in 2003 and is being studied to identify genes and assign functions to them. Other research programs are isolating all the proteins contained in human cells and seeking to understand their structure and function. This new field is called **proteomics.**

9.1 Essentials

- **The Human Genome Project (HGP), which seeks to sequence and map the human genome, began in 1990.**
- **The project will also identify all the proteins encoded by human genes.**

9.2 What Is Gene Mapping?

There are several ways to map genes. The most basic method, developed in the 1930s, involved searching for genetic evidence that two genes are near each other on the same chromosome. Genes close together on the same chromosome are said to show **linkage** because they tend to be inherited together. The closer the genes are, the more likely they are to be inherited together. In 1936, using pedigree analysis, scientists showed that the genes for hemophilia and color blindness are both on the X chromosome and are therefore linked. Because these two conditions are X-linked, it was fairly easy to show that their genes are both found on the X chromosome, as shown at right. (See Chapter 4.)

It is more difficult to map genes to individual autosomes (chromosomes 1–22). This task requires very large families with two specific genetic traits. These families are very rare.

In 1955, researchers found linkage between the B allele of the ABO blood type and an autosomal dominant condition called **nail-patella syndrome,** which causes deformities in the nails and kneecaps. Below is a pedigree of a family with this linkage.

In the pedigree below, notice how the two traits are inherited together (that is, they are linked). However, the pedigree also shows that individuals II-8 and III-3 (starred) inherited only the nail-patella allele or the B blood type allele, but not both. This shows that the two alleles can be inherited separately.

The separation of the two alleles, or the fact that people in the family inherited only one, is the result of *crossing over* between the two genes on chromosome 9. The figure on the next page shows chromosomes crossing over.

Crossing over, which takes place during meiosis, occurs only a few times on any one pair of chromosomes. These crossover events occur randomly.

The closer two genes are on a chromosome, the less likely there will be crossing over between them.

Because the frequency of crossing over is related to the distance between the two genes, one can tell how close together these genes are by looking at how often they are inherited together. This information is used to construct a **linkage map** of a chromosome.

A linkage map (also called a **genetic map**) shows the order of genes on a chromosome and the distance between them. The units of distance are expressed as a percentage of the number of times crossing over occurs between two genes. This means that two genes that undergo crossing over 10% of the time are 10 map units apart on the map; genes that undergo crossing over 1% of the time are 1 map unit apart (■ Figure 9.3). These map units are called **centimorgans (cM).**

We can use this method to calculate the distance between the genes for nail-patella syndrome and the gene for the ABO blood group by measuring how often they undergo crossing over. Looking at the pedigree below, we can see that two family members (II-8 and III-3) out of sixteen inherited only one of the alleles in question, either the allele for nail-patella syndrome or the B blood type. In other words, the frequency of crossing over is 2/16 = 12.5%. From this pedigree, we can calculate the distance between these two genes is 12.5 cM.

In order to refine this measurement, many pedigrees of families with nail-patella syndrome have been studied and combined, giving a more accurate estimate of the distance between these two genes as 10 cM.

— Location of hemophilia A gene

Location of color blindness gene

X Chromosome

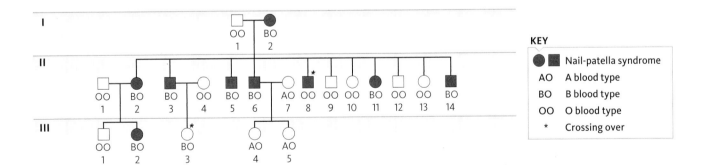

KEY

	Nail-patella syndrome
AO	A blood type
BO	B blood type
OO	O blood type
*	Crossing over

■ **Figure 9.3 A Human Linkage Map**
This chromosome map indicates the location of three genes: *a*, *b*, and *c*. Notice that each gene is a different distance from the others. This distance indicates how often these genes are inherited together.

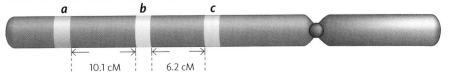

a   *b*   *c*

10.1 cM   6.2 cM

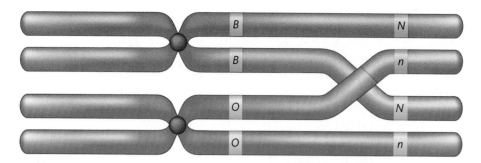

It took about 20 years to find the first example of linkage between autosomal genes, and by 1969 only five cases of linkage had been discovered. It was apparent that the effort to map all human genes by linkage analysis using pedigrees was progressing slowly.

PHOTO GALLERY
These sequencing machines are high-speed computers and are extremely expensive.
Sam Ogden/Photo Researchers

What changed genetic mapping?

In the 1980s, a method called **positional cloning** was used to map the genes for cystic fibrosis (CF), neurofibromatosis (NF), Huntington disease (HD), and dozens of other genetic conditions and to produce genetic maps of most human chromosomes. Some of these genetic disorders are discussed in Chapter 4.

In positional cloning, **markers** are identified that show differences in restriction enzyme cutting sites (see Chapter 6) or differences in the number of repeated DNA sequences (such as short tandem repeats [STRs]). Once these markers are assigned to specific chromosomes, they were used to follow the inheritance of a genetic disorder in pedigrees, and establish linkage between the marker and the mutant allele for that disorder. Although positional cloning identifies one gene at a time, by the late 1980s more than 3500 genes and markers had been assigned to human chromosomes. Some of the genes mapped by positional cloning are listed in Table 9.1.

How is mapping done today?

The development of methods to rapidly sequence DNA and the creation of computer programs that analyzed the

sequence was a springboard for launching the Human Genome Project (HGP). The idea behind the project was that instead of finding and mapping markers and disease genes one at a time, the HGP would sequence and map the entire human genome using high-speed DNA sequencers and computers.

9.2 Essentials

- **Efforts are now focused on identifying genes and their functions.**

- **Genomics is an extension of earlier methods of gene mapping, such as linkage studies.**

9.3 How Are Whole Genomes Sequenced?

Geneticists have developed a rapid method for determining the sequence of a genome. In this method, the DNA from an organism is cut with restriction enzymes to construct a collection of DNA fragments that contains all the sequences in a genome. This collection is called a **genomic library.** Individual fragments from the library are placed in a **DNA sequencer** that generates the order of nucleotides list of the As, Cs, Gs, and Ts we are familiar with. Using specialized software called **assemblers,** sequences of the fragments are put together to form the entire genome's sequences.

In 2003, the sequence of the gene-coding region of the human genome was published. The sequence is stored in a public database so it can be accessed and studied by scientists. New advances in this technology are being developed to sequence the remaining portion of our genome. It will be several years before the entire human genome is sequenced.

What happens after a genome is sequenced?

After a genome is sequenced, the information is organized and checked for accuracy. Then the next task is to find all the genes in that sequence that encode for proteins. Looking at a DNA sequence (■ Figure 9.4) does not reveal whether or not it contains genes (only 5% of human DNA is actually genes).

Software programs scan sequences, searching for certain combinations of base pairs. These include regions at the beginning of a gene called **promoter sequences** that are

Table 9.1 Genes Mapped by Positional Cloning	
Gene	**Chromosome Number**
Huntington disease	4
Familial polyposis	5
Cystic fibrosis	7
Wilm's tumor	11
Retinoblastoma	16
Breast cancer	17
Myotonic dystrophy	19
Amyotrophic lateral sclerosis (ALS)	21

■ **Figure 9.4**

A DNA Sequence

Within this sequence, it is difficult to tell where the genes are without knowing how to search for the coding sequences.

```
AAGTCTCAGG  ATCGTTTTAG  TTTCTTTTAT  TTGCTGTTCA  TAACAATTGT  TTTCTTTTGT
TTAATTCTTG  CTTTCTTTTT  TTTTCTTCTC  CGCAATTTTT  ACTATTATAC  TTAATGCCTT
AACATTGTGT  ATAACAAAAG  GAAATATCTC  TGAGATACAT  TAAGTAACTT  AAAAAAAAAC
TTTACACAGT  CTGCCTAGTA  CATTACTATT  TGGAATATAT  GTGTGCTTAT  TTGCATATTC
ATAATCTCCC  TACTTTATTT  TCTTTTATTT  TTAATTGATA  CATAATCATT  ATACATATTT
ATGGGTTAAA  GTGTAATGTT  TTAATATGTG  TACACATATT  GACCAAATCA  GGGTAATTTT
GCATTTGTAA  TTTTAAAAAA  TGCTTTCTTC  TTTTAATATA  CTTTTTTGTT  TATCTTATTT
CTAATACTTT  CCCTAATCTC  TTTCTTTCAG  GGCAATAATG  ATACAATGTA  TCATGCCTCT
TTGCACCATT  CTAAAGAATA  ACAGTGATAA  TTTCTGGGTT  AAGGCAATAG  CAATATTTCT
GCATATAAAT  ATTTCTGCAT  ATAAATTGTA  ACTGATGTAA  GAGGTTTCAT  ATTGCTAATA
GCAGCTACAA  TCCAGCTACC  ATTCTGCTTT  TATTTTATGG  TTGGGATAAG  GCTGGATTAT
TCTGAGTCCA  AGCTAGGCCC  TTTTGCTAAT  CATGTTCATA  CCTCTTATCT  TCCTCCCACA
```

already known. When the computer finds these regions, it is clear that the sequences that follow are genes.

After a gene has been identified, by matching the nucleotide triplets to the corresponding amino acids, the protein encoded by this gene can be derived.

9.3 Essentials

- **The development of new technology such as sequencers and computer programs helped the HGP move forward.**

9.4 What Have We Learned about the Human Genome?

The content of the human genome has proven to be a big surprise. In some ways, the genome is like many basements or attics, filled with old things having no apparent value, mixed in with small clusters of useful items.

In the genome, these useful items are the genes that code for proteins. Much of the rest is DNA that represents remnants of our genome's evolutionary history.

Although the human genome contains over 3 billion nucleotides of DNA, only about 5% of this DNA contains information for proteins. More than half the genome is made up of sequences that are repeated thousands or hundreds of thousands of times, called repetitive DNA or repeats.

What are the repeats and where did they come from?

There are several different types of repeats in our genome. One of these types is called **transposons** and they make up 45% of our genome. If transposons replicate, the new copies may move (or transpose) and insert themselves into another region of the genome. Fortunately, almost all transposons do not replicate and move around.

The second type of repeat is called a LINE 1 sequence; it makes up 17% of the genome. A third type, the **Alu sequence,** contributes 10% of our genome. Alu repeats first appeared in primate genomes about 65 million years ago and have played important roles in the evolution of our own genome.

Could any of these types of repeats produce a genetic disorder or inactivate a gene? Yes, in fact, copies of Alu sequences inserted in or near genes account for about 0.1% of all known genetic disorders, including hemophilia, neurofibromatosis, and some forms of breast cancer.

On the other hand, an Alu sequence may have played an important role in human evolution. About 2.8 million years ago, an Alu sequence inserted itself and turned off a gene that regulates cell growth in one of our ancestral species. At about the same time, brain size began to increase in the line leading to our species. Scientists think that having this gene switched off caused this evolutionary event.

Other repetitive sequences are found as clusters throughout the genome. They include the short tandem repeats (STRs) used to identify individuals we discussed in Chapter 8.

Are there other surprising findings about our genome?

While cataloging the proteins found in human cells, scientists were surprised to find that the number of proteins our cells can make far outnumbers the genes we carry. It is estimated that our ge-nome has about 20,000 to 25,000 genes, but the catalog of human proteins now includes over 500,000 and may exceed 2 million. How can 20,000 to 25,000 genes encode the information for so many proteins? The difference is the result of several mechanisms that can work independently or together. During processing of

---

BY THE NUMBERS

20,000–25,000

Number of protein-coding genes in humans.

20,000

Number of protein coding genes in *Drosophila* (fruit flies).

50%

Percentage of human genes that are similar to those of a banana.

John A Rizzo/Getty Images

540+

Number of genomes that have been sequenced.

1089

Number of bacterial genomes being sequenced.

Millions

Number of bacterial genomes still to be sequenced.

messenger RNA in the nucleus (see Chapter 5 to review this), the coding segments in the RNA can be rearranged, increasing the number of different proteins that can be produced from a single gene. In addition, once they are synthesized, proteins can be chemically modified. These modifications can change the structure and function of the protein, again multiplying the number of different proteins that can be de-rived from a single gene. An international effort, called the Human Proteome Project (HUPO) is working to identify all the proteins humans can make and to establish their functions and interactions with other proteins.

9.4 Essentials

- While completing the HGP scientists found a number of surprises.

9.5 How Can Information from the Human Genome Project Be Used?

Human geneticists are interested in studying human genetic disorders. Usually they want to know the following:

1. Where is the gene for this condition located?
2. What is the function of the protein encoded by this gene?
3. How does the mutated gene or its protein product result in a disorder?

If we can gather this information, we can learn how a gene and its product work and then develop diagnostic tests and treatments for specific genetic disorders. Although this seems like a difficult task, some progress has been made in several disorders, including cystic fibrosis (CF). As you remember from Chapter 4, CF is a condition inherited as a recessive trait. It causes mucus to accumulate in certain organs of the body.

Using positional cloning, the CF gene was mapped to a region on the long arm of chromosome 7, as shown in the figure above.

— Location of CF gene

Chromosome 7

Soon afterward, the gene itself was isolated and sequenced. The nucleotide sequence was converted into the amino acid sequence of the CF protein. When this sequence was compared with sequences in protein databases, it revealed that the CF protein is similar to proteins found in a cell's plasma membrane. Some of these proteins control the flow of molecules into and out of a cell. This finding helped establish that CF results from a mutation that affects a cell membrane protein.

Once the amino acid sequences of the CF gene were identified, attention turned to the development of drugs to treat CF. Individuals with CF have a defective protein that cannot regulate the flow of chloride ions into and out of the cell. This causes the mucus secreted by these cells to thicken and clog the ducts and lungs of the affected individuals. To treat CF, scientists are developing drugs that partially or completely restore the CF protein's function. At least five drugs are in development or testing to repair the protein.

9.5 Essentials

- Information from the HGP is being used to develop drugs to treat genetic disorders.

9.6 What Is the Future of Genome Sequencing?

At the present time, genome sequencing is very expensive and time-consuming. New technologies are being developed to reduce the cost and the time needed to sequence areas of the genome, or an entire genome. This advance may make sequencing a routine part of medical care, and many couples will have the experience that the Colemans are having at Dr. Rudin's office. Eventually genome sequencing will cost $1000 or less and take less than a day. Having the sequence of your genome will make it possible for doctors to monitor your health, provide information on how to reduce risks for certain diseases, and to make early diagnosis of conditions such as cancer, when the disease is more treatable.

CASE A QUESTIONS

Now that we understand how the future of the HGP might affect our lives, we can look at the case of the Colemans in a different light.

1. What should Dr. Rudin have told the Colemans before he sequenced the genome of their unborn child?
2. Should the Colemans learn about all the genetic information about their unborn

child that Dr. Rudin has? Why or why not?

3. If they do not take the opportunity to learn all the information, which pieces of genetic information should they take?
4. If given the opportunity to learn all the genetic information about your fetus, list five things you'd want to know.
5. In the future, when we will be able to

learn more about genetic information than we can now, how will this change how children are treated? Explain your answer.

6. Knowing whether your fetus carries a gene for a genetic disorder is one thing, but finding out about things like its height as an adult or intelligence level is quite another. Will society want this information when it is available? Why?

- In the near future, genome sequencing will probably be a routine part of medical care.

9.7 Is Gene Therapy Part of the Human Genome Project's Future?

As we saw in Chapter 6, using recombinant DNA techniques, scientists can transfer genes from human cells to bacteria to create drugs and other human proteins. Is it possible to transfer genes *into* humans? Imagine if a gene is not functioning correctly, and it is making an incorrect protein as a result. If the normal gene were transferred into a person with the genetic disease, the correct protein could be produced and the condition cured.

Beginning just before 1990, researchers began to use recombinant DNA technology to develop a method called **gene therapy** to treat genetic disorders. The idea behind gene therapy is to transfer copies of normal genes into cells (or people) that carry defective copies of these genes. Once transferred, the normal genes should direct the synthesis of the normal gene product, which in turn produces a normal phenotype.

How are genes transferred to people in gene therapy?

In most cases, cells are removed from the body of the person affected with a specific genetic disorder and modified in the lab. Normal copies of the gene are inserted using a virus to carry these genes. The virus, called a **vector**, enters the cell and inserts the normal gene into a human chromosome. The gene then becomes part of the genetic information carried by that cell.

After gene transfer, the cells are grown in the laboratory and checked to ensure that the normal gene is active and making its correct protein. Then the cells are transferred back into the body.

Have genetic diseases been treated by gene therapy?

In 1990, Ashanti DeSilva, a young girl with a genetic disorder called **severe combined immunodeficiency disorder (SCID)**, was treated with gene therapy. Children with this disorder have no functional immune system and usually die from an infection that would not be serious in most people. White blood cells were removed from her body, and a gene for the enzyme **adenosine deaminase (ADA)** was inserted into them. The treated cells were injected into her body, and Ashanti went on to develop an immune system and have a normal life.

Are there any problems with gene therapy?

In spite of its promising start, gene therapy has not yet fulfilled its promise as a treatment for genetic diseases. In many cases, the therapy has not worked. In a few cases, patients developed leukemia after gene therapy, and at least two people have died as a result of gene therapy. Scientists are working to correct problems with gene therapy and to

CASE B Ownership of Gene Questioned

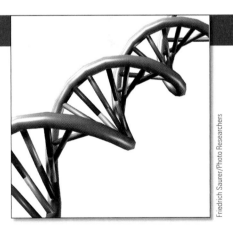

Friedrich Saurer/Photo Researchers

Technogene has been in the biotech business for quite a long time. During those years it has worked in the secondary market supplying materials (DNA and cells) to other biotech companies and has never patented anything.

During some of Technogene's newest work looking for a gene for hypertension, its scientists found an interesting area of the genome that they thought might be part of this important gene. When searching databases for this sequence, they discovered that another company, Markogene, had already patented it.

Technogene's head scientist, Julie Mersiant, was shocked. How could Markogene patent a section of a gene? At the time it was patented, no one even knew what the function of this sequence was or even if it was useful.

Dr. Mersiant wanted to use this sequence in her experiments, but she knew that Technogene would have to pay a substantial amount to have the right to use it. This would increase the cost of any discovery she made by possibly millions of dollars.

1. What might Dr. Mersiant or Technogene do now?

2. Many companies have patented human genes, their products, and the methods by which they were identified. Do you think this helps or hinders scientific knowledge? Why or why not?

3. If you were the attorney for Technogene, how would you argue for the right to use the information?

4. If you were the attorney for Markogene, how would you argue for the right to use the information?

5. Many people think that all information gleaned from the Human Genome Project should be open, and everyone should be allowed access. Do you agree? Why or why not?

develop new approaches to using genes as a treatment for genetic diseases.

9.7 Essentials

- Gene therapy is available only experimentally.

9.8 What Are the Legal and Ethical Issues Associated with the Human Genome Project?

Many of the legal and ethical questions regarding the HGP have been addressed as these technologies become available. Some issues will arise in the future, such as those surrounding the testing of the Colemans' fetus.

Those who originally funded the HGP foresaw some of these problems and allotted a percentage of its funding to study the ethical, legal, and social issues (ELSI) related to the human genome sequence and its many applications.

The chart below addresses some of the ethical and legal questions generated by the HGP.

Question	How are these questions decided today?	How might these questions be decided in the future?
Who will have control of the information gleaned from the HGP?	Confidentiality of medical records is controlled by law. Some of the scientific information has been made available to labs all over the world through computer databases.	If there were more understanding of genetics and its importance, courts could decide to diminish the right to confidentiality. This could make it easier for scientists to see records and use the material in them.
Will we have more genetic counselors as more information becomes available?	The academic community is developing more programs in genetic counseling.	Doctors will look for experts to work with their patients to explain genetic testing and the results that follow.
Will the use of this information lead to engineering the genetics of our children?	The first steps in genetic engineering are already happening in the use of PGD and other genetic testing. Many people are afraid that if this technology becomes commonplace, problems may arise.	The evolution of humans may be affected by genetic engineering, and those with disabilities may be selected against.
Could employers use genetic information to deny people jobs?	Employers are kept from discriminating against employees or job seekers by legislation.	If more genetic information is available, employers may require prospective employees to have genetic tests.
Can people patent human genes?	Genetic sequences, genes, and entire genomes can be patented in the U.S. and other countries. These patents allow companies to work on areas of the genome without another lab doing the same research.	Some scientists feel that the information gleaned from the HGP should be made public so all labs can access it. This is still a contentious issue today.

9.8 Essentials

- Some of the legal and ethical questions surrounding the HGP are being addressed.

Moore v. Regents of the University of California
(249 Cal.Rep. 494 (1988), 793 P.2d 479 Cal.Lex. 2858 (1990))

In 1976, John Moore entered the University of California Hospital in Los Angeles, California, for treatment of **hairy cell leukemia.** Leukemia is a condition in which white blood cells grow at an advanced rate; in hairy cell leukemia, specific white blood cells with hairlike projections proliferate and invade the spleen, causing it to enlarge. It usually attacks young males. *Splenectomy* (removal of the spleen) is the generally accepted initial treatment of hairy cell leukemia; improvement often lasts months or even several years.

Moore's spleen was removed in October 1976. His leukemia went into remission, and he was thrilled with the cure. From November 1976 through September 1983, Moore periodically returned to California from Washington state to have blood drawn and other materials removed (serum, bone marrow, and sperm). He was told that these procedures were part of his treatment, and all of his expenses were paid by UCLA. On April 11, 1983, Moore was asked to sign a consent form to allow the doctors to conduct research on his cells. Moore signed the form.

The doctors at UCLA had determined that the cells of Moore's spleen were unique because they created a protein called **lymphokine** that had been used to treat certain cancers. Using genetic engineering, the doctors created a continuous culture of his cells (a cell line) that would last forever; then they joined sections of his spleen cells' DNA to the DNA of *E. coli.* The resulting cell line produced a pharmaceutical product that had tremendous medical uses as well as high commercial value. Moore was not informed about the use of his cells, the cancer-fighting product derived from them, or their potential commercial value.

On March 20, 1984, the U.S. Patent Office issued Moore's physicians a patent for the cell lines with Dr. David Golde and Dr. Shirley Quan listed as inventors. The cell line was originally named the "Moore Cell Line," but the name was later changed to "RLC Cell Line." Dr. Golde and Dr. Quan made agreements with Sandoz Corporation and Genetics Institute to commercially develop and investigate this cell line. For this, they received the option to buy 75,000 shares of Genetics Institute stock at a reduced price and $330,000 cash.

John Moore filed a lawsuit on September 9, 1984, charging Dr. Golde, Dr. Quan, the Regents of the University of California, Sandoz Corporation, and Genetics Institute with **conversion,** the removal of another's property. In order to win the case, Moore's attorneys had to prove that his cells and their products were his property.

ISSUES:

Two major issues are involved in this case: informed consent and the doctor/patient relationship.

Informed Consent:

Moore's attorneys claimed that Moore was not allowed **informed consent.** All patients who undergo surgery or any medical procedure must give consent. The doctor is responsible for telling the patient all the information needed to make informed consent. A doctor who does not provide enough information or misrepresents what is to be done can be sued. Moore did sign a consent form for the surgery, and later for the blood tests, but he was not informed that his cells could be used and might be sold.

In other cases, doctors have done experimental work on prisoners and others without telling them how serious the consequences might be. These cases, such as the Tuskegee Institute case in the 1940s, had serious results. In the Tuskegee case, men who had syphilis were given experimental drugs to see if the syphilis could be cured. When the experimental drugs did not work, the patients were sent home. Twenty years later, many died from the disease. There were several ethical problems in this case: The men were never told they had syphilis, nor were they treated with antibiotics as these drugs became available. Cases such as this paved the way for *informed-consent* laws. As the years went by, these laws were refined, and more and more information is now necessary to ensure informed consent.

The Doctor/Patient Relationship:

Because of the lack of informed consent, Moore's attorneys argued that his relationship with his physicians was breached. Because of this breach, Moore was treated fraudulently and experienced extreme emotional distress.

RESULTS:

Trial Court:

The trial court did not allow the claim of conversion because Moore did not specifically state when he had his surgery that he did not want his tissue used for research, and he failed to attach a copy of the release for the splenectomy to the court documents. The court did not address the question of ownership of the cells, essentially forcing this decision to a higher court. John Moore appealed this ruling to the California Court of Appeals.

Appellate Court:

On July 21, 1988, after a long discussion, the California Court of Appeals stated in a two-to-one decision that Moore's allegation of a property right to his own tissues "is sufficient as a matter of law," and that he owned his own cells. Because the rights to the cell line had already been sold, it became only a matter of the court deciding who should share in the proceeds. The defendants appealed to the Supreme Court of California.

State Supreme Court:

On July 1, 1990, the Supreme Court of California reversed the appellate court's decision that Moore owned his cells and sent the case back for a new trial on the grounds of breach of doctor/patient relationship, because Dr. Golde should have informed Moore of the possible future use of his cells.

In considering the issue of ownership, the court stated that allowing Moore an ownership interest in his cells could impose "a duty on scientists to investigate the consensual pedigree of each human cell sample used in research." This, the court reasoned, was too difficult and confining. It might also increase litigation in these areas. The court was afraid that scientific experimentation, a very necessary thing, would become more difficult if doctors were not allowed to work with human cells freely. Such work is extremely important because the effects of drugs and other treatments must be tested on living human cells before they can be tested on human beings. Checking into the ownership of cells before experimental use would take a great deal of time and expense, and scientific study would be slowed.

The court would not allow Moore to stop the companies from using his cells or have a part in the profits. But the quality of the doctor/patient relationship was important; the court said Moore could go back to court and sue the doctors because they did not inform him that his cells were being used for this purpose.

No such lawsuit has been filed on behalf of Moore, as suggested by the court. Therefore, it is assumed that a settlement was reached for an undisclosed amount, probably with a gag order.

QUESTIONS:
1. Do you think the doctors owed some of their earnings to John Moore? If so, why and how much? If not, why?
2. If you were Moore's attorney, what arguments would you use to prove that his cells were his property? List three or more.
3. If you were the doctors' attorney, what arguments would you use to plead your case? List three or more.
4. Did the doctors do something wrong? If so, what?
5. Do you agree with the California Supreme Court's ruling? Why or why not?

Read the following cases and answer the accompanying questions.

CASE #1

Recently in Illinois, a man was diagnosed with leukemia and was told he needed a bone marrow transplant. A computer found a perfect match for his rare tissue type. The woman, a stranger, was tested while in the hospital for other treatment. She refused to donate her bone marrow, even when the seriousness of the illness was explained and the leukemia patient called her and pleaded. The patient took her to court to get an injunction to force her to give him the bone marrow, but the judge told the man he could no longer bother her. They were *her* bone marrow cells.

CASE #2

A woman entered the hospital to have an appendectomy. She asked the surgeon for her appendix back. He said no, it was going to be examined in pathology and then destroyed. She went to court to try to force the hospital to release her appendix to her.

[continued]

CASE #3

Recently a California couple had a second child in order to obtain bone marrow for their older child, who had leukemia. An organization tried to stop this use of the infant's cells by asking the court to stop the transplantation.

6. Ethically, one of our most important concerns is autonomy, the right to control our destiny and what happens to us. How does autonomy play a part in the Moore case? How would you use an autonomy argument if you were one of Moore's attorneys?

7. As more information becomes available about our genetics and testing becomes available, we can test our unborn children, our families, and ourselves. If we do not own our cells, as the California Supreme Court has ruled, do we own the results of tests on those cells? What is your opinion?

8. People are allowed to donate organs for transplantation. Courts have allowed some body parts (such as sperm and eyes) to be willed to someone. Does this mean that these parts are property? Would this argument have changed the results of Moore?

9. In Texas, a woman's eye had been removed because of disease and was lost down the drain. She sued for emotional trauma because her parts were treated so badly. How could you use the Moore case here?

10. Which of these cases are most closely related to Moore? Place them in order from most to least.

11. How would you use the Moore case in arguing each of these cases as the attorney for the defendant?

12. How would you use the Moore case in arguing each of these cases as the attorney for the plaintiff?

✪ THE ESSENTIAL TEN

1. The Human Genome Project (HGP), which seeks to sequence and map the human genome, began in 1990. [Section 9.1]
   ▮ Scientists are working on identifying all human genes and their uses.

2. This project will also identify all the proteins encoded by human genes. [Section 9.1]
   ▮ This new field is called proteomics.

3. Efforts are now focused on identifying genes and their functions. [Section 9.2]
   ▮ This information will help us discover what happens when genes function incorrectly.

4. Genomics is an extension of earlier methods of gene mapping, such as linkage studies. [Section 9.2]
   ▮ As scientists found faster and more sophisticated ways to work with DNA , genomics has advanced rapidly.

5. The development of new technology such as sequencers and computer programs helped the HGP move more quickly. [Section 9.3]
   ▮ The sequencers are coupled to high speed computers that analyze and store the data.

6. While completing the HGP scientists found a number of surprises. [Section 9.4]
   ▮ There are many more proteins produced than there are genes in the genome.

7. Information from the HGP is being used to develop drugs to treat genetic disorders. [Section 9.5]
   ▮ The main reason to decode the human genome is to help us treat and possibly cure these conditions.

8. In the near future, genome sequencing will probably be a routine part of medical care. [Section 9.6]
   ▮ Scientists working on the HGP see changes in health care as one of the ultimate uses of their work.

9. Gene therapy is available only experimentally. [Section 9.7]
   ▮ Research is directed at making gene therapy a safe and effective procedure to treat genetic disorders.

10. Some of the legal and ethical questions surrounding the HGP are being addressed. [Section 9.8]
    ▮ These legal and ethical questions are being discussed by the Ethical, Legal and Societal (ELSI) arm of the HGP.

KEY TERMS

adenosine deaminase (ADA) (p. 151)
Alu sequences (p. 149)
assemblers (p. 148)
bioinformatics (p. 146)
centimorgans (cM) (p.147)
conversion (p. 153)
DNA sequencer (p. 148)
expressed genes (p. 146)
gene therapy (p. 151)
genetic map (p. 147)
genome (p. 145)
genomic library (p. 148)
genomics (p. 146)
hairy cell leukemia (p. 153)
Human Genome Project (HGP) (p. 145)
informed consent (p. 153)

linkage (p. 147)
linkage map (p. 147)
lymphokine (p. 153)
map (p. 146)
markers (p. 148)
nail-patella syndrome (p. 147)
nonexpressed sequences (p. 146)
positional cloning (p. 148)
promoter sequence (p. 148)
proteomics (p. 146)
sequence (p. 145)
severe combined immunodeficiency disorder (SCID) (p. 151)
transposons (p. 149)
vector (p. 151)

ONLINE RESOURCES

Preparing for an exam? Log on at academic.cengage.com/login for a pre-test, a personalized learning plan, and a post-test in CengageNOW's Study Tools to help you assess your understanding.

If assigned by your instructor, the Case A and Spotlight on Law activities for this chapter, "The Future Tells All" and "Moore v. Regents of University of California," will be available in CengageNOW's Assignments.

REVIEW QUESTIONS

1. What is the difference between a sequence and a map of DNA?
2. If a sequence is unexpressed, what does this mean in terms of phenotype?
3. What is the combination of software, computers, and genetics called?
4. Why is crossing over important in gene mapping?
5. How is a linkage map constructed?
6. What is the next step after a gene for a genetic condition is located and isolated?
7. What is positional cloning?
8. Who were the leading scientists in the HGP?
9. List three possible consequences of finishing the HGP?
10. What are the goals of the Project?

APPLICATION QUESTIONS

1. Some have compared the Human Genome Project to the development of the atomic bomb. They call it "big science." Research the atomic bomb project and list three similarities and three differences.
2. As the HGP progresses, large numbers of jobs will be created. What types of jobs are they? List three.
3. How is the science of the HGP and DNA forensics similar? Go back to Chapter 8 and reread the sections on how forensic testing is done.
4. After all the human genes are mapped and cloned, we will have a great deal of information about ourselves, our children, and where we come from. Do we want to know this?
5. Research the ELSI program and see what progress it is making in studying ethical, legal, and society issues.
6. One of the most interesting cases involving gene therapy was the Jesse Gelsinger case. Jesse was a member of a clinical trial for gene therapy and died. Research this case and discuss what happened as a result of his death.
7. If there are so many areas of the genome that do not function, why do you think they are still in the human genome?
8. Research some of the animal genomes that have been mapped and sequenced. Why are they important?

What Would You Do If . . . ?

1. You were deciding how far the Human Genome Project should go?
2. Your state wanted to collect samples of DNA from everyone to study?
3. Your child's school began collecting DNA samples to submit to an organization looking for certain genes in the population?
4. You were asked to vote to allot more funds to work on the Human Genome Project?

Science Creative Quarterly, Sept 07–April 08

Entire Genomes Now Analyzed Quickly

K. J. Shelswell

Thanks to ever-improving methods to sequence DNA, scientists can now analyze the genomes of entire communities of microbes, a field known as *metagenomics*. By comparing these bacterial communities in the intestines of different people, scientists can study various health statuses. They hope to find out how microorganisms prevent or increase risk for certain diseases and whether they can be manipulated to improve health.

To access this article online, go to academic.cengage.com/biology/yashon/hgs1.

QUESTIONS:
1. Do you think there might be differences the bacteria in the intestines of people of different ages? Why?
2. Usually microbes make us ill; do we want to study their genomes for cures?

Biology News Net, February 7, 2007

Announcement of DNA Areas with Risk of Type 2 Diabetes

Scientists from a number of companies and universities have announced the discovery of three regions of human DNA that contain clear genetic risk factors for type 2 diabetes. Type 2 diabetes is increasing in the United States as the population ages.

The findings stem from the work of the Diabetes Genetics Initiative (DGI), a public-private partnership established in 2004 between Novartis, a drug company, the Broad Institute of Harvard and MIT, and Lund University. This information is directly from the Human Genome Project as MIT and Harvard are working on it. This information will be used directly in clinical research with patients with diabetes.

To access this article online, go to academic.cengage.com/biology/yashon/hgs1.

QUESTIONS:
1. How can knowing where the gene for higher risk is located help patients with diabetes?
2. How can knowing about this gene help patients who don't have diabetes yet?

New Scientist, October 12, 2002

Race for the $1000 Genome Is On

Sylvia Pagán Westphal

In 2001, Craig Venter, CEO of the Craig Venter Institute, challenged scientists to come up with the $1000 genome. He will award a prize of $500,000 to any company or individual who can sequence 10 genomes in 10 days with each genome costing less than $1000. Venter told *New Scientist* that he is not "taking orders" from people or setting this up as a commercial service, as some reports have stated. Rather, his goal is to collect sequence information from lots of people.

To access this article online, go to academic.cengage.com/biology/yashon/hgs1.

QUESTIONS:
1. Do you think people will have their genomes sequenced if it costs only $1000? Why or why not?
2. Will insurance companies pay for genome sequencing? Why or why not?

New York Times, June 1, 2007

Genome of DNA Discoverer Is Deciphered

Nicholas Wade

The full genome of James D. Watson, who jointly discovered the structure of DNA in 1953, has been deciphered, marking what some scientists believe is the gateway to an impending era of personalized genomic medicine.

A copy of his genome, recorded on two DVDs, was presented to Dr. Watson yesterday in a ceremony in Houston by Richard A. Gibbs, director of the Human Genome Sequencing Center at the Baylor College of Medicine, and by Jonathan M. Rothberg, founder of the company 454 Life Sciences.

"I am thrilled to see my genome," Dr. Watson said.

Source: To access this article online, go to academic.cengage.com/biology/yashon/hgs1.

PHOTO GALLERY
Recently James Watson (of DNA fame) became the first person to have his genome sequenced.
Markus Schreiber/AP Photos

QUESTIONS:

1. What do you think was the reason for picking Dr. Watson?
2. Dr. Watson is a winner of the Nobel Prize; will his genome be different from that of an ordinary person? Explain your answer.

Biology Basics: Genes, Populations, and the Environment

Tim Graham/Getty Images

BB3.1 How do genes and the environment interact?

Our phenotypes are affected by more than our genomes. Our environment also plays a part in who we are. Environment can include things we don't usually think about, such as our prenatal environment, what we eat, what we breathe, and where we live. If someone near us has a cold, for example, he or she may pass it on to us or our immune system may fight it off. The outcome of interactions between both genetics and the environment make us what we are. Traits that are affected by both genetics and the environment are called **multifactorial traits.** We will discuss these in detail in Chapter 10.

BB3.2 Does genetics affect our perception of and response to the environment?

Beginning in the 1930s, researchers observed that people react differently to things in the environment. In 1933, a chemist found differences in how people taste a chemical called **phenylthiocarbamide (PTC)**. Some are tasters and others are not. Tasters find PTC to have a bitter and very strong taste, while nontasters cannot taste it at all. Later, it was shown that the ability to taste PTC has a genetic basis. Nontasters are homozygous for a recessive allele (*aa*), and tasters are heterozygous or homozygous for a tasting allele (*Aa* or *AA*). There is a distribution of tasters around the world. In the map below, darker shades show a high percentage of tasters (up to 85%), while lighter shades have less tasters (about 5%). What do you think might cause this?

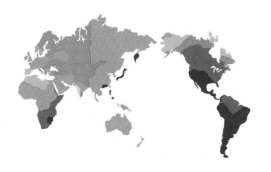

In the 1970s, Linda Bartoshuk and her colleagues at Yale University wondered whether the genetics of tasting might have an impact on diet and food choices. She discovered that tasters can be divided into two groups: tasters and supertasters, who have intensely negative reactions to PTC. In her surveys she found that about 25% of people are nontasters, 50% are tasters, and 25% are supertasters. This seems to correlate with the *aa*, *Aa*, and *AA* genotypes. She also discovered that supertasters can have more than ten times as many taste buds as nontasters and are very sensitive to tastes.

It seems as though food preferences might be related to taster status. Supertasters often find that high-sugar foods are too sweet, coffee is too bitter, and hot peppers and spices have a more intense and often unpleasant taste. As a result of their highly sensitive taste buds, supertasters are less likely to include certain foods and beverages in their diet, including Brussels sprouts, cabbage, spinach, soy products, grapefruit juice, and alcohol.

These findings raise questions about the relationships among our genotypes, taste preferences, and diet. Nutritionists are interested in studying supertasters and seeing if they can answer some of these questions: Do supertasters choose fruits and vegetables that are higher in cancer-fighting compounds? How does our taster status affect our food choices? Can food additives designed to block the sensation of bitterness be used to improve nutrition?

BB3.3 Are there population variations in responses to drugs and medicines?

In recent years, thousands of new drugs have been developed and are being used to treat diseases. Some of these drugs produce a wide range of reactions in different people that may be genetically influenced. For example, some people break down certain drugs very rapidly, making them less effective unless a higher dose is used. Others metabolize the same drug more slowly, and high levels of the drug remain in the body. This can prove toxic. Drug companies have responded to this problem in many ways. Some drugs are put into capsules that delay the release of the drug, spreading the drug level over a longer time period, while other drugs are produced in a variety of dosages.

BB3.4 What is the relationship between cancer and the environment?

There is evidence to show that the environment plays an important role in the development of cancer. Many environmental factors have been implicated in cancer, including natural radiation (x-rays, radon gas), occupational exposure to chemicals (polyvinyl chloride), virus infections (hepatitis B, human papillomavirus), and personal choices such as exposure to ultraviolet light (sunlight and/or tanning lamps), smoking, and diet. Some of this evidence comes from studying workers with years of on-the-job exposure to synthetic pesticides, asbestos, and industrial chemicals such as benzene, polyvinyl chloride, and others.

Naturally occurring environmental agents may also play a role in cancer. These include viral infections which account for almost 15% of cancers worldwide.

Table BB3.1 lists some human viruses that are associated with cancer.

In most cases viral infection alone is not enough to convert a normal cell to a cancer cell; other factors are involved. Some of these include DNA damage, viral infection of actively growing cells, and the ability of the virus to disrupt cell cycle control.

Table BB3.1 Viruses and Cancer	
Virus	**Associated Cancer(s)**
Epstein-Barr virus	Burkitt's lymphoma, nasopharyngeal cancer
Hepatitis B virus	Liver cancer
Human herpes virus 8	AIDS-related Kaposi sarcoma
Human papillomavirus	Cervical cancer

BB3.5 Is there a relationship between genetics and cancer therapy?

More than 70% of breast cancer cases are classified as *estrogen sensitive.* This means that the tumor cells have estrogen receptors on their surface, and that the presence of estrogen in the body promotes growth of the cancer. One of the most widely used drugs to treat estrogen-sensitive breast cancer is **tamoxifen.** Once in the body, this inactive form of the drug is converted into a powerful anti-estrogen compound called **endoxifen.** The conversion of tamoxifen to endoxifen is controlled by an enzyme called CYP2D6 as shown below.

The gene that encodes for this enzyme has several alleles. Four groups of women have been identified: One group metabolizes tamoxifen very slowly, while members of the other three groups rapidly convert it to endoxifen. Analysis of recurrence rates shows that women who metabolize tamoxifen slowly have a much higher risk of recurrence than those with faster metabolism rates. When developing drugs for cancer therapy, it is important to take genetics into account. In some cases, genetic testing is done before a drug is prescribed, to ensure maximum benefits from the therapy.

The genetic and environmental interactions in cancer will be discussed in Chapter 11.

BB3.6 How is behavior affected by the interaction of genes and the environment?

We all know that stress can cause people to change their behavior. In certain situations, one person might become depressed, but others may cope better. Scientists have found that a variation in one gene can cause changes in these differences. Two alleles, long and short, in a gene called *5-HTT,* seem to control these differences. In researching the genetics of *5-HTT,* one study analyzed people between the ages of 21 and 26 who had experienced four or more stressful life events (such as divorce, debt, unemployment, or death of a loved one). Those who had two copies of the short *5-HTT* allele were more likely to become depressed than those carrying two long alleles. This finding adds to the evidence that both genetics (the type of *5-HTT* allele) and environmental factors (the stressful event) can influence behavior and how we react to changes in our lives. The role of genes, behavior and the environment will be discussed in Chapter 12.

BB3.7 How do environmental factors and genes interact with the immune system?

Autoimmune diseases are the third most common form of disease in the United States (after heart disease and cancer). There are 15–80 such diseases that affect some 5–8% of the population. Autoimmune diseases occur when the body's immune system produces antibodies that attack specific cells and tissues in the body. Almost all tissues and organs can be affected by autoimmune attacks, including skin, connective tissue, the digestive system, and the nervous system. Some of these include diabetes, multiple sclerosis, and rheumatoid arthritis. As shown in the chart on the next page, more than 75% of people who have autoimmune diseases are women. The reason for this is unknown.

Identical twin studies have clearly demonstrated that both genetic factors (the presence of specific alleles) and environmental factors (infection) are involved in triggering these diseases. Often a patient has no symptoms of the autoimmune condition until they contract a specific virus. This infection causes the condition to start or get worse. Table BB3.2 lists some autoimmune diseases that are linked to infections.

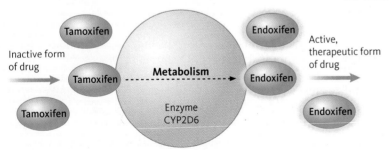

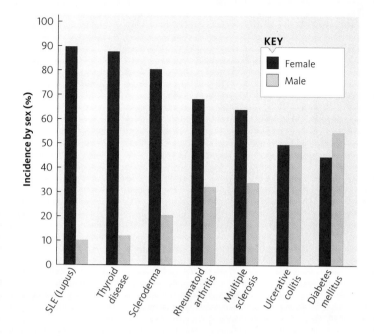

The distribution of autoimmune diseases among women and men. For reasons not yet known, more than 75% of those affected are women.

In addition, bacteria, fungi, and some parasites can cause an autoimmune disease or increase an autoimmune response already in progress. The connection between infection and autoimmune disease is still being studied. It is sometimes difficult to establish a connection between an infection and an autoimmune response, because sometimes the infection has been over for a long time before the abnormal immune response begins. Oddly enough, recent studies also show the opposite effect. Some infections can actually offer protection from autoimmune responses. This happens when the immune system stops attacking a patient's own tissues and starts to defend the body against the infection. This may someday lead to a treatment for autoimmune conditions.

Table BB3.2	Autoimmune Diseases and Infections	
Autoimmune Disease	**Associated Infection(s)**
Multiple sclerosis	Measles virus, Epstein-Barr virus (EBV)
Rheumatoid arthritis	Hepatitis C virus, Lyme disease
Type 1 diabetes	Mumps and rubella virus

A gene complex called the HLA system controls a person's immune system and may determine who will develop an autoimmune disease. The genetics of the immune system and the HLA complex will be discussed in Chapter 13.

BB3.8 How does the environment affect where and how often a genetic disorder occurs?

Sometimes both populations and the environment play a part in genetics. One example is sickle cell anemia. It is most common among people who live in West Africa, in the lowland areas around the Mediterranean Sea, and in some regions of the Middle East and India, but is almost nonexistent in most other populations.

In Chapter 5, Marcia Johnson and her husband both had ancestors that came from West Africa, which would put them at higher risk for carrying the sickle cell gene. But why do people in this region of Africa have a higher frequency of sickle cell anemia? Geneticists often look to the environment for factors that may be responsible. In the case of sickle cell anemia, the relationship between a gene and the environment has been clearly established; in other cases, the link is elusive and still unknown.

As discussed in Chapter 4, sickle cell anemia is an autosomal recessive disorder caused by a single amino acid substitution in the oxygen-carrying protein, hemoglobin. The gene for this disorder is present in very high frequencies in some populations where malaria, an infectious disease, is present. Scientists discovered that carrying the mutant allele for sickle cell anemia gives one resistance to malaria, making sickle cell anemia and resistance to malaria interrelated. The reasons for this relationship will be discussed in Chapter 14.

Glossary

5-HTT A gene that regulates nerve impulses.

autoimmune disease A condition in which a person's immune system attacks his or her own tissue.

endoxifen An anti-estrogen compound used in cancer chemotherapy.

multifactorial traits Traits that are caused by both genetics and environment.

phenylthiocarbamide (PTC) A chemical that only certain members of a population can taste.

tamoxifen A drug given to breast cancer patients that controls the amount of estrogen in the bloodstream.

10 Polygenic and Multifactorial Inheritance

✪ CENTRAL POINTS

▪ Polygenic traits are controlled by two or more genes.

▪ Multifactorial traits are polygenic with an environmental component.

▪ Spina bifida is an example of a multifactorial trait.

▪ There are many other multifactorial traits.

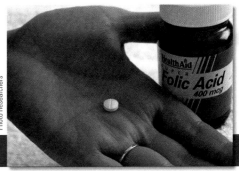

Faye Norman/Science Photo Library/ Photo Researchers

CASE A Prenatal Pills

Just after her friends Brian and Laura went to see Dr. Franco (see Case A in Chapter 2), Vera Smith found out she was pregnant. She was 33 and single. She knew that she should see a doctor right away, but her schedule was hectic. Her friends told her to eat right and not to take any medication, not even an aspirin, because it might hurt her fetus.

When Vera was almost three months pregnant, her mother called from California. She mentioned that Vera should see a doc-

tor for prenatal care. Vera called Laura and got Dr. Franco's number. At her appointment, the nurse first asked whether she was taking folic acid. Vera had never heard of folic acid and said proudly, "I am not taking anything; I don't want to hurt the baby." The nurse pointed to a sign that said all women should take folic acid when pregnant to reduce the chance that the baby would be born with a birth defect called spina bifida. Vera wondered what that meant; the nurse

gave her a pamphlet about nutrition, folic acid, and pregnancy.

Some questions come to mind when reading about Vera and her pregnancy. Before we can address those questions, let's look at how genes and environmental factors work together to affect phenotypes and the outcomes of pregnancy.

10.1 What Is a Polygenic Trait?

Have you ever wondered why a large group of people in a room, on a train, or in a coffee shop all look somewhat different but yet still pretty much the same? This is probably because of **polygenic traits,** which are traits that are determined by two or more pairs of genes.

An example of a polygenic trait is the human immune system, which is responsible for protecting us against infection. This system can have at least 30 million different genotypes, which are a combination of many alleles of multiple genes. The immune system determines blood types, the acceptance or rejection of transplanted organs, and susceptibility to a number of disorders (the immune system will be discussed in detail in Chapter 13). The color of skin, hair, and eyes are polygenic traits because they are influenced by more than one gene, often located on different chromosomes. The result is the slight and often variable range of differences we see throughout a population or in a family.

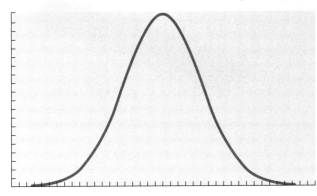

Polygenic traits usually can be measured in some way. This includes height, skin color, and sizes of body parts. This measurable aspect of the phenotype is called the **trait value.**

The graph on the left shows the distribution of a typical polygenic trait in a population. The curve shows that most people have an average trait value. In this distribution, called a **bell curve,** the values for most individuals are clustered around the average, and there are very few individuals at the extremes of the phenotype. Skin color is a polygenic trait with wide phenotypic differences; some people are very pale, while others are very dark. These differences, which are caused by several different genes and environmental factors, can occur in a large population or in a large family.

Polygenic traits by definition involve a number of genes, each of which contributes only a small amount toward the phenotype. Imagine a polygenic trait controlled by 10 genes. There would be many more than 10 possible combinations of phenotypes from these genes. But the situation is more complicated because some genes contribute more than others, and, in addition, the environment often strongly influences the phenotype of a polygenic trait.

10.1 Essentials

- A polygenic trait is one that is caused by two or more genes.
- Polygenic traits usually have a bell-shaped distribution curve in populations.

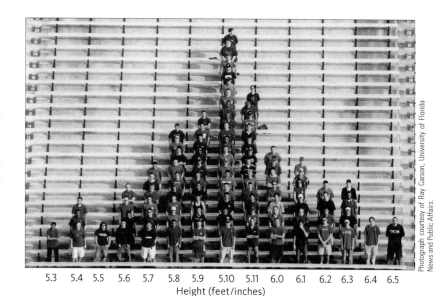

Photograph courtesy of Ray Carson, University of Florida News and Public Affairs.

5.3 5.4 5.5 5.6 5.7 5.8 5.9 5.10 5.11 6.0 6.1 6.2 6.3 6.4 6.5
Height (feet/inches)

■ **Figure 10.1 Height as a Multifactorial Trait**
The distribution of height among males in a biology class.

10.2 What Is a Multifactorial Trait?

A **multifactorial trait** is controlled by two or more genes *and* is affected by environmental factors. Although each gene controlling a multifactorial trait is inherited in Mendelian fashion (see Chapter 4), the interaction of genes with the environment can produce many different phenotypes. Height is a good example (■ Figure 10.1).

If you look at the photo above, you can see that there is a wide range in the distribution of height among the students. Height is a multifactorial trait because it is determined by more than one gene *and* because environmental factors such as nutrition can contribute to adult height. The graph below shows how variation in height can be distributed in a group of men:

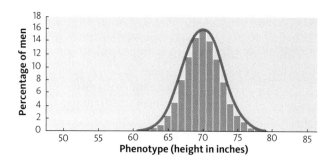

What are the characteristics of multifactorial traits?

Multifactorial traits have several important characteristics:

- Several genes control the trait.
- The trait is not inherited as dominant or recessive.

- Each gene controlling the trait contributes a small amount to the phenotype.
- Environmental factors interact with the genes to produce the phenotype.
- There are many phenotypic differences in the trait within a population; when these distributions are graphed, they form a bell-shaped curve.

Height is clearly a multifactorial trait; it fulfills all of these requirements.

One of the first questions to consider about a multifactorial trait is, how many genes control it? Then we need to determine if there are environmental influences on the trait. For height, environmental factors include hormones and diet. Determining the amount of interaction between genes and specific environmental factors can be a difficult task.

Which diseases are multifactorial traits?

Many common diseases are multifactorial traits, and people with these diseases make up the largest part of patients hospitalized with genetic disorders. The photos on the next page show some of these conditions.

Some other diseases that show multifactorial influences are cancer (discussed in Chapter 11), spina bifida (discussed shortly), hypertension, and cardiovascular disease.

10.2 Essentials

- When a polygenic trait has a significant environmental component, it is called a multifactorial trait.
- The environment can interact with a multifactorial trait in many ways.

Diabetes

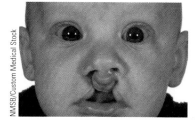

Cleft lip and palate

Club foot

10.3 What Is Spina Bifida?

Spina bifida (SB) is a birth defect involving the nervous system. It appears in the first month of embryonic development when the spinal column forms. SB belongs to a group of conditions called **neural tube defects.** These disorders get their name because they involve problems with the development of the embryo's nervous system (■ Figure 10.2).

The neural tube gives rise to the brain and the spinal cord. Other cells form the meninges, the membranes that cover and protect these structures. The drawing on the right shows the major parts of the nervous system. Notice that the meninges surround the brain and spinal cord.

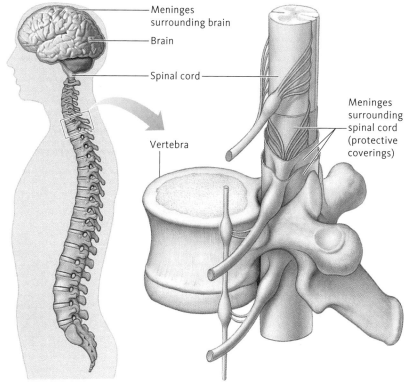

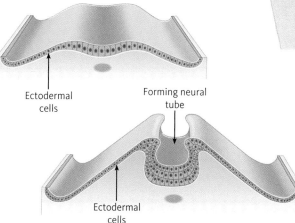

■ **Figure 10.2 Formation of the Neural Tube**
A cell layer called the ectoderm normally folds over the cells that will eventually form the spinal cord, and forms the neural tube.

SB is one type of neural tube defect. The three basic forms of SB are shown in the figure on the next page.

The most extreme form of a neural tube defect is called **anencephaly.** In this condition, the head end of the neural tube does not close. As a result, major portions of the brain and skull do not form, and the remaining portions of the brain may not be enclosed in the skull. (See photo in Case B on page 171.) Fetuses with this condition can survive only within the mother and most are not born alive. Those that are born usually die within a few hours or days due to heart and breathing problems.

In many cases where the lower end of the neural tube does not close, the spine can be surgically repaired, but damage to the nervous system is permanent. Children born with

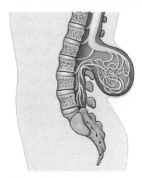

The spinal cord and meninges protrude from the base of the spine.

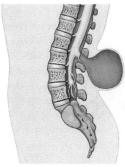

The spinal cord develops normally, but the meninges protrude.

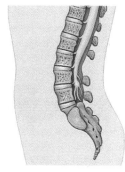

One or more vertebrae are malformed, but nothing protrudes from the body.

SB have varying degrees of paralysis depending on where the opening occurs. In addition, most individuals with SB have learning disabilities and may have bowel and bladder problems.

There is no cure for SB. The damaged nerves cannot be repaired; treatment consists of therapy and assistance in the form of crutches, braces, or wheelchairs. Complications that arise later in childhood may require further surgery or other therapies. Most individuals with SB live into adulthood.

PHOTO GALLERY
This child with spina bifida is functioning well.
Ellen Senisi/PhotoEdit

Spina bifida can be diagnosed by several methods of prenatal testing (see Chapter 7). Vera Smith waited to take folic acid and its preventative effects will not help her. However, maternal blood tests and ultrasound scanning might reveal the presence of a neural tube defect.

Is spina bifida a multifactorial trait?

Spina bifida is a multifactorial trait. It tends to cluster in families; the risk of having a second child with SB or another neural tube defect is significantly increased.

BY THE NUMBERS

10%
Percentage of newborns that will have a diagnosed multifactorial trait during their lifetime.

1 in 1000
Frequency of spina bifida worldwide.

3%
The risk of having a second child with spina bifida.

image100/CORBIS

61%
Percentage of adults in the United States who were overweight according to a 2006 study.

26%
Percentage of adults who were identified as obese in the same study.

A gene associated with neural tube defects has been identified. This gene, called *VANGL1*, is found on chromosome 1, the largest human chromosome. Below is a diagram of chromosome 1 showing the location of this gene.

VANGL1 gene ———

Chromosome 1

In mice, the *VANGL1* gene acts early in the development of the neural tube and mutations in this gene cause conditions similar to SB. Based on the gene's action in mice, a study of more than 100 patients with neural tube defects showed that all affected individuals had mutations in the *VANGL1* gene. One of these mutations, called **V239I**, may cause a partial loss of function in the VANGL1 protein. This mutation, along with other environmental factors, may produce some forms of SB.

What are the environmental risk factors for spina bifida?

Recently, one significant environmental risk factor for SB has been identified. Diets deficient in **folic acid**, a B vitamin, have been shown to be a risk factor for SB. Dietary sources of folic acid include whole grains, green leafy vegetables, and fruit. It is recommended that all women of childbearing age take 0.4 mg of folic acid every day for at least three months before pregnancy, and continue taking it until the 12th week of pregnancy. This is necessary even though foods in the United States have been fortified with folic acid since 1998. For women who have previously had a child with SB or other neural tube defects, the recommended daily dose is 4 mg/day. This dose is ten times higher than for women who have not had a child with a neural tube defect. Taking folic acid beginning three months before pregnancy has been shown to reduce the risk of SB and related conditions by about 70%. How folic acid interacts with genes that control the formation of the neural tube is still unknown. Other environmental factors, if any, remain to be discovered.

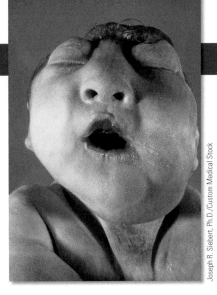

Joseph R. Siebert, Ph.D./Custom Medical Stock

When Samantha found out she was pregnant, she was extremely happy. She already had a wonderful, healthy little girl. At her first ultrasound, Samantha was shocked that the doctor didn't say everything was all right; her face showed worry. When Samantha asked what was wrong, Dr. Chang said the head of the baby looked small, and she was going to send the ultrasound out to an expert.

The expert's analysis confirmed that Samantha's baby had anencephaly, a condition in which the brain and head are abnormally small, sometimes absent.

"Will my baby live?" Samantha asked.

"It will probably live to be born, but it will not live very long after birth," Dr. Chang told her. "In California, parents who had a child with a similar condition donated her organs to babies who needed a transplant. If you might be interested in doing this, it can been done soon after your baby is born."

Dr. Chang had read that the couple in California placed their baby on a respirator at birth. Then, they donated her heart, lungs, and kidneys to babies waiting for organs.

Samantha thought about her other choices. Should she carry a baby to term just to have it die? Could she do this?

1. What should Samantha do? List three options.

2. What reasons could Samantha give for carrying her baby and donating the organs?

3. What reasons could she give to abort the baby?

4. Where could Samantha go for help with this decision?

5. If Samantha finds out that her baby may be born dead and the organs would not be usable, should this make a difference in her decision?

Another approach for studying multifactorial traits uses recombinant DNA techniques and information from the Human Genome Project to identify genes that affect specific polygenic traits, such as reading ability and IQ. Using these techniques, scientists have uncovered genes associated with reading disability (developmental dyslexia). To the right is a diagram of chromosomes 6 and 15 showing the location of some genes for cognitive ability. Cognitive ability goes beyond IQ and tests for verbal and spatial abilities, memory, speed of perception, and reasoning.

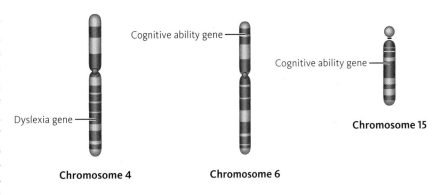

Chromosome 4

Chromosome 6

Chromosome 15

Dyslexia gene

Cognitive ability gene

Cognitive ability gene

More recently, a gene associated with cognitive ability has been identified on chromosome 4. Mutations in this gene are responsible for dyslexia. The location of this gene is shown on chromosome 4, whose illustration is on the right as well.

The accumulated results of research indicates that intelligence is a polygenic and multifactorial trait.

10.4 Essentials

- Obesity is a multifactorial trait.
- Intelligence is another example of a multifactorial trait.
- The interaction of environment and heredity is an active area of research in genetics, psychology, and other fields.

10.5 What are the Legal and Ethical Issues Related to Multifactorial Traits?

Vera's case brings a number of legal issues to mind. For example, if a patient does not want to take folic acid, does she have the right to refuse? If she did refuse, could her doctor force her to take it?

This situation is related to the question of consenting to one's medical treatment. We have discussed informed consent in previous chapters. In Chapters 5 and 7, Marcia and Victoria consented to have testing for a genetic condition.

Marcia had herself tested and Victoria allowed her child to be tested. How does the law handle this and the right to refuse both testing and treatment?

Any patient can consent to treatment but can also refuse treatment for themselves and sometimes their child. Under United States law, any adult can give consent or refuse medical treatment or testing without a problem. Problems often arise when a child is involved.

In the cases in this chapter, both Vera and Samantha are pregnant. A pregnant woman can consent to medical treatment for herself and her fetus. This is based on the fact that the fetus is inside the woman's body and therefore she is the patient. After her baby is born, Samantha can donate its organs because she is the parent and a child cannot consent to donation.

Informed consent and consent to medical treatment have a number of exceptions and one of them is age. If a child is younger than 16 (this varies by state) they cannot consent and their parents or guardians must sign a consent form. No state gives that right to fetuses or newborns.

If a woman is pregnant, no matter what her age, she is considered an adult as far as medical decisions are concerned. Many pregnant women have medical problems (diabetes, etc.) and need special treatment. If they refuse, doctors will try to convince them to accept treatment but they cannot be forced.

The chart below addresses some of the ethical and legal questions regarding consent to treatment:

Question	How are these questions decided?	Related case or legal issue
What might happen if a pregnant woman takes drugs that would harm her fetus?	Usually, not much can be done to force a woman to stop taking drugs, even if those drugs are illegal.	Some judges have tried to put pregnant women in jail to stop their drug intake but this has been shown to be unconstitutional. In one case, a judge tried to put a pregnant woman in jail when it became known she had abused her previous children. She was released and he was taken off the bench.
Can a woman be forced to take medication that is beneficial to her fetus but not to her?	States do not force a woman, or any other patient, to have medical treatment because they are considered adults and can, therefore, make their own decisions.	No cases have tried to give a woman medication or vitamins if she refuses. Judges would probably not allow this. Some religions do not allow medical treatment and the right to refuse has been upheld only for adults.
If a newborn or fetus needs special treatment but the mother refuses, what can be done?	Laws allow physicians to go to court and file an injunction for treatment of critically ill newborns or children, overriding the parents refusal. This has not been extended to fetuses.	Members of the Jehovah's Witnesses have refused blood transfusions for their children and in life and death situations courts have issued injunctions forcing treatment.

Scientists have been trying to measure intelligence for quite a long time. Francis Galton (Darwin's first cousin) thought that all inherited traits, including intelligence, could be measured. But, as mentioned in this chapter, it was a Frenchman, Alfred Binet, who developed an intelligence test when the French government paid him to find a way to distinguish normal children and what were then termed inferior children.

Binet's test came to the United States in 1917 and was used to test soldiers in World War I to see what jobs they would be suited for. Soon after, a Binet test was administered to a prisoner on trial for murder in Wyoming. Because the prisoner fared so poorly on the test, the jury acquitted him by reason of his mental condition. Schools around the world began using the Binet test to place children into learning groups. The tested IQ value followed a person throughout his or her life.

Originally, IQ was calculated as the following ratio: 100 times the person's mental age (as determined by the Binet test) divided by his or her chronological age.

An average IQ, as shown by the bell curve in ■ Figure 10.4, was considered to be 100.

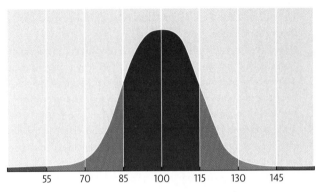

| 55 | 70 | 85 | 100 | 115 | 130 | 145 |

■ **Figure 10.4 IQ and the Bell Curve**
This distribution of adult IQ scores is a typical bell curve. The average score is 100; approximately half of the scores are below 100 and half are above.

This type of calculation was originally applied only to children; but in 1939 David Wechsler published the first intelligence test explicitly designed for an adult population: the Wechsler Adult Intelligence Scale (WAIS).

Although the inheritance of IQ has been investigated for nearly a century, controversy remains as to how much is genetic, what genes control it, and the mechanisms for its heritability. Many studies and educators asked, "How can we test for intelligence?" and "What should be included in an intelligence test?" In educational settings, children were being placed in classes using these tests, which often turned out to be biased against certain groups because of the questions asked. So, the IQ test slowly became discredited.

In 1994, the publication of *The Bell Curve,* written by psychologist Richard J. Herrnstein and political scientist Charles Murray, changed perception of IQ tests yet again. In their book, the authors tried to analyze IQ testing of U.S. children to promote social change. However, to this day, many of their conclusions are controversial, such as the following:

1. Low measured intelligence and antisocial behavior are linked.
2. Low test scores of African Americans (compared to those of whites and Asians) were said to be caused by genetic factors.
3. The bell curve of IQ test scores was said to prove that some groups were inferior and linked poverty with low intelligence.

This book caused an uproar and induced many people to rethink the use of IQs as a measurement of intelligence.

PHOTO GALLERY
An IQ test.
Stockdisc/Alamy Limited

QUESTIONS:

1. If we could determine that intelligence (as measured by IQ) was caused by certain genes, what would this mean to our society?
2. Research what happened after *The Bell Curve* was published.
3. The following question is commonly used to show bias in an IQ test for young children: After three drawings are shown, the child is asked to pick out the handbag. Why do you think this could be considered biased?
4. Many schools do not use IQ to place children in special-education classrooms. Do some research to find three reasons why.
5. Research Mensa, a group that claims to have members with very high IQs. Who are some members of this organization?

✪ THE ESSENTIAL TEN

1. A polygenic trait is one that is controlled by two or more genes. [Section 10.1]
   ❚ The term polygenic means many genes.

2. Polygenic traits usually have a bell-shaped distribution curve in populations. [Section 10.1]
   ❚ This means that the traits or conditions have varying degrees of expression, from slight to severe.

3. Polygenic traits with an environmental component are called multifactorial traits. [Section 10.2]
   ❚ This means that both genetic and environmental factors affect expression of the condition.

4. The environment can interact with a multifactorial trait in many ways. [Section 10.2]
   ❚ Environmental factors can alter the expression of a trait.

5. Spina bifida is one example of a multifactorial trait. [Section 10.3]
   ❚ There are many other multifactorial traits.

6. Spina bifida is one member of a class of genetic conditions called neural tube defects. [Section 10.3]
   ❚ Spina bifida affects the formation of the spinal cord and brain.

7. Taking folic acid reduces a woman's chance of having a child with spina bifida. [Section 10.3]
   ❚ Adding folic acid to the diet is one way of altering the environment that can affect the risk and outcome of a multifactorial condition.

8. Obesity is a multifactorial trait. [Section 10.4]
   ❚ Genes and environmental factors such as diet affect this phenotype.

9. Intelligence is another example of a multifactorial trait. [Section 10.4]
   ❚ Defining intelligence is the first step in researching its inheritance and the effects of environment on its phenotype.

10. The interaction of environment and heredity is an active area of research in genetics, psychology, and other fields. [Section 10.4]
    ❚ As more multifactorial traits are identified, investigators work to understand how each affect the body.

KEY TERMS

anencephaly (p. 167)
bell curve (p. 165)
concordance (p. 170)
dermatoglyphics (p. 169)
Drosophila (p. 170)
folic acid (p. 168)
intelligence quotient (IQ) (p. 170)
meninges (p. 167)
monozygotic (MZ) twins (p. 170)

multifactorial trait (p. 166)
neural tube defects (p. 167)
ob gene (p. 170)
polygenic trait (p. 165)
spina bifida (SB) (p. 167)
trait value (p. 165)
V239I (p. 168)
VANGL1 (p. 168)

REVIEW QUESTIONS

1. List two environmental factors that may influence cancer and heart conditions.
2. If we haven't yet identified a gene for a condition, can the condition still be considered to be multifactorial? Why or why not?
3. Think of one other human genetic condition that you think might be multifactorial. Then identify the genetic and environmental influences.
4. Make a chart of all the multifactorial conditions listed in the chapter and include their genetic and environmental components. Do some research and add to this information.
5. What can a pregnant woman do to reduce her risk of having a child with spina bifida?
6. Explain what is meant by the term trait value and give an example.
7. Is it harder or easier to find the genes for polygenic traits than for traits controlled by single genes? Why?
8. Research information on club foot. What is done to treat this condition?
9. Explain why fingerprints are considered a multifactorial trait.

APPLICATION QUESTIONS

1. What if a gene whose mutant allele caused a decrease in intelligence was identified? What might be done if a baby was shown to carry this gene at birth?
2. Pregnant women have been jailed to keep them from harming their children. Usually this occurs, albeit rarely, with drug addicts who will not stop taking drugs while they are pregnant. Is this a solution for a woman who will not take folic acid? Why or why not?

3. Consider a family in which one parent has very dark skin and the other very light skin. Do some research and discuss what their children will look like and why. If you can, find a photo of such a family.

4. As we know, environmental factors influence multifactorial traits. What might happen if the environmental conditions were altered? For example, suppose a parent learns that her newborn has a mutation in the leptin gene and therefore has a greater risk for obesity. She then feeds him only the right foods and keeps an eye on him. Would this make a big difference? Why or why not?

5. Is it possible that most human traits are multifactorial? Explain your answer.

6. Do some research on conjoined twins. Are they dizyogtic or monozygotic? Why?

7. If educators study SAT scores in a certain group of Americans and find that the group had especially high scores, could the conclusion be that it is hereditary? Why or why not?

8. A man who is 7 feet tall is attracted to women who are short (5 feet). He really wants to have tall children. If he marries a woman who is short, does he have a chance to have a tall son or daughter? Explain your answer.

ONLINE RESOURCES

Preparing for an exam? Log on at academic.cengage.com/login for a pre-test, a personalized learning plan, and a post-test in CengageNOW's Study Tools to help you assess your understanding.

If assigned by your instructor, the Case B and Spotlight on Society activities for this chapter, "Donation of a Baby's Organs" and "The Bell Curve," will be available in CengageNOW's Assignments.

What Would You Do If . . . ?

1. Your child was asked to take an IQ test?

2. Based on your child's IQ test, he or she was going to be placed in a class for students with low IQ?

3. Based on your child's IQ test, he or she was going to be placed in a class for students with high IQ?

4. Your prenatal diagnosis showed that your child has spina bifida?

5. You were given supplements with folic acid for your pregnancy?

dBusinessNews.com, October 1, 2007

Spina Bifida Rates Are Falling in North Carolina

In the last ten years, the rates of spina bifida in North Carolina have been greatly reduced. The March of Dimes is recognizing North Carolina for its program to increase women's consumption of folic acid. Television and print media have been used to stress the importance of this prenatal vitamin. The following statistics demonstrate the program's success:

- Between 1995 and 2005, the number of children born with neural tube defects has decreased 40%.
- Between 1995 and 2005, the number of children born with spina bifida has decreased 45%.
- In western North Carolina, where the program is the oldest, the numbers have dropped even more (80%).

To access this article online, go to academic.cengage.com/biology/yashon/hgs1.

QUESTIONS:

1. If this program works so well, how might the people in North Carolina expand it across the country?
2. Would this work as well for other multifactorial conditions? Why or why not?

Evansville (Indiana) Courier & Press, October 9, 2007

Taking Aim to Reduce SIDS

Libby Keeling

SIDS is the sudden and unexplained death of an apparently healthy live baby aged one month to one year.

Looking at the risk for a specific infant, doctors see that a triple-risk model may be the culprit. Doctors feel that the risks combine three things: first, an infant's genetic predisposition, such as a brain abnormality; then, an unstable period of growth; and finally, an environmental trigger, such as tobacco smoke.

The program, called "Back to Sleep," urged parents to make sure their babies were sleeping on their backs. For some reason, this has caused a drop in SIDS deaths.

Specialists say it's a multifactorial thing, but police still investigate every death due to SIDS.

To access this article online, go to academic.cengage.com/biology/yashon/hgs1.

QUESTIONS:

1. Twenty years ago a specialist in childhood death wrote a book saying that SIDS was genetic, and this was the primary thinking for 10 years. A woman who had five babies die of SIDS was found not guilty of murder based on his books. Research this case and write a short report on the surprising result.

2. How do you think sleeping on their backs keeps babies safe?

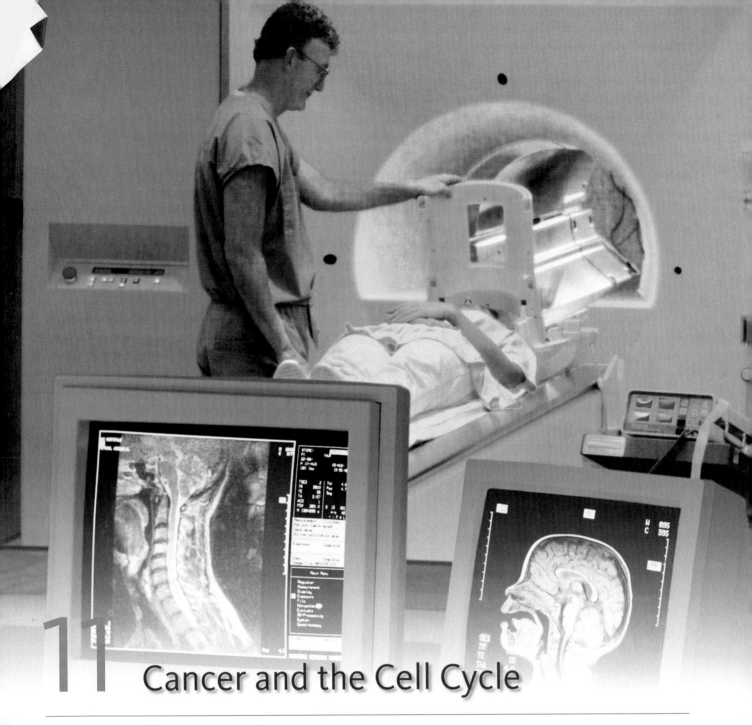

11 Cancer and the Cell Cycle

✪ CENTRAL POINTS

- Cancer involves uncontrolled cell division.
- Mutations in certain types of genes may lead to cancer.
- Cancer is a disease of the cell cycle.
- Breast cancer is a common form of cancer.
- Chromosomal changes are often a cause of cancer.
- Environmental causes of cancer are being studied.
- Some lawsuits have addressed smoking as a cause of cancer.

CASE A Patient Offered New Cancer Treatment

Harriet Abeline had faced bad news before. She was diagnosed with breast cancer five years ago and underwent a lumpectomy and radiation treatment. Two years later, a routine mammogram revealed another lump. She faced this challenge with similar treatment.

Harriet understood medicine; for years she had worked as a recruiter for a major drug company interviewing and hiring scientists and physicians. The company was a reputable and honest business and developed drugs that were used all over the world.

So with each recurrence of cancer, she researched her situation and the available drugs being used for treatment. With each Google search, she read about the experimental treatments that were available for end-stage breast cancer, keeping them in the back of her mind.

Now, as her doctor told her that the cancer had spread to her bones, Harriet was frightened. Dr. Hill understood her fear. She had been working with cancer patients for years and was involved in some of the newer treatments.

She told Harriet that she could enter phase 2 FDA clinical trial for a new genetically engineered drug that would target her specific type of tumor. Harriet knew that phase 2 trials are designed to test for side effects and risks of the treatment. There was no guarantee of a cure.

Some questions come to mind when reading about Harriet's dilemma and the decision she faced. Before we can address these questions, let's look at the biology behind cancer.

11.1 What Is Cancer?

Cancer is a complex disease that affects many different cells and tissues in the body. It is characterized by uncontrolled cell division and by the ability of these cells to spread, or *metastasize,* to other sites in the body. These cells are generally referred to as **malignant.** This type of unchecked growth may result in death, making cancer a devastating and feared disease. The illustrations below show some of the differences between normal and malignant cells.

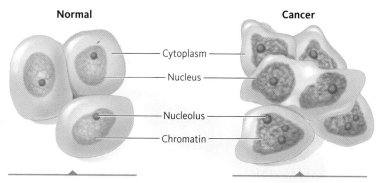

Normal	Cancer
• Large cytoplasm	• Small cytoplasm
• Single nucleus	• Multiple nuclei
• Single nucleolus	• Multiple and large nucleoli
• Fine chromatin	• Coarse chromatin

Improvements in medical care have reduced deaths from infectious disease and have led to increases in life span, but as a person lives longer, cancer becomes a major cause of illness and death. The risk of many cancers is age-related, and because more Americans are living longer, they are at increased risk of developing cancer.

The link between cancer and genetics was discovered early in the 20th century by Theodore Boveri. He proposed that normal cells become malignant because of changes in their chromosomes.

Studying the pedigrees of families with cancer reveals that in some cases, the disease has a genetic component. The pedigree below is of a family with breast cancer.

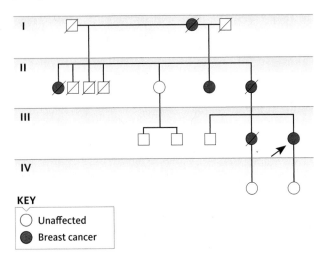

KEY

○ Unaffected
● Breast cancer

A change in the genetic makeup of a cell, called a **mutation,** can cause the cell to become cancerous and to divide uncontrollably. Cancer, then, is a genetic disorder that acts at the cellular level.

Some agents in our environment, called **carcinogens** (such as ultraviolet light, chemicals, and viruses), and certain behaviors (such as diet and smoking) can increase the rate of these mutations, thereby playing a part in cancer risks. These risks are usually expressed as a frequency (such as one in a million). This frequency describes the possibility that someone will develop cancer after exposure to a toxic agent or an increase with age.

Not all tumors are cancerous. Noncancerous growths, called **benign** tumors, increase in size, but do not spread (metastasize) to other tissues. Some of these tumors, including cysts, usually cause problems only when they become large enough to interfere with the function of neighboring organs.

Three things identify growths as cancerous:

1. The tumors begin with a single cell that reproduces by *mitosis.*
2. The cells in these tumors divide continuously.
3. These cells are invasive and move to other sites in the body, a process called **metastasis.**

11.1 Essentials

- Cancer is defined as uncontrolled cell division.

11.2 How Are Genes Involved with Cancer?

Two classes of genes cause cells to become cancerous when they mutate: **oncogenes** and **tumor suppressor genes.** The normal alleles of these genes work in opposite ways to contol cell division.

Normal cells carry genes that turn on cell division, usually in response to signals from outside the cell. These genes are called **proto-oncogenes.** Mutated alleles of these genes are called *oncogenes.* When activated, oncogenes cause cells to divide uncontrollably.

The normal alleles of the other class, called *tumor suppressor genes* turn off cell division and keep cells from forming tumors. When they are mutated, control over cells division is lost, and cancerous tumors form.

How does a normal proto-oncogene become an oncogene?

As discussed earlier (see Chapter 5), mutations can be caused by errors in DNA replication, exposure to environmental agents, or lifestyle choices. When a proto-oncogene is exposed to these agents, it may mutate and be changed to an oncogene, causing a normal cell to become cancerous.

Certain viruses can also play a part in causing cancer-causing mutations. In these cases, after a virus enters a normal cell, it can trigger the activation of events that transform the cell into a cancerous cell. An example is infection by the **human papillomavirus (HPV).** Interaction between the proteins made by HPV and and those made by the cell cause uncontrolled division of cells in the cervix. The result is cervical cancer. Almost 12,000 women will be diagnosed with this form of cancer in 2008.

Mistakes in DNA replication can also cause mutations, and some of these may lead to cancer.

Earlier chapters described how mutation at a single base pair can cause a serious condition such as sickle cell anemia. In cancer, the mutation may be at one or many sites on a gene. A cancer-causing mutation can occur in one or both copies of a gene.

11.2 Essentials

- Cancer appears to run in some families.
- Proto-oncogenes, oncogenes, and tumor suppressor genes control cell division.
- When these genes are mutated, cancerous tumors form.

11.3 What Is the Cell Cycle?

Cells in the body alternate between two basic states: division and nondivision. The time between divisions varies from minutes to months or even years. The sequence of events from division to division is called the **cell cycle.**

The cell cycle consists of two parts: **interphase** and **mitosis** and **cytokinesis.**

The first part of the cell cycle, called *interphase,* is the time between divisions. It has three stages: **G1, S,** and **G2.** Division occurs during the other part of the cell cycle: *mitosis* (division of the chromosomes), and *cytokinesis* (division of the cytoplasm).

Cancer is a disease of the cell cycle. Certain checkpoints in the cell cycle regulate cell division. Cancerous cells bypass these checkpoints, resulting in uncontrolled cell division. Since this fact was discovered, studies of the cell cycle have become an important part of cancer research. If we can find ways to restore control of the cell cycle, we might be able to stop cancerous cells from dividing.

Follow the drawing below as we discuss the cell cycle, beginning with a cell that has just finished division. After division, the two daughter cells are about half the size of the parent cell. Before they can divide again, they must undergo a period of growth. Most of this growth takes place during the first of the three stages of interphase: *G1, S,* and *G2.*

1. G1 begins immediately after division, and is a time when new organelles (including membranes and ribosomes) and cytoplasm are formed. By the end of G1, the cell has doubled in size.

2. During the S phase, a duplicate copy of each chromosome is made (in humans, this means that all 46 chromosomes are copied).

3. During the G2 phase, the cell prepares to begin mitosis. By the end of G2, the cell is ready to divide.

4. During the M phase, the cell divides by mitosis. Following mitosis, the cytoplasm divides and two cells are formed.

STAGES OF THE CELL CYCLE

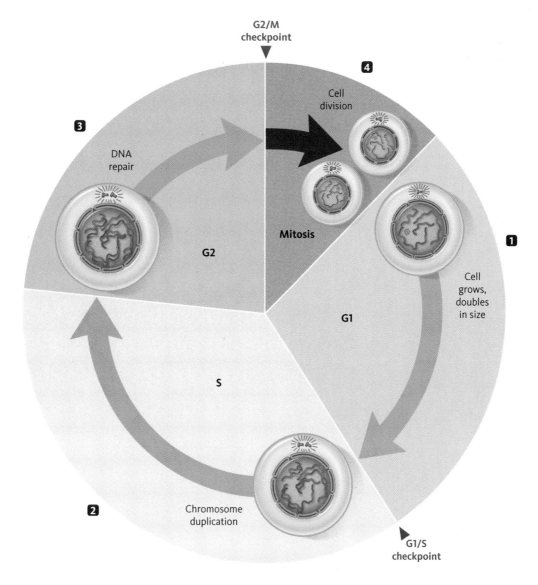

What are the stages of mitosis?

When a human cell completes G2, it enters mitosis, the second major part of the cell cycle. During mitosis, a complete set of chromosomes is distributed to each daughter cell, and the cytoplasm is divided between the new cells. Although mitosis is a continuous process, for the sake of discussion, it is divided into four stages: prophase, metaphase, anaphase, and telophase.

A. Prophase: At the beginning of prophase, the replicated chromosomes (remember, they were copied during the S stage of interphase) condense and become visible. Each of the 46 duplicated chromosomes consists of two sister chromatids joined by a centromere. Near the end of prophase, the nuclear membrane breaks down, and a network of specialized fibers called spindle fibers form in the cytoplasm and stretch across the cell.

B. and C. Early Metaphase and **Metaphase:** Metaphase begins when the chromosomes, with spindle fibers attached, move to the middle of the cell.

D. Anaphase: During anaphase, the centromeres divide, converting each sister chromatid into a chromosome. After this, the newly formed chromosomes migrate toward opposite ends of the cell. At the end of anaphase, there is a complete set of 46 chromosomes at each end of the cell.

E. Telophase: In the final stage of mitosis, the chromosomes begin to unwind, the spindle fibers break down, and the nuclear membrane begins to re-form.

What happens in cytokinesis ?

In this last stage of the cell cycle, the cell membrane constricts and gradually divides the cell into two daughter cells, each of which contains a set of 46 chromosomes.

STAGES OF MITOSIS

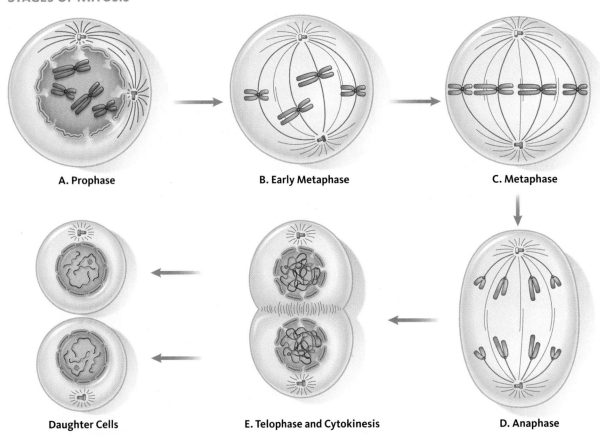

A. Prophase

B. Early Metaphase

C. Metaphase

D. Anaphase

E. Telophase and Cytokinesis

Daughter Cells

What helps regulate the cell cycle?

The cell cycle is regulated at a number of points (see the diagram on the previous page):

1. A point in G1 just before cells enter S, known as the **G1/S checkpoint**
2. The transition between G2 and mitosis, called the **G2/M checkpoint**

The genes we spoke of earlier, the tumor suppressor genes and proto-oncogenes, control these checkpoints. Proteins encoded by the tumor suppressor genes turn off or decrease the rate of cell division. The proto-oncogenes encode proteins that turn on or increase the rate of cell division. Products of these genes normally act at either the G1/S or G2/M control points to regulate the movement of the cell through the cell cycle.

In normal cells, signals from outside the cell can activate tumor suppressor genes or the proteins they produce to turn off cell division. This signaling can also activate proto-oncogenes to turn on cell division. This process is called **signal transduction.**

These signals originate outside the cell and can be proteins, hormones, or nerve impulses. Signals can also come from the environment and may include steroids, pollutants, and other molecules.

As shown in the figure below, signal transduction begins when a signal molecule binds to a receptor in the cell's plasma membrane. This binding sets off a series of interactions inside the cell among protein molecules. In many cases, the signal molecule remains outside the cell. The binding of the molecule changes the shape of the receptor and allows it to transmit a signal to other proteins in the cell. These proteins in turn alter other proteins, activating them to generate a cellular response.

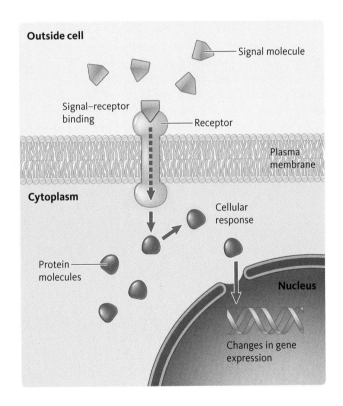

Generally, the basic process of signal transduction involves protein-protein interaction. As the signal is passed along, it is changed, or *transduced,* from one form to another. For example, a signal may bind to and change the shape of a receptor; this may in turn cause a protein within the cell to become chemically modified and interact with and activate a second protein.

This activated protein may bind to yet another protein and this pair of proteins may interact with a third protein,

causing a domino effect, eventually resulting in a new pattern of gene expression.

How is signal transduction related to cancer?

Cancer is related to a loss of cell cycle control. This often involves mutations or change in a cell's signal transduction pathway or the cell cycle control machinery.

Let's look at an example of how a mutation in a proto-oncogene called *RAS* might lead to cancer. The RAS protein (produced by the proto-oncogene) attaches to the inside of the plasma membrane and is part of a signal pathway that turns on cell division. When activated, the RAS protein changes shape and becomes switched on, transferring its signal to another protein in the pathway. Once the signal has been transmitted, the RAS protein changes shape again, switches off, and becomes inactive.

When there is a mutation in the *RAS* gene, the protein becomes stuck in the "on" position, calling for cell division. This causes the cell to lose control of the cell cycle and divide in an uncontrolled fashion, becoming cancerous. *RAS* gene mutations are found in many types of cancer, including colon cancer, pancreatic cancer, and stomach cancer.

11.3 Essentials

- The cell cycle is important to understanding cancer and the control of cell division.

11.4 How Does a Woman Develop Breast Cancer?

Breast cancer is the most common form of cancer in women in the United States. Each year, more than 40,000 women die from breast cancer. An estimated 178,000 new cases are expected in the United States in any given year. Although environmental factors may be involved in breast cancer, geneticists struggled for years with the question, is there a genetic predisposition to breast cancer? After more than 20 years of work, the answer is clearly yes. Mutations in two different genes can predispose women to breast cancer and ovarian cancer. They are called *BRCA1* and *BRCA2.*

The *BRCA1* gene is located on chromosome 17.

In the 1970s, Mary-Claire King and her colleagues analyzed the pedigrees of 1500 families with breast cancer. They found that approximately 15% of the families had multiple cases of breast cancer.

Dr. King decided to test the blood of as many families as possible, trying to locate a genetic marker for breast cancer. Even it was a long shot because most cases of breast cancer, like other cancers, occur

Chromosome 17

at random. In the 1980s, as recombinant DNA techniques became widely available (see Chapter 6), her team began using DNA markers and the polymerase chain reaction (PCR) for genetic screening of blood samples.

Finally, in 1990, after testing hundreds of families using hundreds of markers, the team found a link between breast cancer and a genetic marker called D17S74 on chromosome 17. As discussed in Chapter 9, a *genetic marker* is a small segment of DNA on a chromosome that can be used to find the location of a gene. Many of the original families in Dr. King's study were tested for this marker, and a correlation was found between the marker and breast cancer. This meant that a gene for a predisposition to breast cancer had been located somewhere on chromosome 17, near the genetic marker found by Dr. King. This gene is now called *BRCA1*.

Not all women who inherit a mutation in the *BRCA1* gene will develop breast cancer; what they inherit is an increased risk of developing breast cancer. Women who carry one mutant copy of the *BRCA1* gene will develop breast cancer if their other copy mutates in a breast cell. Approximately 82% of women who inherit one mutant *BRCA1* allele will develop a mutation in the other *BRCA1* allele. Women with a *BRCA1* mutation are also at higher risk (a 44% chance) for ovarian cancer.

A second breast cancer predisposition gene, *BRCA2*, was discovered in 1995. As with *BRCA1*, when mutated, *BRCA2* increases the risk of developing breast cancer. *BRCA2* is found on chromosome 13.

Even though these mutations are rare (less than 1%), in some populations the frequency is much higher. One such population is women of eastern Euro-

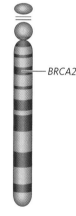

— BRCA2

Chromosome 13

pean Jewish ancestry (**Ashkenazi Jews**); their combined frequency of *BRCA1* and *BRCA2* is 1 in 40.

Many questions about the protein products and the functions of *BRCA1* and *BRCA2* remain unanswered, as do questions about how mutations in these genes lead to breast cancer.

Can men get breast cancer?

Although most people think of breast cancer as a women's disease, men can also develop breast cancer. Men inherit the breast cancer genes mentioned earlier from their parents. Because they have small amounts of breast tissue, it isn't thought of as a male condition. Men are frequently diagnosed at later stages than women because they mistakenly believe that they cannot get breast cancer. Because such cases are detected later, they are often more difficult to treat.

In the United States, about 1% of breast cancers occur in males, with an estimated 2,000 new cases reported each year. In parts of Africa, the rates are significantly higher. In Egypt, males account for 6% of all cases, and in Zambia, male breast cancer represents 15% of all cases.

Risk factors for male breast cancer include age, family history of breast cancer (usually in female family members), and occupational exposure to heat, gasoline, or estrogen-containing creams in the soap and perfume industry. Males of Eastern European Jewish ancestry, also called Ashkenazi Jews, and black males have higher rates of breast cancer. Researchers are working to discover the reasons for increased rates of breast cancer in certain populations.

11.4 Essentials

- **Researchers have discovered two mutant genes, *BRCA1* and *BRCA2*, that increase a woman's risk of developing breast cancer.**

CASE A QUESTIONS

Now that we understand that some cases of breast cancer are caused by a genetic predisposition, let's look at some of the issues raised in Harriet's case.

1. What should Harriet do?

2. If Dr. Hill told Harriet that no other treatment could help her, would that make a difference?

3. Many trials need end-stage patients to see how the drugs work with individuals who are very sick. Can these patients really understand what the trials are about and give informed consent?

4. What should Dr. Hill tell Harriet about the proposed treatment before she agrees? List five items.

5. Informed consent is a very important part of any clinical trial. Is there anything that a researcher or physician shouldn't tell a patient?

6. One serious problem in clinical trials is that researchers often cannot persuade people to participate. Why do you think that is?

11.5 What Are the Other Genetic Causes of Cancer?

Changes in the number and structure of chromosomes are often seen in cancer cells. For example, as discussed in Chapter 3, Down syndrome is caused by the presence of an extra copy of chromosome 21. In addition to most of the symptoms discussed in Chapter 3, people with Down syndrome are 18 to 20 times more likely to develop leukemia (a cancer of the white blood cells) than those in the general population. This was a clue to scientists that they might look on chromosome 21 for a gene associated with leukemia. That connection has not yet been discovered.

The relationship between chromosome rearrangements, or *translocations,* and cancer was first discovered in leukemia. In leukemia there is uncontrolled division of white blood cells. Specific chromosome changes in these cells are easy to identify and are used to diagnose this type of cancer.

In one form of leukemia, a translocation occurs between parts of chromosomes 9 and 22 and they exchange parts. This unusual chromosome rearrangement, (called the Philadelphia chromosome after the city where it was discovered), is common in patients with **chronic myelogenous leukemia (CML).**

Other cancers, including **acute myeloblastic leukemia, Burkitt's lymphoma,** and **multiple myeloma,** are associated with other, specific translocations. The finding that certain forms of cancer are consistently associated with specific types of chromosomal abnormalities is significant in the study of these cancers.

The figure below shows a number of genes associated with cancer and the chromosomes on which they are located.

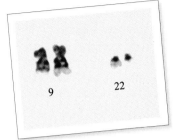

PHOTO GALLERY
The Philadelphia chromosome is commonly found in patients with chronic myelogenous leukemia (CML).
Dept. of Clinical Cytogenetics, Addenbrookes Hospital/ Science Photo Library/Photo Researchers

11.5 Essentials

- Chromosomal abnormalities can also cause cancer.

CANCER GENES ON OTHER CHROMOSOMES

Chromosome	Condition	Gene	Type of Gene
APC *K-RAS* **Chromosome 5** **Chromosome 18**	Familial adenomatous polyposis (FAP; a form of colon cancer)	On chromosome 5 and a mutation in *K-RAS* on chromosome 18	Oncogene
Wilms tumor **Chromosome 11**	Wilms tumor (a tumor of the kidney)	On the short arm of chromosome 11	Tumor suppressor gene
Lung cancer gene **Chromosome 13**	Small-cell lung cancer	On the long arm of chromosome 13	Tumor suppressor gene

11.6 What Is the Relationship between the Environment and Cancer?

A solid relationship exists between environmental factors and cancer. *Epidemiologists* are scientists who study factors related to disease. For example, they may study the geographical areas in which certain diseases are found most often. A large number of cancer cases found in a restricted area is called a **cancer cluster.** Epidemiologists examine the environment surrounding the cancer cluster to look for a cause-and-effect link. The 1986 book *A Civil Action* by Jonathan Harr, described a cancer cluster in Woburn, Massachusetts. The environmental trigger that caused the cancer turned out to be industrial solvents that entered the town's water supply.

Another way epidemiologists study the relationship between disease and environmental factors is to look at populations to determine which types of cancer they develop. Because many forms of cancer in the United States are related to our physical surroundings, personal behavior, or both, it is estimated that at least 50% of all cancer can be attributed to some type of environmental factor.

What environmental factors are associated with cancer?

Smoking is the number one environmental factor implicated in cancer—either directly, in the smoker, or indirectly, through secondhand smoke. Smoking is related to cancers of the oral cavity, larynx, esophagus, and lungs and accounts for 30% of all cancer deaths. Most of these cancers

BY THE NUMBERS

5%
Percentage of women with breast cancer who have the mutant *BRCA1* allele.

1 in 200
Number of women who will inherit the mutant breast cancer gene.

82%
Percentage of women with the mutant *BRCA1* allele who will develop breast cancer.

0.12%
Frequency of the mutant *BRCA1* and *BRCA2* alleles in the general population.

2.5%
Frequency of the mutant *BRCA1* and *BRCA2* alleles in Ashkenazi Jews.

85%
Percentage of men whose lung cancer is caused by smoking.

Nick Hawkes/Alamy Limited

75%
Percentage of women whose lung cancer is caused by smoking.

1 in 5
Chance that an American will get skin cancer.

have very low survival rates. Lung cancer, for example, has a five-year survival rate of 13%, meaning that on average, only 13% of those diagnosed with lung cancer will live for five years. Cancer risks associated with tobacco are not limited to smoking; people who use snuff or chewing tobacco are 50 times as likely to develop cancer of the mouth compared to non-users.

An estimated 1 million new cases of skin cancer are reported in the United States every year, and almost all cases are related to ultraviolet (UV) light exposure from the sun or tanning lamps. Skin cancer cases are increasing rapidly in most populations, presumably as a result of an increase in outdoor recreation and people moving to regions of the United States with more sun exposure. Epidemiological surveys show that lightly pigmented people are at much higher risk for skin cancer than heavily pigmented individuals. This supports the idea that genetic characteristics can affect the susceptibility of individuals or groups to environmental agents.

PHOTO GALLERY
Melanoma, a form of skin cancer.
Dr. P. Marazzi/Photo Researchers

Ozone depletion of the atmosphere in certain geographic regions also contributes to increased levels of UV radiation exposure, which in turn is associated with increases in skin cancer frequency. This is because the ozone layer around the planet blocks a great deal of UV radiation. More than 80% of lifetime skin damage occurs by age 18. In spite of this, many Americans think suntans are healthy looking, and only 25% consistently use sunscreen lotions or oils.

11.6 Essentials

- Smoking is only one lifestyle factor that may affect cancer risks.
- Environmental agents often cause normal cells to become cancerous.

PHOTO GALLERY
The appearance of lungs in a non-smoker.
James Stevenson/Science Photo Library/ Photo Researchers

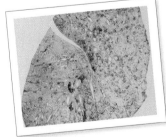
PHOTO GALLERY
The appearance of lungs in a smoker.
James Stevenson/Science Photo Library/ Photo Researchers

Michael Newman/PhotoEdit

CASE B Government Experiments on Children

Even though it happened in 1969, Robin Matchison still had nightmares about it. Early one morning, her mother told her to get up and say good-bye to her brother—he was going away for a long time. Robin was only 10 years old, but she remembered it clearly. A big car was parked in the driveway, and her 5-year-old brother, Jack, was being carried into the car.

"Where is he going?" she asked her mother.

"To a place where they can take care of him," her mother said.

"But we can take care of him, can't we, Mommy?" Robin asked.

Six years later, Robin's mother told her that Jack had been taken to a home for children with mental retardation in Massachusetts. They went to visit Jack, who at that time was quite ill. The doctors told her mother that Jack had leukemia, which is often associated with Down syndrome (the cause of Jack's retardation).

Recently, Robin read an article in the newspaper about experiments conducted by the United States government laboratories. Some of these involved exposing people to radiation when they were not aware of it. One of the places where this happened was in the home in Massachusetts where Jack had lived until he died of leukemia in 1977. Robin didn't know much about leukemia, but she went to the library and learned that one of the causes of leukemia is exposure to radiation, especially in children. She wondered if this was what caused her brother to die.

It seemed awfully unfair that the U.S. government had used children like her brother as guinea pigs. A friend at work told Robin she should see a lawyer about it.

1. Should Robin see a lawyer? Why or why not?

2. Who might the lawyer sue as part of this lawsuit?

3. Give three arguments why experimentation on people with mental retardation would be wrong.

4. Give three arguments why such experimentation might be considered all right.

5. What problems do you think may arise because this occurred so many years ago?

6. What might the government have done in 1969 to make this type of experimentation more acceptable?

7. Should experimentation be done on humans? Why or why not?

11.7 What Are the Legal and Ethical Issues Associated with Cancer?

In our society, some people with cancer look for someone to blame for their illness. Certain kinds of lawsuits, called **tort** cases, address the issue of who has been legally wronged. These cases affect the individuals involved, but they are actually intended to change society.

Some of the most famous of these lawsuits have involved tobacco companies. In these suits, people with cancer, and sometimes the families of those who have died of cancer, sue tobacco companies for negligence in causing the cancer.

The primary evidence presented in these cases was that the tobacco companies knew two things: that tobacco is harmful and possibly fatal, and that it is addictive.

At first, these lawsuits were not successful, but later, memos found in the files of the companies showed that they did know that tobacco is both harmful and addictive, and they did not warn smokers. Today, because of these lawsuits, warnings appear on all packages of tobacco products and that may stop many people from smoking.

An interesting ethical problem arises with smokers as well. One of the ethical principles that most societies hold dear is autonomy. This means that we have the personal right to make decisions for ourselves, no matter what the decision. Laws have been passed saying that people cannot make independent decisions until they are over 18, or in certain cases over 21. After this age, their decisions are their own. This applies to smoking because many feel they should not have any restrictions on their choice to smoke. In law suits to stop smoking inside buildings, another argument was made. This was that some who chose not to smoke were being affected by the smokers.

The following chart addresses some of the ethical and legal questions related to smoking.

11.7 Essentials

- Successful lawsuits have been brought against cigarette companies.

Question	How are these questions decided?	Related case or legal issue
If a person wants to smoke, can a law say he or she can't?	Most states have a minimum legal age (16 or 18) for purchasing cigarettes.	People who are old enough can decide for themselves. Autonomy is upheld.
Can laws control where a person can smoke?	Most states have laws about smoking in either public buildings or all indoor space, with the exception of private homes.	These laws protect not the smoker, but nonsmokers inhaling secondhand smoke. Employees of bars who were exposed to a great deal of secondhand smoke first introduced this legal issue by suing their employers when they became ill.
Could laws be passed that would control other environmental causes of cancer?	Both the federal government and state legislatures have regulated the control of dangerous chemicals such as pesticides and pollution.	Even with these laws in place, lawsuits have been brought to demand payment for treatment of patients who became ill from these exposures.

SPOTLIGHT ON ETHICS

Henrietta Lacks and Her Immortal Cells

The *HeLa* cell line is an immortal cell line used in medical research. Most cell lines divide only a limited number of times and then die, but the HeLa cell line is an exception. The HeLa cell line was derived from cervical cancer cells taken from Henrietta Lacks, shown in the photo at right, who died of her cancer in 1951. The cells were removed during the biopsy taken from a visible lesion on the uterine cervix as part of Lacks's diagnosis of cancer.

A cell line is made from a single cell or a group of cells that is kept alive *in vitro* (in the lab) and used for scientific study.

HeLa cells are termed *immortal* because they can divide an unlimited number of times in a laboratory cell culture plate as long as they are given the right nutrients. Medical students and scientists all over the world use the HeLa cell line to study these cervical cancer cells.

These cells are shown in the photograph below:

Obstetrics & Gynaecology/Science Photo Library/Photo Researchers

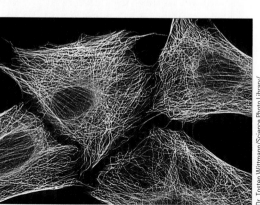

Dr. T. Tosten Wittmann/Science Photo Library/Photo Researchers

The cells in this cell line were originally grown by George Otto Gey without Lacks's knowledge or permission and later sold to medical schools. This gave Dr. Gey a nice income. However, he never patented the cell line or the cells in their original form.

Remember from Chapter 9 the case of Moore v. Regents of University of California, in which the court said that the doctors did not have to inform Moore of their use of his cells after they were removed. The same was true for the HeLa cell line and Henrietta Lacks.

But even if no law forces physicians or scientists to inform patients of the use of parts removed, are they somehow ethically obligated? In the transcript of the Moore case, Henrietta's case was discussed. It is difficult to know whether this might have influenced the California Supreme Court decision.

Initially, the HeLa cell line was said to be named after a "Helen Lane" or "Helen Larson" in order to preserve Lacks's anonymity. But the press used her real name within a few years of her death. Later her daughter asked that the cell line be named after her mother to show her contribution.

Recently, these cells were used to find a connection between certain strains of the human papillomavirus, discussed earlier, and human cervical cancer. A vaccine has been developed for this form of human papillomavirus and is being marketed to 9- to 14-year-olds and their parents.

QUESTIONS:

1. Consider the following hypothetical case based on the HeLa cell line:

   Heidi Hagawa remembered her mother, Frieda, very well. Heidi was 20 when her mother was diagnosed with cancer and underwent many tests at the hospital, and Heidi stayed by her side. Heidi's parents were divorced, and Heidi was the only family member her mother had.

   Heidi remembered when the doctors told them that Frieda's blood cells made special molecules called *monoclonal antibodies*. These cells might be used to fight off cancer and some viral infections in other patients. They asked for samples of her blood, which would be used to help her and others. Frieda signed a consent form stating that it was all right to use her blood to treat others with cancer. She died eight months later.

   Heidi didn't really think about the blood samples after her mother died, but a few years later she heard about a man named Moore in California who had his spleen removed. His spleen cells produced lymphokines, which could be used to treat others. He was suing his doctors for part of the millions of dollars they had made from selling his cells (along with their DNA) to a biotechnology company to make a cell line. A cell line keeps reproducing cells and their products for many, many years, and these products can then be sold.

   Heidi wondered if this was the same situation as her mother's case. She called Irv Kutler, an attorney friend of hers, to ask about it. Irv said he thought they had a case. The doctors had sold Frieda's cell line, and a great deal of money and stock had changed hands. The rights to Frieda's cell line were sold to a company called Montansic, and even though Frieda did sign a consent form, she was never told what they would do with the cells. Irv said they should sue.

   Could they win? Why or why not?

2. The company that marketed the vaccine for the human papillomavirus used an interesting advertising strategy. It ran television ads that appeared to be educational. They were telling women that a virus causes cancer. Soon after these ads ran, an ad for the vaccine appeared, with girls saying they wanted to be "one less" woman with cervical cancer. What do you think of advertising the product this way?

3. Next, the company began to lobby state legislatures to make vaccination of 9- to 14-year-old girls mandatory for entering school? Is this ethically right? Why or why not?

4. Make a checklist comparing and contrasting the Moore case with Henrietta's case.

5. After all these years, would there be any way to determine what Henrietta Lacks would have wanted to be done with her cervical cancer cells? How?

★ THE ESSENTIAL TEN

1. Cancer is defined as uncontrolled cell division. [Section 11.1]
   ▌Uncontrolled division can occur in any cell.

2. Cancer appears to run in some families. [Section 11.2]
   ▌Pedigrees of families often show a number of members with different forms of cancer.

3. Proto-oncogenes, oncogenes, and tumor suppressor genes control cell division. [Section 11.2]
   ▌Depending on which type of gene is mutated, uncontrolled cell division may result.

4. When these genes are mutated, cancerous tumors form. [Section 11.2]
   ▌Some of these genes turn on cell division, while others turn it off.

5. The cell cycle is important to understanding cancer and the control of cell division. [Section 11.3]
   ▌The normal cell cycle gives us clues as to where uncontrolled growth can occur.

6. Researchers have discovered two mutant genes, *BRCA1* and *BRCA2*, that increase a woman's risk of developing breast cancer. [Section 11.4]
   ▌Testing for these two genes is becoming popular and marketed directly to the public via the Internet and other media.

7. Chromosomal abnormalities can also cause cancer. [Section 11.5]
   ▌An extra chromosome 21 has been shown to increase the risk of certain forms of leukemia.

8. Smoking is only one lifestyle factor that may affect cancer risks. [Section 11.6]
   ▌Warnings are now present on cigarette packages to inform consumers of this risk.

9. Environmental agents often cause normal cells to become cancerous. [Section 11.6]
   ▌Scientists are studying which environmental agents are related to cancer.

10. Successful lawsuits have been brought against cigarette companies. [Section 11.7]
    ▌The cigarette companies knew that their product caused cancer.

KEY TERMS

acute myeloblastic leukemia (p. 185)
Ashkenazi Jews (p. 184)
benign (p. 180)
BRCA1 (p. 183)
BRCA2 (p. 183)
Burkitt's lymphoma (p. 185)
cancer (p. 179)
cancer cluster (p. 186)
carcinogens (p. 180)
cell cycle (p. 180)
chronic myelogenous leukemia (CML) (p. 185)
cytokinesis (p. 181)
G1 (p. 181)
G1/S checkpoint (p. 182)
G2 (p. 181)
G2/M checkpoint (p. 182)

human papillomavirus (HPV) (p. 180)
interphase (p. 181)
malignant (p. 179)
metastasis (p. 180)
mitosis (p. 181)
multiple myeloma (p. 185)
mutation (p. 180)
oncogenes (p. 180)
Philadelphia chromosome (p. 185)
proto-oncogene (p. 180)
RAS (p.183)
S (p. 181)
signal transduction (p.183)
tort (p. 187)
tumor suppressor genes (p. 180)

REVIEW QUESTIONS

1. Why is cancer so difficult to treat?
2. What is the difference between benign and cancerous tumors?
3. Not so long ago, it was thought that cancer did not have a genetic component. Now cancer's genetic causes are discussed in a human genetics book. What has changed?
4. What are the differences between tumor suppressor genes and oncogenes? List three.
5. How much should we do to keep carcinogens out of our environment? What do you do?
6. List the parts of the cell cycle.
7. What is signal transduction and how might it relate to cancer?
8. How did scientists find the markers for breast cancer?
9. What is the Philadelphia chromosome and why was it called this?
10. If we know what environmental agents might cause cancer, why can't we just get rid of them?

APPLICATION QUESTIONS

1. Give three reasons why people are so afraid of cancer. Apply the science that you have learned.
2. Recently, a vaccine has been developed to protect women from cervical cancer. Research this vaccine and how it was marketed.
3. It is easy to understand how older people's cells can become cancerous, but many children, even babies, get cancer. Do some research on this topic to see why it occurs, and write a short report.

What Would You Do If . . . ?

1. You were offered a space in an experimental study to cure cancer?

2. You were advising your mother about whether she should join such a study?

3. You were advising young people not to smoke?

4. You were voting on a bill to take cigarettes off the market?

5. You were working on marketing strategies to keep people out of the sun and had to reach people who have to work outside (construction workers, letter carriers, etc.)?

4. A large number of cancer cases in a certain geographic area is called a cancer cluster. Research a cancer cluster in your area and write a short report.
5. Research the problems with plastics in our environment and cancer and write a short report.
6. Now that you know about signal transduction, go back and read over the example of the *RAS* gene. How might you use this knowledge to prevent cancer from occurring?
7. Do some research on environmental causes of cancer. Make a list of how you might change what you do to cut your cancer risk.
8. Do some research on Mary Claire King and how she became a scientist.

ONLINE RESOURCES

Preparing for an exam? Log on at academic.cengage.com/login for a pre-test, a personalized learning plan, and a post-test in CengageNOW's Study Tools to help you assess your understanding.

If assigned by your instructor, the Case A and Spotlight on Ethics activities for this chapter, "New Cancer Treatment" and "Henrietta Lacks," will be available in CengageNOW's Assignments.

Learn by Writing

In Chapter 1 we suggested that you start a blog with members of your class or others who are interested in our topics. Now is the time to revisit your blog and consider some of the questions in Chapters 9 and 10. Email others you think might be interested and invite them to contribute.

Here are some ideas to address in your blog:

- Should there be changes in the way patients are treated with cancer, now that we know that there is a genetic component?
- Is it important that patients like Harriet know about the clinical trials being done on cancer?
- Do some research on clinical trials being done in your area and share the web sites in your blog.
- Should the law actually get involved in any of the questions raised in Chapters 9–11?
- Discuss any other questions or comments you want to make or address with your fellow students.

In The Media

ABC News, September 6, 2007

Breast Cancer More Deadly in Black Women

Marilynn Marchione

A new study explains why breast cancer seems to be more deadly in black women: their tumors did not respond to hormone-based treatments that succeed in other women. Previously, differences in recovery were blamed on the fact that black women seemed to get fewer mammograms and less aggressive treatment. Even though the women studied were diagnosed at an earlier age and had tumors in earlier stages, they were more likely to die of their disease.

To access this article online, go to academic.cengage.com/biology/yashon/hgs1.

QUESTIONS:
1. What might doctors do to help patients in this situation?
2. Could laws be passed to help lower the number of black women dying of cancer?

Times of India, September 8, 2007

Blood Test to Spot Cancer Early

British scientists tracked over 11,000 women over a 30-year time frame. The central premise of the study was to see which of the women developed breast cancer and develop a blood test that might spot the cancer before symptoms appear.

The women all gave blood samples that were tested for biomarkers that are proteins created by cancer cells. Many of these biomarkers have already been linked to certain cancers.

After studying the women's blood samples, markers were found in those that developed cancer. The test has been used to detect signs of breast cancer in some women and it is hoped that this will help in early screening and better prevention.

To access this article online, go to academic.cengage.com/biology/yashon/hgs1.

QUESTIONS:
1. Would you take the blood test described in the article? Why or why not?
2. Sometimes researchers take samples and test them but do not tell study participants about the results. Do you think the scientists should have told the women of their possible diagnosis?

12 Behavior

⊛ CENTRAL POINTS

▪ Behavior is a reaction to our environment.

▪ Animals and humans have similar behaviors.

▪ Brain chemicals play an important part in human behavior.

▪ A single gene or groups of genes can affect a person's behavior.

▪ Twin studies are an important part of behavioral genetics.

▪ Courts are unclear on how to address the issue of genetics and behavior.

CASE A Twins Found Strangely Alike

Jim Lewis didn't learn he had a twin until he was 5, but it never sank in. Jim Springer learned he had a twin when he was a little older, but thought he was dead.

When the two men were 39, researchers at the University of Minnesota Twin Study Group, who had been studying twins reared apart, reunited them.

The "two Jims" are famous because they share so many similarities. It even amazed the doctors that they weighed exactly the same (180 lb. [81.6 kg]), had the same first name, and were both 6 ft. (1.8 m) tall.

In addition to their similar physical traits:

- Both had a dog named "Toy."
- Each had been married twice.
- Each one's first wife was named Linda.
- Each one's second wife was named Betty.
- Both had a son named James Allan.
- Both smoked Salem cigarettes.
- Both drank Miller Lite beer.

- Both had worked as sheriffs, part time.
- Both had recurring migraine headaches.
- Both bit their nails.
- Many of their mannerisms and speech patterns were alike.

Some questions come to mind when reading about the two Jims. Before we can address those questions, let's look at the genetics of behavior.

PHOTO GALLERY
Jim Springer and Jim Lewis, the "two Jims" twins
Enrico Ferorelli Photography

12.1 How Is Behavior Defined?

In its simplest form, behavior can be defined as a reaction to **stimuli.** Most behavioral studies began with animals. We still look at animals as having a simpler form of behavior, which makes them easier to study, but human behavior has many similarities to animal behavior. However, human behavioral response can be more variable, and people do react differently to similar situations, which makes their behavior more difficult to study.

Take, for example, alcoholism. It can be defined as an excessive consumption of alcohol. But is this a clear enough definition to identify people who are alcoholics? Many people react differently to alcohol, and that makes alcoholism difficult to define.

Behavioral genetics is the study of the influence of genes on behavior. One of the big questions in behavioral genetics is, which is more important: genetics or environment? How we react to certain situations may be influenced by genetics, parents, siblings, education, and many more variables. The interactions among all these factors makes the study of our behavior somewhat complicated.

12.1 Essentials

- Behavior is response to stimuli.
- A person behaves in a certain way based on his or her heredity and environment.

12.2 How Does the Body Work to Cause Behavior?

Brain chemicals, or **neurotransmitters,** can change our moods and our actions. These chemicals are released when certain nerve cells in the brain become active, in a process called **neurotransmission.** The speed or frequency of neurotransmission can alter how we react in a certain situation.

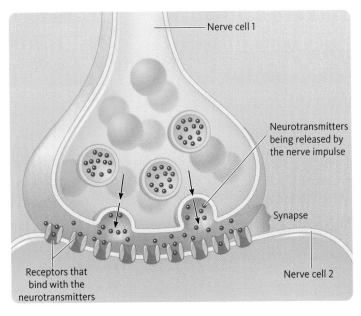

Nerve cell 1

Neurotransmitters being released by the nerve impulse

Synapse

Nerve cell 2

Receptors that bind with the neurotransmitters

As shown in the drawing above, two nerve cells are often close to each other. When an impulse travels to the end of nerve cell 1, it must jump to nerve cell 2 to continue its journey. Follow the arrows to the **synapse,** a gap between nerve cells. A neurotransmitter (or chemical) is released to expedite the jump across the synapse. If anything increases or decreases the release of the neurotransmitter, a person's behavior may change.

One prominent example is the group of brain chemicals known as **endorphins.** When endorphins are released, often during exercise, stress, or excitement, we feel a rush of exhilaration. Many drugs that change our mood or reactions, such as cocaine, mimic the naturally occurring neurotransmitters and create a similar feeling (Table 12.1).

Because brain chemicals are produced by nerve cells, DNA carries the genetic information for their compounds.

Table 12.1 Selected Recreational Drugs and the Neurotransmitters They Mimic	
Drug	**Brain Chemical Equivalent**
Cocaine, morphine, and heroin	Endorphins
Mescaline	Dopamine
Nicotine	Endorphins
Amphetamines	Norepinephrine

As discussed in Chapter 5, proteins are synthesized in cells. In the case of neurotransmitters, proteins are involved in making these molecules—and feelings and behaviors result. Therefore it is clear that at least some of our behavior is genetic.

12.2 Essentials

- **The release of neurotransmitters in the brain causes us to act and react.**

12.3 How Can Changes in Genes Cause Changes in Behavior?

A variety of genetic changes including chromosomal abnormalities, single-gene defects, and multigenic conditions can alter our behavior.

An example of a single-gene defect that changes behavior is Huntington disease (HD). (Review Alan's case in Chapter 4.) Some of the symptoms of HD are involuntary movements and a progression of personality changes. The HD gene codes for a large protein called huntingtin. This protein is necessary for the survival of certain brain cells. When a person has HD, the mutant gene produces an altered version of the protein that breaks apart and accumulates in the brain cells. Cells with accumulations of huntingtin die, and as more cells die, behavioral changes appear. The photo below shows the difference between a normal brain and one with HD.

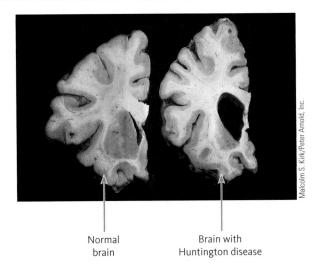

Malcolm S. Kirk/Peter Arnold, Inc.

Normal brain

Brain with Huntington disease

Changes in behavior can also be caused by chromosomal abnormalities. People with a mutation in a specific region of the X chromosome have specific behavioral changes. This condition, called **fragile X syndrome,** causes autistic-type behaviors as well as problems with aggression. The photo of the X chromosome on the next page shows the location of the fragile site, which causes chromosome breaks.

A multifactorial condition (multifactorial traits are discussed in Chapter 10) that affects behavior is **schizophrenia.** Schizophrenia is a collection of mental disorders that can have many symptoms. Some of these are hallucinations, delusions, disordered thinking, and changed behavior.

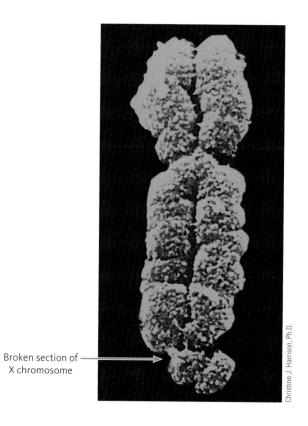

Broken section of X chromosome

Christine J. Harrison, Ph.D.

Genes associated with schizophrenia have been found on the X chromosome and autosomes. There also seems to be an environmental component to schizophrenia. Genetic and environmental links to schizophrenia are being studied by many scientists.

12.3 Essentials

- Huntington disease is an example of a genetic condition that causes changes in behavior.

12.4 How Do We Study the Genetics of Behavior?

Behavioral geneticists search for genetic influences on human behaviors, an area that is also studied by **psychologists** and **psychiatrists**. Medical geneticists study the genetic basis for mental illnesses that can be treated by physicians. However, the differences between these fields are blurring.

Take, for example, the study of depression. At first, depression was treated by psychologists and was considered to be based only in behavior. Now, with information from pedigree analysis, molecular genetics, and neurobiology, it is clear that depression is a medical condition with genetic components. Physicians as well as specialists in psychology now treat depression as a physical illness.

One way that scientists study the genetic basis of behavior is to study cases like that of the two Jims. The Jims are special because they are identical twins adopted by different families and raised in separate environments. As we know, identical twins have identical genomes and, therefore, each

has a set of genes that are exactly the same as that of their twin. This means that if a specific trait or behavior is the same in identical twins, it most likely has a significant genetic component. If the two twins are raised in separate environments and have different behavioral traits, those traits have a significant environmental component.

For example, if one identical twin is alcoholic, there is a 55% chance that his or her identical twin will be alcoholic regardless of their environment. This statistic was determined by studying groups of identical twins raised apart. This relationship is called **concordance.** Concordance is a way of calculating how often a trait occurs in both members of a pair of twins. Because identical twins are genetically identical, it follows that any genetic mutation carried by one would also be carried by the other and, if this is true, the concordance rate would be 1.0.

Take, for example, cystic fibrosis (genotype *cc*). If one identical twin has the condition, the other will also have it (genotype *cc*) and the concordance will be 1.0.

Differences occur when a phenotype shows up in one twin but not the other. One example is schizophrenia. The concordance in identical twins has been found to be 0.69. This means when studying schizophrenia in identical twins, 69% of the time, both twins had schizophrenia, and in 31% of the cases, only one twin had the condition.

Thus, the concordance for a given trait helps establish whether or not it is has a genetic basis. Concordance is also studied in *fraternal twins.* Fraternal twins do not have identical DNA, but like all brothers and sisters, share half their genes.

Table 12.2 shows the concordance of schizophrenia and a few other traits.

If a specific behavioral trait is to be studied in a family, a pedigree is constructed and the trait is traced through as many generations as possible.

In alcoholism, many members of a family may have the condition, which will show up clearly in a family's pedigree. As discussed above, however, there are often problems in defining behavior. In the case of alcoholism, is it someone who drinks everyday? Someone who can't stop drinking?

Table 12.2 Concordance for Selected Traits

Trait	Concordance in Identical Twins	Concordance in Fraternal Twins
Blood type	1.0	0.66
Eye color	0.99	0.28
Schizophrenia	0.69	0.28
Diabetes	0.65	0.19
Alcoholism (males)	0.41	0.17
Attention deficit disorder/ hyperactivity	0.58	0.31
IQ score	0.69	0.88

The following figures show pedigrees of some traits thought to affect behavior.

PEDIGREES OF SOME BEHAVIORAL TRAITS

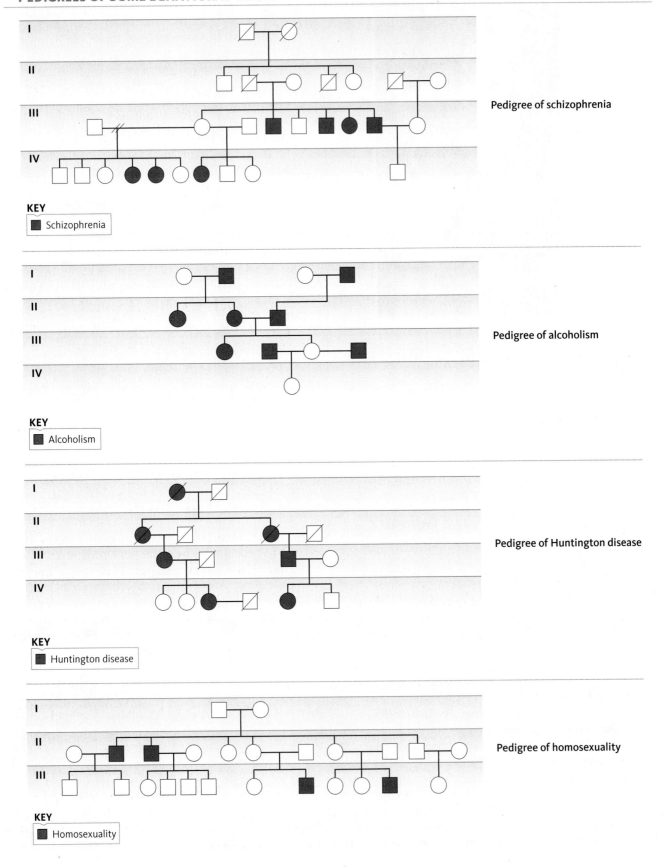

Pedigree of schizophrenia

KEY
■ Schizophrenia

Pedigree of alcoholism

KEY
■ Alcoholism

Pedigree of Huntington disease

KEY
■ Huntington disease

Pedigree of homosexuality

KEY
■ Homosexuality

Animal studies are another way scientists study behavior. Because mice reproduce quickly and their behaviors are observable, many behavioral geneticists study them. If a gene is suspected to be involved in a behavior, researchers can use recombinant DNA techniques to insert a human gene, mutate a mouse gene, or change its pattern of expression and study the effects on the behavior of these genetically modified mice and their offspring.

In 1999, scientists working at Princeton University published the results of one such study. They inserted a human gene into a mouse. This gene codes for a protein in brain cells known to be associated with memory. The transgenic animals performed better in mazes and other tests used to measure intelligence in mice.

How can all of the methods just discussed be used to study one behavioral trait?

The genetic causes of sexual orientation have been studied using combined methods that include twin studies, chromosomal analysis, and pedigree analysis. In one twin study, 56 sets of male identical twins and 54 sets of male fraternal twins were studied. Table 12.3 shows the results.

This study and others like it have shown that there is a strong genetic component to homosexual behavior.

Other researchers used linkage and pedigree studies to identify an area on the X chromosome that is associated with homosexual behavior. To do this, they first identified 38 families that included two homosexual brothers. After finding a number of molecular markers on the X chromosome of the brothers, they then searched for these markers on the X chromosomes of family members. The gene associated with these markers is shown below.

Does this definitely prove that homosexual behavior is genetic? No; but this work has provided a foundation for further work on genetics and homosexuality.

12.4 Essentials

- Behavioral genetics is studied in both animals and humans.
- Because of the identical genomes of identical twins, many scientists study twins reared apart to assess the role of genetics and environment in behavior.
- Recombinant DNA techniques have been used to produce changes in behavior in mice.
- Scientists are still studying the roles of heredity or environment in influencing behavior.

Table 12.3 Concordance of Homosexuality		
Trait	Concordance in Identical Twins	Concordance in Fraternal Twins
Homosexuality	0.52	0.22

Possible gene for male homosexual behavior

X chromosome

CASE A QUESTIONS

Now that we know more about genetic control of behavior, let's answer some questions about the case of the "two Jims."

1. Why were these two men a perfect set of twins to study behavior?

2. Do you think the evidence shows that the behaviors listed in the case are 100% genetic?

3. Do you know any identical twins? Are they alike or different? List four things about them that are alike and four that are different.

4. Can you see any problems that might arise from studying twins raised apart? List three.

12.5 Can a Single-Gene Defect Cause Aggressive Behavior?

In 1993, scientists began to study a large European family that showed aggressive and violent behavior. The pedigree of this family is shown below. Many of the men in this family had been jailed for committing violent offenses. Notice that the condition is present only in men. The gene for this condition was mapped to the short arm of the X chromosome and is, therefore, an X linked trait (see Chapter 4).

This gene encodes for an enzyme called **monoamine oxidase type A (MAOA)** that breaks down neurotransmitters in the brain. When mutated, this gene causes a condition called *MAOA deficiency*. Failure to rapidly break down neurotransmitters such as **serotonin** disrupts the normal transmission of impulses. This changes normal functions in the nervous system and can cause abnormal behavior.

To study the role of neurotransmitters in behavior, researchers knocked out one gene for serotonin receptors on brain cells in mice. This reduced the ability of nerve cells to bind and process signals. When an unfamiliar mouse was placed in a cage with a knockout mouse, the knockout mice were very aggressive, as shown below.

Jay Van Rensselaer/ Homewood Photo Labs

12.5 Essentials

- MAOA deficiency is another genetic condition that is studied in behavioral genetics.

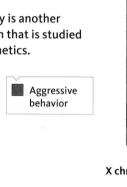

Gene for *MAOA* deficiency

X chromosome

■ Aggressive behavior

CASE B Important Conference on Hold

Dr. David Wasserman first envisioned his planned meeting as a gathering of an academically diverse group to discuss cutting-edge information about genetics and behavior. He had organized other symposia, and as in the past, he carefully planned whom to invite as speakers and guests, and even what food to serve.

Not so long ago, an academic meeting would not have attracted the attention of the national press, but in this case, the meeting was in the headlines. Dr. Wasserman was really surprised when the first reporter called.

He gave the reporter all the information she asked for, including the title of the symposium, "Genetic Factors in Crime: Findings, Uses, and Implications." Of course,

Dr. Wasserman and his colleagues knew that the topic might raise some eyebrows because of the history of genetics and behavior, but they didn't think of it as a sensational topic. After all, they were scientists, and approached the topic in a scientific way.

Then the problems began.

Dr. Wasserman and his colleagues had no idea that this meeting would cause such a stir. Protests against the meeting came from many directions, and the controversy caused the meeting to be cancelled.

1. What should Dr. Wasserman and his colleagues do?

2. How could they have the conference without causing any problems?

Richard Levine/Alamy Limited

3. Years ago the public would never have known about such academic conferences; why has this changed?

4. Some people see a racial component to these kinds of studies. Research this topic and give your opinion.

5. The National Institutes of Health revoked the funding for this meeting; do you think it should have done so? Why or why not?

6. Research what actually happened with this conference.

12.6 What Are the Legal and Ethical Issues Associated with Behavior?

Scientists are not sure what causes many aspects of human behavior. As a result, courts are finding it difficult to deal with this topic. The most important question in the <u>Frye</u> case (see "Spotlight on Law: <u>Frye v. U.S.</u>" in Chapter 1) asks if a specific scientific theory is acceptable among members of the scientific community. Behavioral genetics is a science still in its infancy; therefore, experts do not agree on findings and, as a result, individual courts cannot decide how it will be used to determine guilt or innocence.

In our society we often ask the question, why does someone commit a crime? In legal circles, this is referred to as *motive* and is often a component of a criminal trial. If we could determine that certain genes or groups of genes cause someone to commit a crime, motive would no longer be relevant.

In the 1970s, some scientists studied individuals with XYY syndrome (also called Jacobs syndrome; see Chapter 3). In the United Kingdom, a study of prisoners incarcerated for violent crimes showed a higher percentage of men with XYY syndrome than there is in the general population. Initially, this finding sparked similar studies all over the globe, including the United States, to see if there is a relationship between criminality and XYY individuals. Generally, XYY males are taller, and often more aggressive, but this does not mean they are criminals. Studies in the U.S. have been halted, so information on this possible relationship is still incomplete.

PHOTO GALLERY
Richard Speck claimed an extra Y chromosome as part of his defense.
Bettmann/CORBIS

In 1966, Richard Speck was tried and convicted of killing eight student nurses in Chicago, Illinois. He was identified by the one living witness to the crime. During his sentencing, his attorney prepared a defense trying to use the XYY karyotype as an argument against the death penalty. It was not used at the hearing because tests showed that Speck did not have the XYY karyotype. However, since that time, many unanswered questions about genetics and behavior remain in the minds of both prosecutors and defense attorneys. For example, can a genetic cause of criminality be established and used in court (see "Spotlight on Law: <u>Mobley v. Georgia</u>" later in this chapter)?

Another question is whether a person can be rehabilitated after committing a crime. The idea of rehabilitation is the basis for setting up our penal system. But to change a person's behavior, it is necessary to control that behavior. If criminal or antisocial behavior is genetically controlled, how can we expect a person with a genetic behavioral disorder to control or change his or her behavior?

In addition, the questions raised in the <u>Frye</u> case were set up to separate **junk science** from "real" science. To date, most judges and juries have considered a genetic predisposition to crime to be in the junk science category.

PHOTO GALLERY
Was Christopher Simmons, at age 17, unable to understand the consequences of his actions?
Missouri Dept. of Corrections/AP Photos

An interesting case that addressed these issues was <u>Roper v. Simmons</u>. In 1993, Christopher Simmons and two other boys were convicted of pushing a woman off a railroad trestle into a river, killing her. At the hearing to decide whether Simmons would be given the death penalty, his attorney argued that because he was only 17 when the crime was committed, he was too young to be able to realize the consequences of his actions. To prove this, the attorney presented an **amicus curiae** ("friend of the court") brief written by scientists who had done studies on teenage brains. MRI brain scans from thousands of teens showed significant differences from those of adults. The scans showed that area of the brain that control impulsive behavior (the frontal lobe) was underdeveloped in teens when compared with adults. In addition, the emotional responses of teenagers were different than those of adults (■ Figure 12.1).

In <u>Simmons</u>, the court decided that this scientific information was not strong enough to stop an execution. The case was taken to the U.S. Supreme Court and argued in October, 2004. The Court determined that the execution of offenders who were 18 years of age or younger when their crimes were committed, was unconstitutional. Therefore, Simmons was not executed and is serving life in prison. In the written opinion Justice Kennedy wrote, "Retribution is not proportional if the law's most severe penalty is imposed on one whose culpability or blameworthiness is diminished, to a substantial degree, by reason of youth and immaturity."

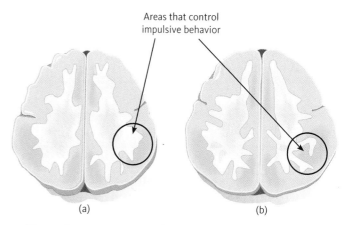

Areas that control impulsive behavior

(a)   (b)

■ **Figure 12.1 Cross-sections of Adult and Teenage Brains**
These illustrations of a teenage brain (a) and an adult brain (b) show large differences in areas that control impulsive behavior.

One of the first behavioral issues to be addressed in criminal trials was insanity. If a person could be proven to be insane, the courts asked, would this make him or her not responsible? For the most part, a medical diagnosis of schizophrenia can allow a person to be found not guilty by reason of insanity. But what actually is insanity? That has been asked in court cases throughout the United States and the answers vary.

In general, courts have held that only if a defendant cannot tell right from wrong and has a definite diagnosis of a medical condition, such as schizophrenia, can he or she be found not guilty by reason of insanity. Schizophrenia has been shown to have a genetic basis. Does this mean a person with this condition cannot control this behavior?

The chart below addresses some of the ethical and legal questions related to genetics and behavior.

12.6 Essentials

- **The question of behavioral genetics in courts of law has not been decided.**

Question	How are these questions decided?	Related case or legal issue
Is a person who has been diagnosed as being insane responsible for his or her actions?	Courts make their decisions based on legislation passed on a state-by-state basis. Some states allow pleas of "not guilty by reason of insanity"; others use "guilty but insane."	When John Hinckley tried to kill President Reagan (U.S. v. Hinckley), he argued that because he had schizophrenia, he was not guilty by reason of insanity. His attorney showed brain scans that indicated he had a smaller than normal brain. He was found not guilty by reason of insanity.
If a person has a genetic condition that makes him or her aggressive, such as MAOA deficiency, can he or she be considered guilty of a crime?	In general, courts today in the U.S. are not likely to accept genetic defenses.	See "Spotlight on Law: Mobley v. Georgia" later in this chapter.
Can a person who was under the influence of drugs or alcohol argue that he or she is not responsible for a crime?	The courts have held that voluntarily taking drugs and alcohol cannot mitigate a person's responsibility for a crime.	
If a person has a genetic condition that causes aggression, could this be used in his or her defense?	This has not been addressed in U.S. courts; however, some scientists believe that if genes control certain behaviors, people may be unable to control them.	The question of what should be done with people who have this type of mutated gene may become important. See "Spotlight on Law: Mobley v. Georgia" on the following page.
If a behavior is obviously "insane," can that be used in a case?	Many feel that women who kill their children should be considered insane.	Andrea Yates drowned her five children in the family's bathtub in 2001. In 2003, Deanna Laney beat her two young sons to death and injured a third with stones, and Lisa Ann Diaz drowned her two daughters in a bathtub. Dena Schlosser fatally severed her 10-month-old daughter's arms with a kitchen knife in 2004. All four of these women were found innocent by reason of insanity. Yates initially was convicted of capital murder, but that verdict was overturned on appeal.

On February 17, 1991, John Collins was shot in the back of the head during a robbery of the Domino's Pizza store where he worked in Oakwood, Georgia. On March 13, after robbing a dry cleaner, Stephen Mobley was arrested and confessed to the murder and armed robbery at Domino's.

During this confession he boasted about how John Collins fell on his knees and begged for mercy. Afterwards, Mobley had himself tattooed with a Domino's logo and plastered his cell with Domino's boxes. His criminal history included rape, robbery, assault, and burglary. The prosecution said, "Mobley is evil, a cold-blooded, heartless killer."

Daniel Summer, Mobley's court-appointed attorney, tried to enter a guilty plea and arrange a deal for life in prison, but the deal was rejected. The prosecutors wanted the death penalty for Mobley. During questioning of the family, Summer met Mobley's aunt, Joyce Ann Childers. She told Summer that "volcanic, aggressive, physical abuse and violent behavior is prevalent throughout the family tree." Summer then remembered an article he had read in the *Chicago Tribune* in which scientists at Harvard, NIH, and overseas were conducting research on genetic ties to violence. Below is the pedigree of Mobley's family based on his aunt's testimony. Compare it to the pedigree of the family with MAOA deficiency, also below.

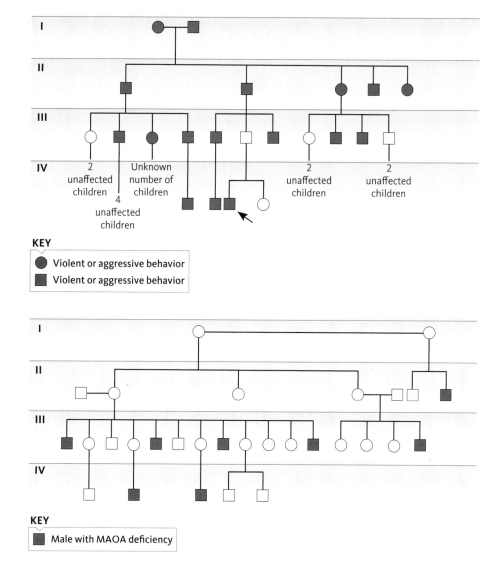

KEY

● Violent or aggressive behavior
■ Violent or aggressive behavior

KEY

■ Male with MAOA deficiency

There are obvious differences between the two families. The most glaring one is that women in Mobley's family are identified with the trait. Could this be MAOA deficiency?

Before the sentencing hearing, Summer contacted Dr. Xandra Breakefield at Harvard Medical School and Dr. Bahjat A. Farat at Emory University. When Mobley's story was revealed, the doctors began to see an emerging pattern of a history of violence through several generations in his family. Both doctors offered their services free of charge; Mobley needed specialized blood and urine tests to determine whether he had a mutation of the MAOA gene.

One might expect Mobley to have come from a poor family and bad surroundings, but the opposite is actually true. Even though many of the people in Mobley's family tree were violent, many were amazingly successful. His father, for example, even though he refused to help Mobley's defense, is a self-made millionaire. He tried sending his son to private school, then to psychiatrists, and finally to jail, but stated, "He never developed a value system or a conscience." In the end he washed his hands of all responsibility.

Mobley could not afford the cost of the MAOA tests (about $1000), and Summer asked the court to cover the expense. This allowed the trial court, and eventually the Supreme Court of Georgia, to weigh the question of validity of genetic causes of criminality.

RESULTS:

The Georgia trial court that was deciding Mobley's fate could find no reason to allow for testing; the decision stated, "The theory of a genetic connection is not at a level of scientific acceptance that would justify its admission." In other words, this did not meet the <u>Frye</u> standards of scientific evidence.

QUESTIONS:

1. Would you have allowed the testing if you were the Georgia Supreme Court?
2. If a gene or group of genes were found to cause violent behavior, how could defense attorneys use this information?
3. How might the prosecution argue for *not* including genetic information in Mobley's case?
4. Is a person who is drunk, taking drugs, or mentally ill responsible for any crimes he or she commits? Why or why not?
5. How does this relate to a genetic cause of crime?

⊛ THE ESSENTIAL TEN

1. Behavior is a response to stimuli. [Section 12.1]
    ▮ This response can be trivial or a major behavioral change.

2. A person behaves in a certain way based on his or her heredity and environment. [Section 12.1]
    ▮ Twins are often used to study the influence of heredity.

3. The release of neurotransmitters in the brain causes us to act and react in certain ways. [Section 12.2]
    ▮ These chemicals cause changes in the nervous system and produce changes in our behavior.

4. Huntington disease is an example of a genetic condition that causes changes in behavior. [Section 12.3]
    ▮ The mutation in Huntington disease affects the brain and nervous system.

5. Behavioral genetics is studied in both animals and humans. [Section 12.4]
    ▮ Studying behavior in animals can teach us something about human behavior.

6. Scientists are still evaluating the roles of heredity and environment in behavior. [Section 12.4]
    ▮ This can be difficult to study because people can react differently to the same situation.

7. Because of the identical genomes in identical twins, many scientists study twins reared apart to study the role of genetics and environment in controlling behavior. [Section 12.4]
    ▮ In this situation, people with identical genomes are exposed to differing environments.

8. Recombinant DNA techniques have been used to generate and study behavioral changes in mice. [Section 12.4]
    ▮ Changing specific genes in mice causes them to act aggressively.

9. MAOA deficiency is a human genetic condition that is being studied in behavioral genetics. [Section 12.5]
    ▮ Men with this condition show uncontrollable aggression and may commit crimes.

10. The question of genetic control of behavior has not been decided in courts of law. [Section 12.6]
    ▮ If our behavior is caused largely by genetics, are we responsible for what we do?

KEY TERMS

amicus curiae (p. 199)
behavioral genetics (p. 193)
concordance (p. 195)
endorphin (p. 194)
fragile X syndrome (p. 194)
junk science (p. 199)
monoamine oxidase type A (MAOA) (p. 198)

neurotransmitters (p. 194)
neurotransmission (p. 194)
psychiatrists (p. 195)
psychologists (p. 195)
schizophrenia (p. 194)
serotonin (p. 198)
stimuli (p. 193)
synapse (p. 194)

REVIEW QUESTIONS

1. How do neurotransmitters work to cause our feelings and behaviors?
2. Research examples that show how humans are similar to animals in their behaviors and write a short report.
3. Is the study of animal behavior a good way to identify human behaviors? Why or why not?
4. Research a few more examples of genetic conditions that change our behavior and write a short report.
5. What would you have included in the amicus curiae brief for the Simmons case? List three things.
6. Behavior seems to have a simple definition. What is it?
7. How do we compare the study of animal behavior to human behavior? Why are some animals better for this than others?
8. List three brain chemicals and their effect on behavior.
9. Schizophrenia is a fairly well known human condition. What external (environmental) factors affect it?
10. What is the difference between a psychiatrist and a psychologist?

APPLICATION QUESTIONS

1. Find an example of another set of twins studied by the Minnesota twin study. How do they compare to the study of the two Jims?
2. If you were going to adopt a child who was an identical twin, do you think it would be better to adopt both or just one? Why?
3. In this chapter we discuss how twins are studied to determine if their have similar behaviors. Design a way to study behaviors in people who are not twins. What trait (such as anger or kindness) would you study? How would you set up the experiment?
4. If sexual orientation were shown to be genetically caused, would this change the way people view homosexuals? How?

5. If MAOA deficiency were shown to cause homicidal behaviors, should we test all newborns for this trait? Why or why not?

6. Look at the chart titled "Pedigrees of Some Behavioral Traits" in Section 12.4. What do you think might occur if we could identify these genes in the future?

7. How does the fact that some drugs mimic naturally occurring brain chemicals explain why people get addicted to them?

8. Research some legal cases where schizophrenia was used as a defense. Does this mean that because schizophrenia has a genetic component, these defendants have a true defense?

9. In the law and ethics section of this chapter there is a quote from the Supreme Court decision in Roper v. Simmons. How does it apply to genetics?

ONLINE RESOURCES

Preparing for an exam? Log on at academic.cengage.com/login for a pre-test, a personalized learning plan, and a post-test in CengageNOW's Study Tools to help you assess your understanding.

If assigned by your instructor, the Case B and Spotlight on Law activities for this chapter, "Important Conference on Hold" and "Mobley v. Georgia," will be available in CengageNOW's Assignments.

What Would You Do If . . . ?

1. You were a juror for a trial in which a defense attorney said her client was not responsible for the crime because he had an extra Y chromosome?

2. You had been asked to speak at Dr. Wasserman's conference?

3. You were on a jury being asked to force a man with schizophrenia to take his medication?

4. Your identical twin girls were asked to be in a twin study?

5. You were a reporter being asked to write about criminal behavior?

In The Media

Baltimore Sun, August 14, 2007

Insanity Defense Muddles Case

Justin Fenton and Andrea F. Siegel

Should a person with schizophrenia be found guilty and placed in prison? This is the question raised in the case of Zachary Thomas Neiman, who was accused of killing his mother with two shotgun blasts as she sat on her sofa.

He has refused to take his medication that would make him competent to stand trial. If he continues to refuse treatment he won't be released from custody. His treatment for schizophrenia allows him to function in society. Without medication, he has severe symptoms.

If he were found not guilty by reason of insanity, he would still have to spend time in a state institution where psychiatrists, psychologists, therapists, nurses and other experts will continually watch, test, and re-evaluate him. If he continued to take his medication he might be allowed parole.

To access this article online, go to academic.cengage.com/biology/yashon/hgs1.

QUESTIONS:
1. If medications can treat mental illness, can the courts force defendants to take them?
2. Is a person insane if he or she is taking medications that allow "normal" behavior?

AScribe Live Newswire, September 18, 2007

Four University of Chicago Scientists Receive $8 Million in Innovation Awards from National Institutes of Health

Eight million dollars has been granted to research the factors that cause adolescents to have problems. The four researchers from the University of Chicago have backgrounds in the study of development and genetics.

One of the studies will work with 7000 sixth to eighth graders in 10 public schools in Chicago. The researcher, Dr. Jacobson, will obtain data on environmental and social factors that may account for problem behavior.

In another study , 800 children will be chosen along with their siblings, half siblings, and unrelated siblings. These students and their guardians will be brought to the University of Chicago Hospital to have neuroscience and behavioral assessments to identify individual differences in problem behavior and brain activity.

To access this article online, go to academic.cengage.com/biology/yashon/hgs1.

QUESTIONS:
1. How might the results of this study translate into information courts could use?
2. Could these studies be harmful to the teens involved in them?
3. How do you think they will pick the students to study? Why?

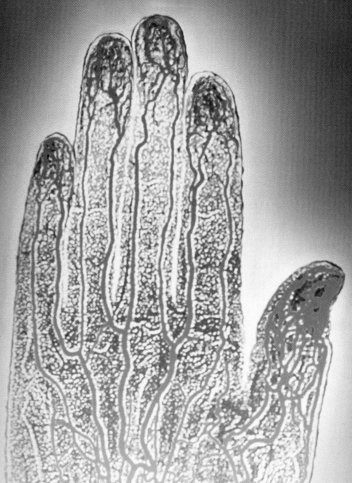

13 Blood Types, Organ Transplants, and HIV

✪ CENTRAL POINTS

- Genetics plays a part in the development of the immune system.

- Immune system compatibility is an important consideration in organ transplantation.

- Human blood types are inherited.

- Problems with the immune system can cause serious disorders.

- Allergies are related to the immune system.

- Carrying organ donor cards makes one's wishes about donation known.

GJLP/CNRI/SPL/Photo Researchers

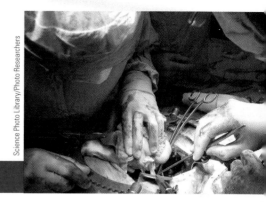

CASE A Sister Wants to Donate Kidney

Kidney disease had devastated the Sanchez family. Mr. Sanchez had failing kidneys; his 15-year-old son, Julio, was already on dialysis, but his older daughter, Maria, was healthy. When her biology class studied heredity, Maria did a family pedigree, using kidney disease as the trait to be studied. She discovered that her great-grandmother had died of kidney disease in Mexico, and so had her grandfather.

Maria desperately wanted to help Julio, who was on a list for a kidney transplant but

his tissue type was rare. She knew from reading her biology book that siblings are often the best match for transplants. She went to see Dr. Tulley, her family doctor, and asked him to test her to see if her tissue type matched Julio's. He was hesitant, but she said, "I am your patient, and I want you to do this—and don't tell my parents." He agreed.

The test results showed that she was a close match. So Maria asked her father for permission to donate a kidney to Julio, but

he became angry. "No," he said, "I want you to stay healthy . . . we'll wait."

Maria still wanted to donate her kidney to Julio, but she was only 17 (not yet legal age) and didn't know what to do.

Some questions come to mind when reading about Maria's family. Before we can address those questions, let's look at the biology of immunity and transplantation.

13.1 What Does the Immune System Do?

The human immune system protects the body from infection caused by bacteria, viruses, and other foreign invaders. It is composed of chemicals and cells that attack and inactivate most things that infect the body. Our body's first line of defense is the skin, which acts as a barrier and blocks these invaders at the entrance to the body. The second line of defense is the inflammatory response which occurs during an infection. The third line of defense is specific white blood cells (*lymphocytes*) called **T cells** and **B cells,** and the molecules they secrete. This stage of defense activates the production of molecules to fight infections.

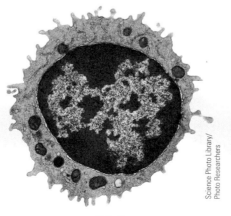

T Cell

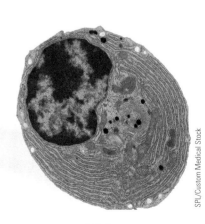

B Cell

Some agents that infect the body carry proteins called **antigens.** These molecules are detected by the immune system and trigger a response that usually involves several stages. Follow these stages in the figure below:

1. Detection of the antigen activates one type of T cell called a **T4 helper cell,** and these cells activate B cells.
2. The activated B cells produce and secrete proteins called **antibodies** that will bind to and inactivate the antigen.
3. Bacteria marked by the antibodies are destroyed by other white blood cells.

Memory cells are the basis of **vaccination** against infectious diseases. A **vaccine** contains an inactivated or weakened disease-causing agent. When injected, these antigens do not cause an infection, but they do stimulate the immune system to produce antibodies and memory cells against that antigen. Later, if the person is exposed to the disease-causing agent (the virus, bacteria, or fungus that was the source of the antigen), he or she will be immune to infection by that agent.

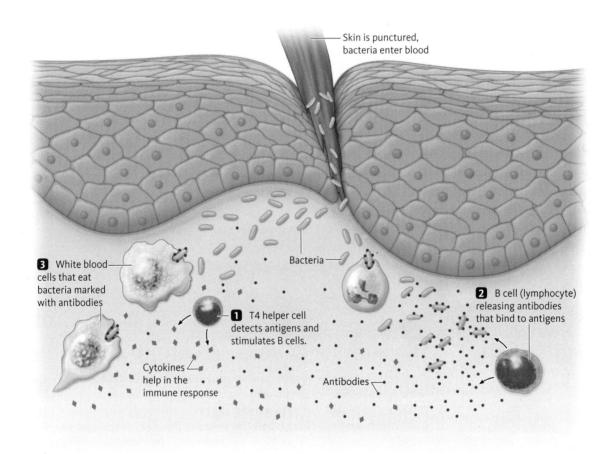

Skin is punctured, bacteria enter blood

3 White blood cells that eat bacteria marked with antibodies

Bacteria

2 B cell (lymphocyte) releasing antibodies that bind to antigens

1 T4 helper cell detects antigens and stimulates B cells.

Cytokines help in the immune response

Antibodies

Antigens may enter the body via a blood transfusion, a cut, or a transplanted organ. They may be part of a disease-causing agent such as a virus, bacteria, or fungus.

B cells are activated and produce specific antibodies that bind to an antigen, marking it for destruction by other cells of the immune system.

Activated B cells also produce **memory cells** that make the body immune to future attacks by the same antigen. If the body is exposed to the same antigen a second time, a massive response is quickly triggered by the immune system, preventing any infection.

Several groups of genes control our immunity by coding for the antibodies that directly attack foreign antigens as well as the antigens on the surface of our cells. Because these genes control immunity, mutations in these genes can cause diseases of the immune system, including **autoimmune disorders,** and *allergies.*

13.1 Essentials

- **The immune system protects the body from infections and foreign invaders.**
- **Two of the most important parts of the immune system are the skin and the T and B cells.**

13.2 How Does Transplantation of Organs or Tissues Affect the Immune System?

When a new organ is placed in the body of a recipient, the cells of the transplanted organ carry antigens that are different from those of the recipient. These antigens, which serve as molecular identification tags, are encoded by a group of genes on chromosome 6 known as the **HLA complex.** This is a cluster of genes close together that tend to be inherited together. The set of HLA alleles carried on each copy of chromosome 6 is called a **haplotype** (■ Figure 13.1). Each of the genes in the HLA haplotype has many different alleles, and the combinations are nearly endless. Therefore, it is difficult to find two people with exactly the same HLA haplotypes . . . except, of course, identical twins. **Rejection** of an organ transplant can occur because of the mismatch of antigens on the cell surface and the immune system's response to them.

Many of the genes in the HLA complex have been identified, and we can test to see if the HLA haplotypes of a potential donor match those of someone in need of an organ transplant.

— HLA complex

■ **Figure 13.1 The HLA Complex**
This map of chromosome 6 shows the position of the HLA gene complex.

Chromosome 6

What determines whether organ transplants are successful?

Successful organ transplants and skin grafts depend on matches between the HLA complex of the donor and the recipient. Because so many allele combinations are possible in the HLA complex, two individuals rarely have a perfect HLA match. That's why finding a compatible donor for those who need organ transplants often takes a long time.

Before an organ transplant is performed, the HLA markers of the donor and the potential recipient are analyzed. If there is at least a 75% match, the transplant will usually be successful.

PHOTO GALLERY
Ronald Herrick donated a kidney to his twin brother, Richard.
AP Photos

After the surgery, the recipient must take **immunosuppressive drugs** to suppress his or her immune system and therefore reduce the possibility that the transplanted organ will be rejected.

Rejection begins when cells of the recipient's immune system, called killer cells, attack and destroy the transplanted organ. Once the rejection process begins, the cells of the transplanted organ are killed rapidly and the organ must be removed. The patient will then need another organ or he or she will die. Closely matching the HLA haplotypes of the donor and the recipient is absolutely necessary to ensure successful transplants.

There is a 25% chance that a sibling will have an organ with matching haplotypes. Maria has already been tested, and her kidney would be a ideal match for Julio. If Maria's father does not allow her to donate a kidney to Julio, he will have to go on a waiting list with over 74,000 others who need a kidney transplant. There are only about 17,000 kidney transplants performed each year, and hundreds of people on the waiting list die each year before receiving a transplant. It is estimated that several thousand lives would be saved each year if enough donor organs of all types were available.

PHOTO GALLERY
Dr. Joseph E. Murray (on right) earned the Nobel Prize for successfully performing the Herrick kidney transplant.
Eric Miller/AP Photos

Can animal organs be transplanted into humans?

One way to increase the supply of organs for transplantation is to use animal donors. Animal–human transplants (called **xenotransplants**) have been attempted many times, but with little success. Some important problems related to rejection currently prevent the use of animal organs. The first problem occurs when an animal organ (e.g., from a pig like the one pictured below) is transplanted into a human.

Sasha Radosavljevich, 2008/used under license from Shutterstock.com

Proteins on the surface of the pig cells (antigens) are so different from those on the human cells that they trigger an immediate and massive immune response, known as a **hyperacute rejection.** This reaction usually destroys the transplanted organ within hours.

To overcome this rejection, several research groups have been using recombinant DNA techniques (see Chapter 6) to modify the antigens of donor pigs. To do this, they isolated and cloned human genes of the HLA complex. These genes were then injected into fertilized pig eggs (or zygotes), and the resulting transgenic pigs carried human antigens on all their cells. Organs from these transgenic pigs would appear as human organs to a recipient's immune system, preventing a hyperacute rejection. Transplants from genetically engineered pigs to monkey hosts have been successful, but the ultimate step will be an organ transplant from a transgenic pig to a human.

Even if the hyperacute rejection can be suppressed, transplanted pig organs may cause other problems. For example, even if genetically modified, pig organs may trigger a stronger immune reaction by the human recipient. Another potential problem is that the cells of the pig organs may carry viruses that are potentially dangerous to humans.

A more radical way to ensure that pig donor organs will be compatible with humans is to transplant bone marrow from a donor pig to the human recipient. The resulting pig–human immune system, called a **chimeric immune system,** would recognize the pig organ as if it were human,

PHOTO GALLERY
Baby Faye received the first baboon heart transplant in 1984.
Duane R. Miller/AP Photos

and still retain normal immunity to fight infectious diseases. As far-fetched as this may sound, animal experiments using this approach have been successful. Bone marrow transplants from donor to recipient have been used in human-to-human heart transplants to increase chances of success.

■ **Figure 13.2 Xenografts**
Jim Finn, shown here before treatment, suffered from Parkinson's disease and couldn't walk, talk, or use his hands. As part of a clinical trial he had fetal pig neural cells injected into his brain. Six months later, he could sit, stand, and walk independently. Today he is an advocate for using animal cells for treatment.

As recently as 10 years ago, the possibility of animal–human organ transplants seemed remote, more suited to science fiction than to medical reality. But today more than 200 people in the United States have already received **xenografts** of animal cells or tissues (■ Figure 13.2).

The advances described here make it likely that xenotransplants of major organs will be attempted in the next few years.

13.2 Essentials

- The antigens on the cells of a donor organ determine whether it will be a match for the recipient.

- By testing HLA genes, matches between recipients and potential organ donors can be made.

- Xenotransplantation may become widespread in the near future.

CASE A QUESTIONS

Now that we understand more about the basics of organ transplantation, let's look at some of the issues raised in Maria Sanchez's case.

1. List four arguments Maria could use to persuade her father to allow her to donate her kidney.

2. List four arguments Maria's father could use to persuade her not to donate her kidney.

3. Does Maria need a lawyer? Why or why not?

4. In the state where the Sanchezes live, the legal age for consent is 18. How does this affect Maria's case?

5. How does family heredity play a part in Maria's argument?

13.3 What Are Blood Types?

Blood types are determined by antigens on a cell's surface. Humans have more than 30 different blood types. Each is defined by the presence of specific antigens on the surface of red blood cells. These antigens are similar to those that are found on the cells of transplanted organs. Antigens serve as markers that identify the cells of the donor and the recipient in a transfusion. We'll focus on two examples of blood types: the ABO system that is important in blood transfusions, and the Rh factor, which plays a role in a disease called **hemolytic disease of newborns (HDN).**

What determines the ABO blood types?

ABO blood types are determined by a gene, **I,** which encodes for cell surface proteins, or antigens, that identify blood type as A, B, AB, or O. This gene has three alleles: I^A, I^B, and I^O, often written as **A, B,** and **O.** The A and B alleles each encode a slightly different version of the antigen found on the cell surface, and the O allele produces no antigen. Those with type A blood carry the A antigen on their red blood cells and make antibodies against the B antigen. Those with type B blood carry the B antigen on their red blood cells and make antibodies against the A antigen. Therefore, type A blood will not accept type B blood and vice versa. Type O cells carry no antigen on their surface and, therefore, do not stimulate an antibody response when used in a transfusion.

How are ABO blood types inherited?

To understand how the ABO blood type is inherited, first look at Table 13.1. It shows that if a person has type AB blood, both the A and B antigens are present on the red blood cells and the body does not make antibodies against either the A or B antigen. The table also shows that those with type O blood have neither the A nor the B antigen on their red blood cells, but do make antibodies against the A and B antigens. This situation helps define how ABO blood types are inherited.

As we stated, the I gene has three alleles that contribute to blood types A, B, and O. People with type O blood have the genotype $I^O I^O$ (or OO). The I^O gene acts as a recessive trait when it is paired with the I^A or I^B allele. Therefore, a person with the genotype $I^A I^O$ would have type A blood, and someone with $I^B I^O$ would have type B blood. However, when the genotype contains I^A and I^B genes, they are not dominant or recessive to one another. Instead, they are said to be **codominant,** and therefore those with the genotype $I^A I^B$ will have an AB blood type.

If you have type A blood, your genotype could be $I^A I^A$ or $I^A I^O$. If you have type B blood, your genotype could be $I^B I^B$ or $I^B I^O$. If you have type AB blood or type O blood, your genotype can be only $I^A I^B$ or $I^O I^O$, respectively.

How do blood transfusions work?

People with type A blood can donate to others who are type A, type B individuals can donate to type B individuals, and so forth. However, the situation is a little more complicated than that. Because people with an AB blood type do not make antibodies against the A or B antigen, they can receive a transfusion using blood of any type. Such individuals are known as **universal recipients.** Those with the O blood type have neither antigen on their blood cells, and as a result they can donate blood to anyone and no reaction will occur. Type O individuals are known as **universal donors.**

Can there be problems with transfusions?

For successful transfusions, the ABO antigens of the donor and recipient must match. If there is a mismatch, the recipient's immune system will make antibodies against the antigens, causing the blood cells from the donor to form clumps (■ Figure 13.3). The clumped blood cells block circulation in small blood vessels, reduce oxygen delivery, and have often fatal results. The clumped blood cells can break down and release large amounts of hemoglobin into the blood. This hemoglobin forms deposits in the kidneys and can cause kidney failure.

Table 13.1 Summary of A, B, and O Blood Types

Blood Type	Antigens on Plasma Membranes of Red Blood Cells	Antibodies in Blood	Safe to Transfuse to	Safe to Transfuse from	Possible Genotypes
A	A	Anti-B	A, AB	A, O	AA or AO
B	B	Anti-A	B, AB	B, O	BB or BO
AB	A + B	None	AB	A, B, AB, O	AB
O	None	Anti-A + anti-B	A, B, AB, O	O	OO

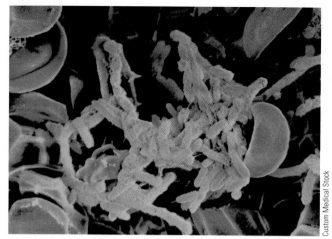

■ **Figure 13.3 Transfusion Reaction**
Red blood cells clumping together after a transfusion reaction.

Custom Medical Stock

How is the Rh factor inherited?

The **Rh blood group** (named for the rhesus monkey in which it was discovered) has a somewhat complex organization. To simplify the genetics, we can say it consists of **Rh positive (Rh⁺)** individuals, who carry the Rh antigen on their blood cells, and **Rh negative (Rh⁻)** individuals, who do not carry this antigen. In this blood type, the Rh⁺ allele is dominant to the Rh⁻ allele.

As a result, if you are Rh⁺, you can carry the allele combination Rh⁺Rh⁺ or Rh⁺Rh⁻. But if you are Rh⁻, you can carry only the Rh⁻Rh⁻ combination.

How does the Rh factor cause problems in newborns?

When a woman is pregnant, a small number of fetal cells may cross the placenta and enter her bloodstream (see

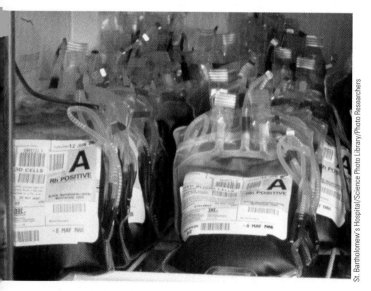

St. Bartholomew's Hospital/Science Photo Library/Photo Researchers

Chapter 2). These blood cells can also enter her bloodstream during childbirth. If she is Rh⁻ and her fetus is Rh⁺, fetal blood cells that cross the placenta will stimulate the mother's immune system to make antibodies against the Rh antigen on the fetal cells. If this is the mother's first pregnancy, this situation will usually not harm either the fetus or the mother.

However, during a second pregnancy, if the mother's second child is also Rh⁺, antibodies from her blood can cross the placenta in the late stages of pregnancy and destroy the fetus's red blood cells. The result is a serious disease called *hemolytic disease of newborns (HDN)*. The drawing below shows HDN and how it affects the mother and fetus.

To prevent HDN, Rh⁻ women are given an Rh-antibody preparation (called RhoGAM) during their first pregnancy if the child they are carrying is Rh⁺. These antibodies move through the mother's circulatory system and destroy any fetal cells that may be present. To be effective, the RhoGam must be given before the mother's immune system has had a chance to make antibodies against the Rh antigen. The drawing below shows this reaction.

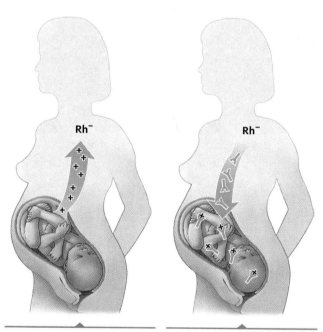

Plus antigens enter the maternal circulation.

Antibodies against the plus antigen attack and destroy fetal blood cells.

13.3 Essentials

- A, B, AB, and O are the major human blood types.
- The Rh blood type can cause problems in newborns when the mother's type is Rh⁻ and the fetus is Rh⁺.

13.4 How Are HIV Infection and AIDS Related to the Immune System?

Acquired immunodeficiency syndrome (AIDS) is a clinical disease that develops after a person is infected with the human immunodeficiency virus (HIV) (see the figure on the right). HIV infects and kills *T4 helper cells.* As we discussed earlier, these cells are extremely important for the onset of an immune reaction. Without them the body would not be able to recognize foreign bacteria or viruses. At the beginning of an immune response, the T4 helper cell recognizes the antigen and then activates the B cells, which in turn form antibodies.

HIV, once inside a cell, copies its genetic information and inserts it into a chromosome in the infected cell. This viral genetic information can remain inactive for months or years. Later, when the infected T4 cell is called upon to participate in an immune response, the viral genes become active. New viral particles are formed in the cell and bud off the surface of the T cell, rupturing and killing it. Over the course of an HIV infection, the number of T4 helper cells gradually decreases, and the body loses its ability to fight infection. Without an active immune system, an infected person can become ill from many diseases that they would otherwise fight off.

PHOTO GALLERY
Many patients with AIDS live in Africa.
Karen Kasmauski/CORBIS

Eventually, by killing T4 cells, the HIV infection disables the immune system, resulting in AIDS. In turn, AIDS causes premature death from one or more infectious diseases that overwhelm the body and its compromised immune system.

The figure at right shows how the genetic information in HIV becomes part of T4 helper cells.

HIV is transmitted from infected to uninfected individuals through body fluids, including blood, semen, vaginal secretions, and breast milk. The virus cannot live for more than 1–2 hours outside the body and cannot be transmitted by food, water, or casual contact.

Are some people naturally resistant to HIV infection?

About 15 years after the first AIDS case was reported in the United States, researchers discovered a small number of people who engaged in high-risk behavior, such as unprotected sex with HIV-positive partners, but did not become infected with HIV. Shortly thereafter, researchers discovered that these individuals carried two copies of a mutant allele of a gene called *CC-CKR5.*

HIV AND HOW IT WORKS

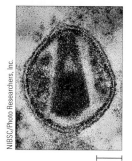

25–30 nm

Photomicrograph of HIV

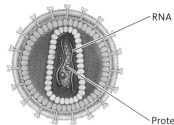

RNA

Protein

Drawing of HIV, showing its RNA genome and the proteins it carries.

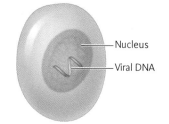

Nucleus

Viral DNA

After HIV injects its RNA into the T cell, the RNA is copied into DNA which is then integrated into the T cell's DNA.

New virus particles

The viral DNA makes new viral RNA and proteins, creating new virus particles that bud off the surface of the infected cell.

When the *CC-CKR5* gene functions normally, it makes a protein on the cell surface that signals the immune system when an infection is present. HIV uses the normal CC-CKR5 protein to infect T4 helper cells. If the *CC-CKR5* gene is mutated and has a small deletion (32 base pairs are missing), the resulting protein is shorter and HIV cannot use this protein to infect and kill T4 helper cells. As a result, individuals who are homozygous for the *CC-CKR5* mutation are resistant to HIV infection.

After this discovery, researchers asked several questions: Which populations carry this mutant allele, and how widespread is it? Can knowledge about how HIV uses the CC-CKR5 protein to infect T4 cells help in the design of anti-HIV drugs?

In which populations are the mutated *CC-CKR5* alleles found?

Population studies show that the mutant allele is present only in Europeans and those of European ancestry. The highest frequency of the mutant allele is found in northern Europe, and the lowest frequency is found in Greece and Sardinia.

Scientists speculate that this mutation is more frequent in certain populations because at some time in the past it may have offered resistance to an unknown but deadly infectious disease (other than HIV/AIDS). Those carrying two copies of the mutant *CC-CKR5* allele were protected from this deadly infection and lived to pass this gene to their offspring.

Other populations carry mutant versions of other genes that confer resistance to HIV infection. These are being identified and studied to find new ways of treating HIV and AIDS.

Are some people naturally susceptible to HIV infection?

Recently, scientists have discovered why HIV infection is highest in sub-Saharan Africa. A gene that evolved to protect people from malaria increases their vulnerability to HIV infection by 40 percent. Variation of this gene may interfere with the ability to fight HIV in its early stages. The study was done over 25 years and involved DNA samples from thousands of participants. African Americans who carry this gene may be more susceptible to HIV infection than the general population.

Can drugs prevent HIV from entering cells?

Many of the drugs now used to treat HIV infection work to prevent the virus from replicating once it is inside the T4 helper cells (■ Figure 13.4). Other drugs block HIV at other stages of its infection and reproduction cycle. Combinations of these drugs are successful in slowing or stopping the

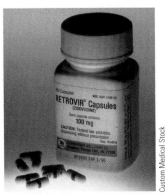

■ **Figure 13.4 AZT**
AZT was one of the first drugs used to treat patients with HIV/AIDS.

Custom Medical Stock

Paul Rapson/Alamy Limited

BY THE NUMBERS

25%
Chance that a sibling will be a match for an organ transplant.

17,000
Number of kidney transplants done each year.

90,000
Number of people on a waiting list for organs on any one day.

10%
Percentage of the U.S. population with at least one allergy.

30,000
Number of cases of anaphylactic shock in emergency rooms in one year.

3 million
Number of people allergic to peanuts in the United States.

progress of HIV. However, these drugs have serious side effects, and, in addition, drug-resistant strains of HIV have developed.

By studying the way HIV enters cells using the protein encoded by *CC-CKR5* gene and other proteins on the cell surface, researchers are developing a new generation of drugs that prevent entry of the virus into its target cells. One of these drugs, **enfuvirtide,** has been approved by the U.S. Food and Drug Administration (FDA) for clinical trials. (See "Spotlight on Society: Drug Development and Public Outcry.") Other drugs are under development and will soon be ready for trials.

13.4 Essentials

- HIV infects cells of the immune system and can lead to AIDS.

13.5 How Are Allergies Related to the Immune System?

Allergies result when the immune system overreacts to antigens that do not cause an immune reaction in most people. These special antigens, called **allergens,** are carried by dust, pollen, and certain foods and medicines. One of the most serious food sensitivities is the allergy to peanuts, which is a growing health concern in the United States.

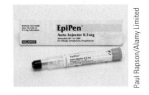

Allergic reactions to peanuts, bee stings, or other proteins may provoke a more serious reaction called **anaphylactic shock.** In this reaction, the bronchial tubes constrict, restricting air flow in the lungs and making breathing difficult. Heart **arrhythmias** and cardiac shock can develop and can cause death within one to two minutes.

Parents of children with these severe allergies often carry an injectable medicine (epinephrine contained in an Epi-Pen®) that can be used as soon as an anaphylactic reaction begins. This drug counteracts the immune response as soon as it is injected, and is a life-saving treatment.

About 80% of all anaphylactic shock cases are caused by allergies to peanuts. Individuals with peanut sensitivities must avoid eating peanuts and any products that include peanuts. Even peanut dust in the air can be dangerous to a person with this allergy. The card shown below explains the dangers of peanut allergies to those who work in a restaurant.

WARNING!
FOOD ALLERGY:
PEANUTS

I have a severe food allergy to peanuts, any foods cooked with peanuts, and anything that has touched peanuts. Please help me dine safely by making sure all utensils, cookware, and surfaces are very clean and have not been previously used. Please review the ingredients on the back of this card and read food labels carefully. If I accidentally ingest even a small amount of peanuts, please dial 911 as I will require immediate emergency medical care.

Courtesy Achoo Allergy & Air Products

The number of children and adults who are allergic to peanuts appears to be increasing. In a 1988–1994 survey, allergic reactions to peanuts in American children were twice as prevalent as those in a 1980–1984 survey.

What is causing the increase in peanut allergies? The answer is still unclear, but environmental factors appear to play a major role. For example, peanut allergies are extremely rare in China, but the children of people who immigrate to the United States from China have about the same frequency of peanut allergies as the children of native-born Americans. This suggests that there is involvement of some environmental factors.

Some researchers feel that because peanuts are now a major part of the diet in the United States, the exposure of newborns and young children (age 1–2) to peanuts is more common. This exposure may occur through breast milk after the mother has eaten peanut butter, or other peanut-containing foods. The immune system of newborns is immature and develops over the first few years of life. As a result, exposure to some antigens is likely to cause food allergies to develop during this time. Until more is known about how peanut allergies develop, some physicians recommend that mothers avoid eating peanuts and peanut

CASE B Prisoners Used as Guinea Pigs

The request had come from the governor's office. It couldn't be ignored. But Warden Wilson had lots of doubts about it. He felt that he had an obligation to his prisoners and that they had the same rights as people in the general population.

A few years ago the state legislature had passed a law requiring that inmates with HIV (which causes AIDS), be isolated in a special wing of the prison. Warden Wilson didn't like this because he was against any form of segregation, but he could understand the safety concerns. Since then, the spread of AIDS had slowed in the prison population, and the prisoners in the special wing got better medical care—he saw to that.

Now he had received a letter explaining that a new study would use inmates to test a vaccine for HIV. The vaccine had been successfully tested in animals, and approved by the FDA for testing on humans. The drug company developing the vaccine needed a population to test the side effects of the vaccine on humans.

The letter went on to explain that prisons were the ideal place for testing on

humans because the population could be carefully monitored, and each person had already been tested and isolated because of his or her HIV status. Each inmate would be paid $1000 a month (a huge salary for inmates) and receive better food and medical care than the other inmates.

Half the participants would be HIV positive and half would be negative. Each inmate would be given an attenuated virus (a weakened, noninfective form of the virus). Even though the virus is supposed to be noninfective, inmates who began the test as HIV negative might contract HIV from the vaccine. This had not happened with the animal tests, but the scientists mentioned this as one of the possible side effects.

Warden Wilson couldn't sleep; he was worried about whether he should allow this study to be done in his prison.

1. What should Warden Wilson do?
2. Give three arguments Warden Wilson should use to persuade the governor's office not to allow the testing in his prison.

picturebox-uk.com/Alamy Limited

3. Give three arguments the scientists might give to persuade the governor's office to allow this kind of testing in the prison.
4. How do you think the prisoners would respond?
5. What reasons might the prisoners use for wanting to participate in the study?
6. Why do you think the researchers picked prisons for this experimentation? Explain your reasoning.
7. Do you think prisoners have the right to refuse such treatment? Why or why not?

products during pregnancy or while nursing, and that children not be exposed to peanuts or other nuts for the first three years of life.

13.5 Essentials

- **Some allergies are becoming a serious problem in our society.**

13.6 What Are the Legal and Ethical Issues Associated with Organ Donation?

The ability to transplant organs has created a number of legal and ethical issues. As the Sanchez family discovered, finding a matching donor isn't easy. Siblings are the best donors, but they cannot always consent.

Most laws require an individual to be 18 years of age to consent to medical treatment. Those below age 18, or those unable to consent for some other reason, are not allowed to donate an organ. In addition, any person who has been identified as incompetent by a court of law, or patients who are comatose or unconscious, cannot consent to donation or any other medical treatment. However, their next of kin can give consent for them. One problem with this type of permission occurs when a family cannot agree on the decision.

Because the number of people on transplant lists has increased over the years, more donors are needed. As a result, many states have passed laws that make organ donation easier for those who want to do it. Most states have organ donor stickers or cards that are issued with driver's licenses. These cards can be used to inform family members and doctors of the donor's wishes. Directions for organ donation can also be spelled out in a living will written by the donor or discussed with friends and family. But many people still do not donate or make their wishes known.

Some states have been thinking of reverting to **assumed consent.** This would mean that everyone would become an automatic organ donor. Only those who carry a card that states their desire not to be donors would be exempt.

The following chart addresses some of the ethical and legal questions associated with organ donation.

Question	How are these questions decided?	Related case or legal issue
What if family members cannot agree about donating the organs of a comatose relative?	Courts can decide whether the comatose patient wanted to donate. They use organ donor cards, living wills, and conversations with friends as a way of determining this.	Often these cases need to be sent to a judge to decide what the actual desires of the patient are.
What happens when a wealthy or famous person needs an organ, but other people are ahead of him or her on the transplant list?	The waiting list is carefully set up so that those whose condition is the most serious get organs first.	The United Network for Organ Sharing (UNOS), run by the U.S. government, sets up the organ donor list, which does not take wealth or celebrity into account.
Some people are afraid they will not be treated as aggressively in an emergency room if they carry an organ donor card. How is this falsehood dealt with?	Public education, including television ads, has been slowly changing the mistaken idea that organ donors receive less aggressive treatment.	Courts often have to decide these questions, but in general this is *not* allowed.
There are still not enough organs; how can the supply be increased?	Many labs have been working on using recombinant DNA technology to create transgenic animals that will be a source of organs.	Courts and legislatures are trying to simplify the method of organ donation by passing laws that make this decision automatic.

13.6 Essentials

- **Those who want to donate organs should make their wishes known to their family.**

Drugs that are developed by scientists or companies must be approved by the FDA through a time-consuming and costly process. Approval involves animal testing and then four stages of human testing (clinical trials) before a drug can be put on the market. These tests show the safety and effectiveness of the drugs.

But what if the condition is life threatening and affects many people? In the 1980s, this was the situation with drugs being developed to treat people with HIV/AIDS. Thousands of people were dying of AIDS, and a treatment or cure did not seem to be in the works. Because of the social stigma associated with AIDS (most people who had it were gay men, poor, or drug users), many organizations and drug companies didn't want to become involved.

The leaders of an organization called ACT UP (AIDS Coalition to Unleash Power) began to use social activism to generate change, bring more drugs to the market, and to increase awareness about AIDS. This small organization of gay men and women used the idea of a grassroots movement to get AIDS into the national spotlight and force the U.S. government to realize the health risks involved. ACT UP staged a number of events and rallies, including the following:

- political rallies
- large demonstrations in front of the White House
- protests at the International AIDS Conference
- sit-ins at hospitals

This focused public attention on ACT UP's cause.

In June 1989, one of the most successful protests was held in front of Sloan-Kettering Hospital in New York City. Protestors, dressed as doctors and patients, sat in front of the hospital for four days while acting out scenarios about people dying of AIDS. The protest was to demand that more people with AIDS be included in clinical trials of AZT (zidovudine or retrovir), a drug now routinely used as a treatment for AIDS.

Andrew Holbrooke/CORBIS

Also in 1989, ACT UP protested the FDA's new protocol for another drug, DHPG (ganciclovir). The new protocol would have denied many current patients access to the drug.

This protest (as shown in the photo to the left) forced the FDA to make HIV drugs available for treatment under an already existing rule called the "compassionate use rule," which made new drugs available more quickly for patients who were terminally ill. This could occur while the FDA was reconsidering a change in its protocols.

Finally, in 1997, the FDA formally introduced the Fast Track Designation and Priority Review. The term *Fast Track* refers to this new FDA system, which was aimed at speeding up the process of development and approval for drugs that are identified as important for treatment of serious diseases and which address unmet medical needs.

To qualify for Fast Track designation, a drug must meet the following criteria:

1. an unmet need
2. a disease with no adequate therapy (such as cancer, heart failure, or Alzheimer's)
3. serious (chronic) conditions

The difference between life-threatening and serious conditions was an important distinction that the FDA considered carefully. The level of seriousness is based on how the disease affects a person's day-to-day living, and by taking into account what would happen if the condition were left untreated. For example, cancer is considered serious because if it is left untreated, it can be fatal. The same is true of Alzheimer disease and heart failure. HIV was included in these conditions and ACT UP had been successful in their quest to speed up the approval of drugs for AIDS.

QUESTIONS:

1. The FDA developed its rules for drug development to protect the public from exposure to untested drugs. But imagine that a member of your family had a fatal condition whose treatment was very close to being approved by the FDA but not close enough. What would you try?
2. ACT UP was criticized for putting AIDS in the face of the public. But it worked. Do you think this was worth upsetting some people? Why or why not?
3. Our government was founded on activism; the ACT UP protests were an example of this. Do some research and find others, and write a short report.
4. Make a short list of some the drugs used to treat HIV/AIDS that have been fast-tracked.
5. Fast-tracking began more than 10 years ago, and it has now been applied to many more drugs than just those used to treat HIV. Do you think this helps us or hurts us?

⊛ THE ESSENTIAL TEN

1. The immune system protects the body from infections and foreign invaders. [Section 13.1]
   ▮ Many of these invaders come in through openings in the skin.

2. Two of the most important parts of the immune system are the skin and the T and B cells. [Section 13.1]
   ▮ These cells, called lymphocytes, function to protect us from infectious agents.

3. The antigens on the cells of a donor organ determine whether it will be a match for the recipient. [Section 13.2]
   ▮ If they are different from our own, the body will reject a transplanted organ.

4. By testing the HLA genes, a suitable match for an organ transplant can be determined. [Section 13.2]
   ▮ The HLA genes encode the antigens on cell surfaces.

5. Xenotransplantation may become widespread in the near future. [Section 13.2]
   ▮ The transplantation of animal organs to humans is being investigated.

6. A, B, AB, and O are some of the major human blood types. [Section 13.3]
   ▮ Blood transfusions are usually based on these blood types.

7. The Rh blood type sometimes causes problems in newborns when the mother is Rh⁻ and the fetus is Rh⁺. [Section 13.3]
   ▮ The two Rh blood types are Rh positive and Rh negative.

8. HIV infects cells of the immune system and can lead to AIDS. [Section 13.4]
   ▮ The virus that causes AIDS attacks and kills cells that trigger the immune response.

9. Allergies are becoming a serious problem in our society. [Section 13.5]
   ▮ Allergens are part of our environment and they can trigger an inappropriate immune response if they enter our bodies.

10. Those who want to donate organs should make their wishes known to their family. [Section 13.6]
   ▮ If one wishes to donate organs, but does not communicate that fact, they may not end up as a donor.

KEY TERMS

ABO blood types (p. 211)
allergens (p. 214)
allergies (p. 214)
anaphylactic shock (p. 214)
antibodies (p. 208)
antigens (p. 208)
arrhythmias (p. 214)
assumed consent (p. 216)
autoimmune disorders (p. 208)
B cells (p. 207)
CC-CKR5 (p. 213)
chimeric immune system (p. 210)
codominant (p. 211)
enfuvirtide (p. 214)
haplotype (p. 209)
hemolytic disease of newborns (HDN) (p. 211)
HLA complex (p. 209)

hyperacute rejection (p. 210)
I (p. 211)
I^A (p. 211)
I^B (p. 211)
I^O (p. 211)
immunosuppressive drugs (p. 209)
memory cells (p. 208)
rejection (p. 209)
Rh blood group (p. 212)
Rh negative (Rh⁻) (p. 212)
Rh positive (Rh⁺) (p. 212)
T cells (p. 207)
T4 helper cells (p. 208)
universal donors (p. 211)
universal recipients (p. 211)
vaccination (p. 208)
vaccine (p. 208)
xenografts (p. 210)
xenotransplants (p. 209)

REVIEW QUESTIONS

1. Name three differences between antigens and antibodies.
2. When a person needs an organ transplant, a plea is often made on television or the Internet asking many people to come and be tested. What test is done, and what happens when a match is found?
3. Who is the best match for an organ transplant?
4. How are blood transfusions and organ donation similar? How are they different? Give three reasons for each.
5. What stops HIV from entering cells?
6. Make a list of the steps our immune system goes through to attack an invader.
7. An HLA match isn't easy to make, but what might happen if an organ didn't match and was transplanted anyway?
8. Anti-rejection drugs are very important in transplantation. How do they work?
9. Blood types were once used to determine paternity. How?
10. The HIV virus is transmitted via bodily fluids. Why is this infection so difficult to stop?

What Would You Do If . . . ?

1. Someone in your family asked you to donate a kidney?

2. Someone you didn't know asked you to donate a kidney?

3. You were asked to donate bone marrow? Would your answer change? Why or why not?

4. You were asked to move to a different seat in an airplane because you were eating a peanut butter and jelly sandwich?

5. You were asked to participate in a trial for an AIDS vaccine?

APPLICATION QUESTIONS

1. We are exposed to many things during a single day. Make a list of some of the things you are exposed to in a single day that your immune system works against.

2. Search the Internet for a story of a donor who gave an organ to a complete stranger. Write a short report and share it with the class.

3. Do you think people would be willing to receive an animal donor organ in a transplant? Why or why not?

4. Can a person who is HIV positive donate blood? Why or why not?

5. Research the most recent work being done on the *CC-CKR5* gene and write a short report.

6. There is a new procedure being advertised on television that freezes the cord blood of newborns. Do some research on this process and write a paragraph discussing who could benefit from the use of cord blood, and what it costs to store cord blood.

7. If a patient rejects the donated kidney after transplant, it becomes more difficult to find a match for this person. Why?

8. Who is the better match for a person needing a transplant:
   a. an identical twin or a fraternal twin?
   b. a fraternal twin or a sibling?
   c. a mother or a father?
   d. a complete stranger or friend?
   Give reasons for your answer.

ONLINE RESOURCES

Preparing for an exam? Log on at academic.cengage.com/login for a pre-test, a personalized learning plan, and a post-test in CengageNOW's Study Tools to help you assess your understanding.

If assigned by your instructor, the Case B and Spotlight on Society activities for this chapter, "Prisoners Used as Guinea Pigs" and "Drug Development and Public Outcry," will be available in CengageNOW's Assignments.

In The Media

(Plattsburgh, New York) Press-Republican, **August 29, 2007**

Baby Makes Strong Recovery after Five-Organ Transplant

Andrea VanValkenburg

In Lewis, New York, Elijah Moulton was born with a condition called intestinal atresia, which means his internal organs were underdeveloped. He was fed through a tube inserted into his stomach until doctors decided he was strong enough to endure one of the first five-organ transplants ever done. On July 10, 2007, a liver, small bowel, pancreas, colon, and stomach were implanted into Elijah's abdomen during the seven-hour procedure at Morgan Stanley Children's Hospital in New York City.

One difficulty with transplants for children is the size of the organs. The organs donated by adults are too big, so a child must be the donor. Elijah received his new organs from an anonymous donor baby.

To access this article online, go to academic.cengage.com/biology/yashon/hgs1.

QUESTIONS:
1. Who would donate the organs of their child? Come up with three examples.
2. Legally, a child can not consent to such a donation. Who can consent for a child?
3. If parents say no to the donation but the doctors know that the child is a perfect match, what can they do?

Associated Press, August 27, 2007

Blood Banks Seek More Latino Donations

Lauran Neergaard

Blood banks seem to be targeting Latinos as blood donors. It appears that Latinos are more likely than other Americans to have type O blood.

Type O blood is the most common type and blood banks want as much of it as they can get.

Less people are donating blood and banks are faced with dwindling supplies. Over 60% of the US population can donate blood, but only 5% does. Hospitals need a great deal of blood everyday (34,000 pints) and the demand is constantly rising.

To access this article online, go to academic.cengage.com/biology/yashon/hgs1.

QUESTIONS:

1. How might blood banks get more Latinos to donate? Name three ways.
2. What does the aging of the population have to do with the scarcity of donors? The amount of blood needed?

Voice of America, August 31, 2007

Scientists Search for Allergy-Free Peanut

Paul Sisco

With peanut allergies increasing, scientists like Mohamed Ahmedna at North Carolina Agricultural and Technical State University are searching for a way to eliminate the allergens found in peanuts. He found a way to process peanuts after they are harvested that removes the allergy-related proteins. This means that farmers can continue to grow the strains they are familiar with, but the peanuts will be processed differently. These processed peanuts have not yet been tested on humans.

In another effort, Maria Gallo at the University of Florida is trying to grow peanuts that do not contain the proteins that cause allergic reactions.

To access this article online, go to academic.cengage.com/biology/yashon/hgs1.

QUESTIONS:

1. If these scientists succeed in eliminating the allergen in peanuts, do you think it would be worth all the money spent in research? Why or why not?
2. One doctor is giving patients small amounts of peanut dust to test subjects who have a known peanut allergy. Knowing what you do about the allergic reaction, explain how this may work.

14 Genetics and Populations

⊛ CENTRAL POINTS

- Genetic conditions often become common in a specific community.

- Huntington disease affects large numbers of people in two villages in Venezuela.

- The frequency of traits can vary from one population to another.

- Calculations can determine how frequently an allele is present in a population.

- Population genetics has been used in cases involving DNA forensics.

CASE A   A Marker in the Blood

It began with a study of one family in Venezuela. This family, and another, much larger one, live in small villages around Lake Maracaibo. These villages have a higher frequency of Huntington disease (see Chapter 4 for a description of this disorder) than anywhere else in the world.

Nancy Wexler is a scientist who studies genetic disorders, including Huntington disease (HD). When she learned of these villages in the 1970s, she began traveling there to learn why HD is so common.

She and her colleagues began by creating a pedigree of the families and collecting blood samples to test their DNA for the gene. The pedigree has grown to include almost 19,000 people, and more than 4,000 blood samples were collected. These blood

samples were used to identify a DNA marker in the people with HD. This marker then led to the identification of the HD gene and, finally, to the development of a genetic test for HD.

For her work in identifying the gene for HD, Nancy Wexler won the Lasker Award, often called the U.S. equivalent to the Nobel Prize. However, she also had a personal reason for being interested in HD: her mother, Lenore, and three of her uncles had died from HD. In the later stages of the disease, Nancy's mother had lost her memory and had to be fed and cared for by Nancy's father, Milton. Later, Milton Wexler started the Hereditary Disease Foundation to study Huntington disease and related disorders, and Nancy made HD her life's work.

Although a genetic test for HD was developed in 1993 and Nancy and her sister Alice are at risk for HD, Nancy has decided not to take the test. Alice was tested and found that she does not carry the HD gene.

Some questions come to mind when reading this case. Before we can address those questions, let's look at how the study of populations can open our eyes to certain genetic conditions.

PHOTO GALLERY
This map shows Venezuela and Lake Maracaibo.

14.1 Why Do Geneticists Study Populations?

In small communities such as those around Lake Maracaibo, people rarely travel very far, and thus they tend to marry others in the same village or nearby villages. Over many generations, inherited conditions can become more common because the populations are so isolated. Often the result is that many people are affected with the condition.

In these situations, usually the mutant allele causing the disorder was brought into the community many generations ago. In the Venezuelan population with HD, the mutant allele came into the village in the beginning of the 19th century. Almost all of the affected individuals in the pedigree can trace their ancestry to one woman, Maria Concepción Soto. She had 10 children and became the founder of the population of more than 18,000 people. Populations whose origins can be traced to a small number of people are an example of a phenomenon geneticists call the **founder effect**.

When constructing pedigrees of large families with a founder effect, geneticists discovered that these populations often have a high frequency of one or more genetic disorders. Using pedigrees, along with blood and tissue samples, scientists can identify, map, and isolate genes responsible for a number of genetic disorders. The gene for Huntington disease (HD) was one of those success stories.

What causes Huntington disease?

Huntington disease is caused by a repeated DNA triplet in the HD gene, located on chromosome 4. This triplet, CAG, may be repeated many times. Normally, people have between 10 and 35 copies of the CAG repeat in this gene. People with fewer than 27 CAG repeats do not get HD. People with 27–35 copies of the repeat do not get HD themselves, but their children are at risk if they inherit an HD allele that expands the number of repeats. Those who carry 36–40 copies of the repeat may or may not get HD, but those with 40 or more repeats almost always get HD.

The number of repeats in this gene is not always stable, and as the HD gene is passed to new generations, the number of repeats can increase, putting more people at risk.

Recall that in Chapter 4, Alan and his family worried about getting HD. If Alan or other members of his family carry the mutant HD allele, the disease will appear in midlife, usually after people have already had children.

The population in Venezuela sparked the curiosity of geneticists. Other populations around the world with high frequencies of other genetic disorders have been studied to discover what they have in common.

14.1 Essentials

- Certain genetic conditions can occur at a high frequency in a single community.
- Pedigrees can be helpful in tracking a condition through a large population.

- Huntington disease (HD) is caused by an increased number of copies of a repeated DNA sequence.

14.2 What Other Genetic Disorders Are Present at High Frequencies in Specific Populations?

When looking at the map of Africa, Europe, and Asia (shown below) investigators saw that the geographic distribution of sickle cell anemia almost completely overlapped with the areas affected by malaria. They wondered if this might be more than a coincidence and whether there was a link between sickle cell anemia and malaria.

KEY

Percentage of population that has sickle cell allele

| 14+ | 10–12 | 6–8 | 2–4 |
| 12–14 | 8–10 | 4–6 | 0–2 |

People with malaria experience recurring episodes of illness throughout life, and often die at a young age. Although malaria may seem like an exotic disease to many of us, it affects more than 500 million people worldwide and kills more than 3 million people each year.

Malaria is caused by a parasite that infects red blood cells as part of its life cycle. Mosquitoes (■ Figure 14.1) that bite humans take in a small amount of their blood. These insects become infected with the malaria parasite and spread the disease when they bite uninfected people.

Researchers discovered that people who either are carriers of sickle cell anemia (heterozygotes, *Ss*) or have the disease (homozygotes, *ss*) are resistant to infection by the malaria parasite. In these individuals, the membrane of red blood cells is altered, making it difficult for the parasite to get into the cells. Although those with sickle cell anemia are resistant to malaria, they still have sickle cell anemia, and they often become very ill.

■ **Figure 14.1 The Anopheles Mosquito**
This mosquito carries malarial parasites and injects them into a human's bloodstream when it bites.

But carriers of the gene survive better because they do not have the serious symptoms of sickle cell anemia *and,* at the same time, they are resistant to malaria. Their children are often carriers and are, therefore, resistant to malaria like their parents. In turn, these children become parents and spread the sickle cell gene as well as malaria resistance to their offspring and the population. Therefore, areas with high frequencies of malaria infections may have high frequencies of sickle cell carriers.

14.2 Essentials

- Sickle cell anemia is an example of a gene that has stayed in a population.

14.3 What Other Populations Have Specific Genetic Traits in Common?

Distributions of genetic disorders among certain human populations show some interesting patterns. Familial hypercholesterolemia (see Chapter 4) has a worldwide frequency of about 1 in 500 people. However, in a few populations, the frequency of this disease is much higher. For example, in descendants of Dutch migrants to South Africa (the Afrikaners), the frequency is about 1 in 70 people.

As seen in Table 14.1, some populations have a much higher frequency of carriers of recessive traits than others;

Table 14.1	Frequencies of Carriers of Tay-Sachs within Populations	
Genetic Condition	**Population Affected**	**Frequency of Carriers in a Population**
Tay-Sachs	Ashkenazi Jewish in the U.S.	1 in 30 people are carriers of Tay-Sachs
	All people in the U.S.	1 in 300 people are carriers
Cystic Fibrosis (CF)	People in U.S. of Northern European descent	1 in 22 are carriers of CF
	African Americans	1 in 66 are carriers of CF
	Asian Americans	1 in 150 are carriers of CF

this is called a **carrier frequency.** You can see the huge differences among different populations. In high risk populations, the chances are greater that members of the population will be carriers.

14.3 Essentials

- A number of conditions have increased frequency in certain populations.
- We can measure the number of carriers of a gene in a population.

14.4 Do Environmental Conditions Change the Frequency of Genetic Traits in Populations?

The frequency of a trait may vary from population to population. Cystic fibrosis (CF) is a good example. As discussed in Chapter 4, CF is an autosomal recessive genetic disorder that is common in some populations but nearly absent in others. Many children with CF need specialized care. Medical centers are now in many areas of the United States and other countries to help families with CF. Looking at ■ Figure 14.2, you can see where these centers are located. Do these centers indicate where populations of CF are located?

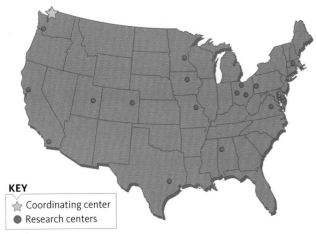

KEY

☆ Coordinating center
● Research centers

■ **Figure 14.2 Cystic Fibrosis Centers**
Cystic fibrosis centers in the United States are shown on this map.

Cystic fibrosis affects the glands that produce mucus, digestive enzymes, and sweat, causing far-reaching symptoms. Most individuals with cystic fibrosis develop obstructive lung disease and infections, leading to premature death.

Recent developments in treating CF now allow affected individuals to live longer. Previously, affected individuals usually died before having children. At that time, scientists thought that the gene might slowly be eliminated from most populations because no one could pass it along. But this hasn't happened.

There is some evidence that people heterozygous for the CF gene (*Cc*) are more resistant to typhoid fever, an infectious disease caused by a bacterium (*Salmonella typhi*). This bacteria infects cells of the intestinal lining. When transgenic mice that were carriers of the human cystic fibrosis gene were injected with typhoid fever bacteria, their intestinal cells were infected by fewer bacteria than those of normal mice.

How this relates to humans with the CF gene has not been determined. If heterozygous carriers of the CF allele (*Cc*) are protected from one or more infectious diseases, the mutant allele may be maintained in the population because of that protection. This relationship is similar to sickle cell anemia and malaria in Africa.

14.4 Essentials

* **The frequency of traits may vary from one population to another.**

CASE A QUESTIONS

Now that we understand how certain genetic disorders interact with the environment and populations, let's look at some of the issues raised in the case of HD in Venezuela.

1. A small part of the Venezuelan population has many individuals affected with HD. As we discussed earlier, the cause of HD is an increase in the number of copies of CAG repeat sequence in the HD gene, carried on chromosome 4.

As members of a population reproduce, the greater the chances they will pass the HD gene on to their children, and the greater the chance that, in some of these alleles, the number of CAG repeats will increase.

Although most cases of HD appear in adults, a person with more than 50 repeats may develop symptoms during his or her teen years. Could you use prenatal diagnosis to determine your child's number of repeats?

2. What might be done to help the Venezuelan families to plan for their future?

3. As discussed in Alan's case in Chapter 4, HD symptoms do not appear until later in life. But Nancy Wexler was born in 1945. Should Nancy Wexler take the test?

4. Draw the pedigree of Nancy and her sister. Include her father, her mother, her mother's three brothers who died of HD, her mother's mother and father, and her father's mother and father.

14.5 How Can We Measure the Frequency of Alleles in a Population?

If a genetic disorder such as sickle cell anemia or cystic fibrosis is caused by a recessive allele, we cannot directly count those who carry that allele in a population. Why? Because we cannot tell the difference between carriers of the mutant allele and those who do not have the allele at all. An example of this is cystic fibrosis. A person who is a carrier of CF has the genotype Cc and has no symptoms. A person who does not carry the CF gene has the genotype CC and also has no symptoms. They both have the same phenotype and cannot be distinguished from each other without a genetic test. If researchers cannot tell who is a heterozygote (Cc) and who is not (CC), there is no direct way of counting how many heterozygotes (Cc) are in the population being studied.

PHOTO GALLERY
Photos of Godfrey Hardy (left) and Wilhelm Weinberg (right)
Hulton-Deutsch Collection/CORBIS; From Stern, C. (1962) 47:1–5/Genetics (Journal)/Genetics Society of America

But early in the 20th century, Godfrey Hardy and Wilhelm Weinberg independently developed a simple formula that allows geneticists to measure the numbers of genotypes and alleles in a population without DNA testing of the population itself.

This formula, now called the **Hardy-Weinberg law,** is widely used by geneticists, clinicians, population biologists, evolutionary biologists, and others to study genes in populations.

To show how the Hardy-Weinberg law works, let's look at a gene that has two alleles, a dominant allele (A) and a recessive allele (a), similar to the situation in CF. In their calculations, Hardy-Weinberg used p to represent the allele A (therefore, $p = A$) and q to represent the allele a (therefore, $q = a$).

Follow these statements below and it will become clear how the Hardy-Weinberg law works:

1. If a genotype has two copies of the dominant allele (AA), it is represented as p^2 ($p \times p = p^2$).

2. If a genotype has one copy of each allele (Aa or aA), it is represented as $2pq$ ($pq + qp$).
3. If a genotype has two copies of the recessive allele (aa), it is represented as q^2 ($q \times q = q^2$).
4. These are the only possible genotypes in a population (aa, Aa, and AA).
5. When added together, these three genotypes equal all the genotypes for this condition in the entire population.
6. Therefore, $p^2 + 2pq + q^2 = 1$. This equation represents 100% of the genotypes AA, Aa, aa in the population.

How can we use the Hardy-Weinberg equation?

Measuring the frequency of alleles and genotypes in a population can be used to provide information about these populations and the people in them. A couple can learn the risk factors for having a child affected with a genetic disorder. When a genetic disorder (such as sickle cell anemia or cystic fibrosis) has a much higher frequency in one population compared to another, the Hardy-Weinberg law can be used to calculate how common the alleles are in these populations and how common a carrier might be. Then a person from one of these populations can know his or her risk of having a child with a specific condition.

How are allele frequencies and heterozygote frequencies calculated?

Suppose we want to discover how common the allele for cystic fibrosis (c) is in a population, such as people with Northern European ancestry in the United States, and also calculate how many heterozygotes (Cc) there are. Here is how we would do it:

1. Count the number of people in the population who have cystic fibrosis. This will tell you how many people have the genotype cc. In the population studied, we find 1 in 2500 people have cystic fibrosis (CF). That is the frequency of CF in this population.
2. Convert the frequency of CF (1 in 2500) in the population to a decimal. So, if 1 in 2500 people in a population have CF, it is expressed as a decimal as 0.0004 (1 divided by 2500). So, the genotype $cc = q^2 = 0.0004$.
3. To calculate q (the frequency of c), take the square root of 0.0004. When you do that, you find that $q = 0.02$ or 2%. Therefore, 2% of all the CF genes in this population

David Young-Wolff/PhotoEdit

Carina Jones was always different. Most of her fellow students could identify their race by looking in the mirror. She never could. On top of that, she was teased about her heritage.

Her father, Dan, an African American artist, had met her mother while he was stationed with the army in Japan. He always said that the first time he saw his wife, Yoshi, he was in love. They married in Japan and came back to live in California, where Dan had grown up. Dan's mother, Carina's grandmother, always told her proudly that her great-grandmother had been a Sioux, one of the first Native Americans to be college-educated.

Right before Carina left for college, her mother's father died and they all went to the funeral in Japan. Her grandmother was very nice to her, and even though she didn't speak English they seemed to bond. But Carina thought the Japanese looked at her strangely. Her father said she was imagining it.

On the return flight, Carina read a magazine article about Oprah Winfrey, who had just returned from a trip to Africa and confessed that she had had her DNA tested to investigate her African origins. The company she used offers tests to African Americans to see which of the 400 ethnic groups in Africa they are related to.

Carina thought she might major in biology, and making it personal seemed to make sense. She looked over at her father and wondered if she should ask him about the test.

1. What should Carina do? Does she need her father's permission to take this test?

2. If Carina asks her father for permission, should he allow her to take the test? Why or why not?

3. Research some of the companies doing these tests and report on their prices and what tests they do.

4. Draw Carina's family's pedigree.

5. When Carina finds out her racial make-up, what might happen?

are the mutant allele c; this means that 98% of the CF alleles in this population are the normal allele C.

Knowing the frequency of the alleles, you can calculate the percentage of carriers (Cc) in the population by substituting the calculated allele values into the Hardy-Weinberg equation. In this equation the carriers (Cc) are written as $2pq$. This is $2 \times p \times q$, or $2 \times 0.98 \times 0.02$, which equals 0.03 or 3%.

In other words, about 3% or 1 in 23 people of Northern European ancestry in the U.S. population carry the gene for cystic fibrosis in the heterozygous condition.

How else can we use the Hardy-Weinberg equation?

Even though malaria is no longer found in the United States, some African Americans with West African ancestry still carry the sickle cell gene as part of their genetic heritage. By counting the number of African American children with sickle cell anemia born in the United States and using the Hardy-Weinberg law, we can calculate that about 8% or 1 in 12 African Americans with West African ancestry are carriers (Ss) of the mutant allele that causes sickle cell anemia (s). This information may be of interest to couples such as the Johnsons before they decide to get genetic testing.

In areas such as West Africa, where malaria is still a serious health problem, we can get useful information about the frequency of sickle cell anemia using this same technique. Studies have shown that in some areas of West Africa, 20–40% of the population are carriers (Ss) for the sickle cell gene, a striking difference from the frequency for African Americans that is only 8%.

14.5 Essentials

- One can measure the frequency of a gene in a population.
- Mathematically determining genetic information about a population can benefit its members.
- Even if the benefit of a specific gene still exists, the environment may change and reduce the need for those benefits.

14.6 What Are the Legal and Ethical Issues Associated with Population Genetics?

The storing of DNA in databases and how those are used present a number of problems. When DNA is collected from a large population it is usually stored in databases for later use. DNA databases raises several legal and ethical questions:

- Should anyone be forced to provide a DNA sample? What about those who are suspects, those who are arrested, or those who are convicted of a crime?

- Who has the authority to order that a DNA sample be taken? A police officer at the scene? A judge based on a subpoena? A trial judge after someone is convicted of a crime?

- Should the DNA profiles of those arrested but found innocent of a crime remain in the database or should they be deleted?
- What crimes should be included in the database? All fifty states in the United States require those convicted of sex crimes to provide DNA samples and most states require convicted felons to provide a sample. Should this be extended to lesser crimes such as fraud, income tax evasion, etc?
- Private information unrelated to crimes such as family relationships and disease susceptibility can often be deduced from the samples. Police, forensic scientists, and researchers have access to people's DNA profile without their permission. Is this a violation of privacy rights?
- In the United Kingdom, the DNA database collects samples from all suspects, and ethnic minorities are overrepresented in the population of arrestees. Does this show that the criminal justice system is racially or ethnically biased?

The Hardy-Weinberg equation can be used to measure the frequency of DNA markers used in profiles after they are placed in databases. This is how they are sorted.

Using allele frequencies in various populations to sort DNA results can cause problems. Defense attorneys can argue that defendants do not have fair testing if their sample was compared to a large population and not to their owm ethnic group.

The opposite can also be argued. An attorney can ask that a larger population be used to match samples if this would give his client a fairer trial.

The following chart addresses some of the ethical and legal questions related to DNA databases and population genetics.

Question	How are these questions decided?	Related case or legal issue
Comparing a defendant's DNA to a database that includes other members of a different ethnic group might not bring about a match. Should a defendant have access to their DNA test results as well as the frequency of database alleles in his or her population?	If this information is available, the defense must be given all of the data. Courts demand that all evidence be disclosed by the prosecution and some by the defense.	This rule, called the **discovery rule,** has been in contention for a number of years. Some legislatures (CT and RI) have passed laws that restrict what the prosecution has to give the defense.
Has the problem of which population databases to use in DNA matching been solved?	The FBI is compiling databases that take into consideration certain population subgroups and geographic backgrounds.	
Should states separate their DNA databases according to ethnicity or race?	Many think that this would be prejudicial to defendants because it narrows the number of suspects for comparison.	
Should DNA samples be taken from anyone suspected of a crime and placed in a state's database?	Some states have tried to institute this type of law.	So far NY and NJ have these laws on the books. Some legal experts believe this will be the trend.

The Heritage of an African Tribe Is Revealed

An interesting combination of history and science developed a few years ago when a sociologist and a geneticist got together to discover the truth behind the oral history of an African tribe.

In England, University of London scholar Dr. Tudor Parfitt was studying the background of the Lemba people, many of whom live in southern Africa. The Lemba have a long-standing oral tradition that tells of their ancestors who were Jews that left the Middle East thousands of years ago. Present-day members of the Lemba tell the story that they are Jews descended from this population. Parfitt was interested in tracing their background to prove or disprove their claims.

He first met members of the Lemba when he gave a lecture at the University of South Africa about another African tribe that had emigrated to Israel. After his talk two men wearing yarmulkes, or Jewish skullcaps, came up to him. They said that they were Jews and their ancestors had emigrated from the Middle East centuries before. Parfitt found this rather intriguing but difficult to believe. He was curious and visited Lemba settlements to study their culture.

He was amazed to find that their religion was very similar to Orthodox Judaism. They had kosher food restrictions, didn't allow marriage outside the faith, and had their sons circumcised.

PHOTO GALLERY
Lemba in worship.
Christian Holland

One night in a London pub, soon after his visit to South Africa, he was introduced to a group of scientists from the Genetic Anthropology Center at the University College of London. They were involved in an ongoing study tracing the inheritance of a genetically unique Y chromosome found in a group of males in Israel. These men were called the *Cohanim*, or Jewish priests. They claimed descent from a single Jewish ancestor, Aaron, the son of Moses, who lived 3000 years ago.

As discussed previously (see Chapter 1), the Y chromosome passes from father to son. It does not engage in crossing over with the X chromosome, but may undergo some change via mutation. In general, however, it is unchanged. The University College group had identified a pattern of molecular markers on the Y chromosome in some Israeli men. They called this pattern the **Cohen Model Haplotype.**

Parfitt suggested they work together to investigate the background of the Lemba and determine whether their oral history about Jewish origins was true by looking at their Y chromosomes. A group of geneticists traveled with him to southern Africa and took DNA samples from the Lemba leaders and their rabbis. Analysis showed that 1 in 10 males of the Lemba carried the Cohen Model Haplotype. The following DNA analysis compares the Y chromosome haplotype of the Lemba and the *Cohanim.*

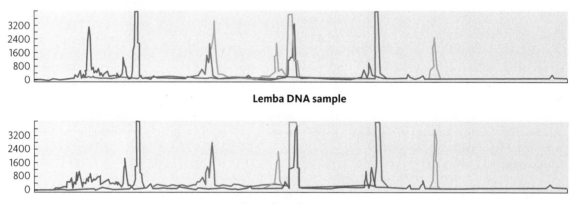

Lemba DNA sample

Cohanim **(Israeli) DNA sample**

QUESTIONS:

1. The analysis seems to prove that the Lemba do have a history of migration from the Middle East. Could they use this to become citizens of Israel? Research this and write a short report on what you find.

2. *NOVA*, the public television show, had a program on this topic. It was titled "The Lost Tribes of Israel" because legend has it that 2 of the original 12 tribes from the Bible have never been found. Could the Lemba be one of the tribes? Why or why not?

3. If you look at Figure 14.3, you might not identify this man as being a Jew, even though he is wearing a traditional Jewish prayer shawl. What does this tell you about the genetics of this group after it migrated to Africa?

■ **Figure 14.3**
A leader of the
Lemba Tribe (Rabbi).

Jay P. Sand

⊛ THE ESSENTIAL TEN

1. Certain genetic conditions can appear at a high frequency in a single community. [Section 14.1]
   ▮ This is because one person many generations ago passed the gene on to his or her offspring.

2. Pedigrees can be helpful in tracking a condition through a large population. [Section 14.1]
   ▮ These pedigrees are often difficult to obtain.

3. Huntington disease (HD) is caused by an increased number of copies of a repeated DNA sequence. [Section 14.1]
   ▮ This sequence is CAG and can be repeated dozens of times in a mutant HD allele.

4. Sickle cell anemia is an example of a gene that has stayed in a population. [Section 14.2]
   ▮ There is a benefit to carrying the gene, which helped it survive.

5. A number of conditions have increased frequency in certain populations. [Section 14.3]
   ▮ One of these conditions involves the amount of cholesterol in one's blood.

6. We can measure the number of carriers of a gene in a population. [Section 14.3]
   ▮ This can help in determining the chances of having a child with a genetic disorder.

7. The frequency of traits may vary from one population to another. [Section 14.4]
   ▮ An example of this is cystic fibrosis.

8. One can measure the frequency of a gene in a population. [Section 14.5]
   ▮ Gene frequency can be mathematically determined using the Hardy-Weinberg law.

9. Mathematically determining genetic information about a population can benefit its members. [Section 14.5]
   ▮ People in such populations can determine their risk of having a child with a genetic condition.

10. Even if the benefit of a specific gene still exists, the environment may change and reduce the need for those benefits. [Section 14.5]
    ▮ If a population moves to a different area, it may not need the benefit anymore.

KEY TERMS

carrier frequency (p. 225)
Cohen Model Haplotype (p. 230)
discovery rule (p. 229)
founder effect (p. 223)
Hardy-Weinberg law (p. 227)

REVIEW QUESTIONS

1. What is the founder effect?
2. Why are large populations studied by geneticists?
3. What do the number of CAG repeats in the HD gene tell us about the people who have them?
4. How are malaria and sickle cell anemia related?
5. What is an allele frequency?
6. Why do we need a mathematical equation to tell us how many carriers there are in a population?
7. The example in Section 14.5 does not reveal what the condition is. Do we need to know that to figure out the allele frequency?
8. Name two ways that using the Hardy-Weinberg equation can help people in a population.
9. What is the cause of malaria?

APPLICATION QUESTIONS

1. Do some research on other communities with a large number of people with HD and write a short report.
2. Large groups of people with leukemia and other cancers have been found all over the world. These are called *cancer clusters*. Do you think this is the same situation as with the people in Venezuela? Why or why not?
3. Do some research to find out how many repeats are found in those who develop HD in their teens and twenties and write a short report.
4. How are the following people affected by malaria (*S* = normal, *s* = sickled): *SS*, *Ss*, and *ss*?
5. How might the information you have learned in this chapter help the following people in the chapter cases?
   a. Martha Lawrence and her husband (Chapter 3)
   b. Chris Crowley and his son, Mike (Chapter 6)
   c. The attorney for William Bern (Chapter 8)
   d. Maria Sanchez's family (Chapter 13)

6. Physicians sometimes use a blood type called MN in their diagnoses. It is similar to the ABO blood type discussed in Chapter 13 but has only two alleles (*M* and *N*). In testing all 1-year-old children in a small population, you find genotype frequencies of *MM* = 0.25, *MN* = 0.5, and *NN* = 0.25. Using the Hardy-Weinberg law, determine the allele frequencies for *M* and *N*.

7. In adults in the population described in question 6, the genotype frequencies are *MM* = 0.3, *MN* = 0.4, and *NN* = 0.3. Using the Hardy-Weinberg law, determine the allele frequencies for *M* and *N*.

8. In Section 14.5, we learned that in some areas of West Africa, 20–40% of the population are carriers for sickle cell (*Ss*), and in African Americans this number is only 8%. Why do you think this is the case?

ONLINE RESOURCES

Preparing for an exam? Log on at academic.cengage.com/login for a pre-test, a personalized learning plan, and a post-test in CengageNOW's Study Tools to help you assess your understanding.

If assigned by your instructor, the Case A and Spotlight on Society activities for this chapter, "A Marker in the Blood" and "The Heritage of an African Tribe Is Revealed," will be available in CengageNOW's Assignments.

What Would You Do If . . . ?

1. You were Nancy Wexler and asked to take a test for HD?

2. You were on a jury deciding a case in which a man's DNA matched that found at a crime scene, and he was Asian but the sample was matched against a mixed population of Asians, African Americans, and Caucasians?

3. You had the opportunity to test your ancestry the way Carina did?

4. You sent your DNA to an ancestry-testing site and found that you were from a different ancestral population than you thought previously?

In The Media

BBC News, April 24, 2008

Human Line Nearly Split in Two

Paul Rincon

The Genographic Project, run by the National Geographic Society, has been tracking human migration by DNA samples. Based on analysis of mitochondrial DNA of Africans living today, researchers found that two different populations existed in isolation between 50,000 and 100,000 years ago. The two populations seemed to have come together in the Stone Age forced by population expansion.

To access this article online, go to academic.cengage.com/biology/yashon/hgs1.

QUESTION:

If the population of early humans had split into two, what do you think might have happened?

Science Daily, April 21, 2008

Breakthrough in Migraine Genetics

Even though studies have found susceptibility of many common diseases among populations, no one had studied migraine headaches before groups in Finland and Australia used a new technique. By comparing the genomes of families that were identified as having migraines, they identified a gene on chromosome 10 that is linked to migraines in women and some men. Four thousand people with migraines carried this mutant allele. These studies were the first to tie a gene to migraine headaches. Because this occurred in two diverse populations from different parts of the world, the results will be studied carefully.

To access this article online, go to academic.cengage.com/biology/yashon/hgs1.

QUESTION:
How might this discovery be used to determine who will have migraines?

15 A Different World: The Past, Present, and Future of Human Genetics

⊛ **CENTRAL POINTS**

▎ We can learn a lot from the eugenics movement of the 1920s.

▎ Genetics is moving ahead with many new applications.

▎ We face a future full of important questions to be decided by society and individuals.

15.1 What Can We Learn from the Past?

Thinking back on the cases discussed in this book, we can make some interesting observations. The Carters, the Crowleys, the Franklins, and others were worried about the genes that had been passed on in their families. In the recent past, some scientists thought they could control which genes were passed on to the next generation and over time, create a "better" human. The following is an incredible, but true story from the early 20th century.

Fitter Families Win Prizes

■ **Figure 15.1 Fitter Families**
These are the winners of a Fitter Family Contest in Kansas in 1927.

In the early part of the 20th century, scientists at the U.S. Eugenics Office, a privately-funded organization, began running contests around the country to pick the families with the "best genetics." These families were given medals and awards and encouraged to have more children.

This program, part of the *eugenics movement,* existed in the United States for more than 40 years. A number of scientists, led by Charles Davenport, began by applying Darwin's theories to humans. Eugenicists first decided which traits were desirable and then encouraged people with those traits to have many children. This is called **positive eugenics.** At the same time, they compiled a list of traits that were not desirable. Laws were passed that forced the sterilization of people with these traits so they could not reproduce. Immigration was limited from certain countries whose people were considered to have undesirable traits. This is called **negative eugenics.**

States began passing laws to sterilize criminals, "imbeciles," and women who were "promiscuous." These laws were up-held by the U.S. Supreme Court in the case of Buck v. Bell, which challenged the sterilization of a woman named Carrie Buck. According to officials in Virginia, Carrie was said to be suffering from "feeblemindedness and promiscuity." She already had an illegitimate child, who was presumed to be "feeble-minded," as was Carrie's mother. The Supreme Court, led by Justice Oliver Wendell Holmes, stated that her sterilization was proper because "It is better for all the world, if instead of waiting to execute degenerate offspring for crime or to let them starve for their imbecility, society can prevent those who are manifestly unfit from continuing their kind . . . Three generations of imbeciles are enough."

Using U.S. eugenics laws as a model, Nazi Germany passed laws that forced the sterilization of people who were regarded as undesirables, including people with epilepsy, physical deformities, and alcoholism. Under the Nazi regime, they later expanded their efforts to include sterilizing Jews, Gypsies, and other groups, using genetics as the basis for these atrocities. These laws were amended to allow mercy killing of newborns who were incurably ill with genetic disorders. Gradually this program of mercy killing was expanded to include adults in mental institutions and from there to include whole groups of people in concentration camps, most of whom were Jews, Gypsies, homosexuals, and political opponents of the Nazi regime. The scientific argument the Nazis used to rationalize this killing was genetics. To rid a population of "bad genes" you must stop them from reproducing. Killing them was the fastest way to do this. After these killings were revealed in Germany, the eugenics movement in the United States rapidly declined. A fear of misuse of genetics still exists today in the United States and other countries.

What might happen if...

1. A law were proposed that would require sterilization of people who carry a "lethal" gene, such as Tay-Sachs?

2. Preimplantation genetic diagnosis (PGD) was available to all parents using IVF?

3. We could manipulate the genes of fetuses so they would all be smart, tall, and good-looking?

15.2 What Are Some of the Newest Technologies of the Present?

Today our society has taken a real interest in science. This may be because discoveries are not only available to scientists. Every newspaper and television network has researchers who read the latest issues of scientific journals and then write articles for the media. Because of this, the general public has access to the newest findings.

In addition, new medical treatments, prescription drugs, medical tests, and genetic tests are advertised directly to the public in the media and on the Internet. This direct-to-consumer marketing is fairly new and assumes a certain knowledge of one's medical condition. Some of the new technologies available to the public are listed below:

Stem Cell Research Explained

There probably isn't anyone who hasn't heard about stem cells. Arguments pro and con about research using stem cells have become a major political and scientific issue. In Chapter 9, we discussed a hypothetical case involving Technogene, a company that wanted to patent genes. In addition to controlling the use of genes for testing or treatment, companies such as Technogene might be able to obtain and control the use of stem cells.

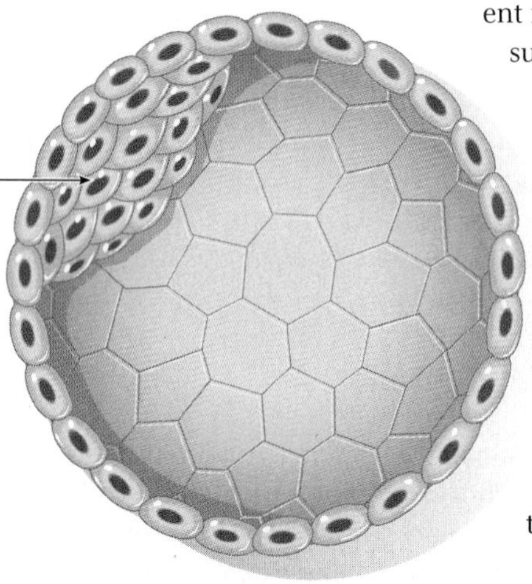

Inner cells mass. (Source of stem cells)

■ **Figure 15.2 Stem Cells**
Embryonic stem cells in a human blastocyst.

There are two types of stem cells. The first type is **embryonic stem cells,** which form in the developing embryo. These stem cells will become all the cells, tissues, and organs of the adult body. Because they can grow into any type of cell, these cells are called **pluripotent stem cells.** If obtained from embryos, these cells can reproduce themselves in the lab for many years, forming a **cell line.**

The second type is called **adult stem cells.** They are present in the adult body. They form only specific types of cells, such as blood or skin. One example of adult stem cells is the cells found in bone marrow. These stem cells divide continuously to form new red and white blood cells.

Because of the ability of embryonic stem cells to grow into multiple types of tissues, scientists have been studying them to see how this happens and how these cells may be manipulated to form specific tissues and organs. Many scientists believe that embryonic stem cells will be the cure for degenerative diseases such as Parkinson's disease, Alzheimer disease, and multiple sclerosis. They may also be used to grow new tissues to repair spinal cord injuries and treat burn patients.

Up until recently, one problem with pluripotent stem cells is that they can be only obtained from human embryos. When the stem cells are removed, the embryo dies. This has created a great deal of controversy that has spread to the political sphere.

In his first year of office, President George W. Bush declared that research using newly created stem cell lines could no longer be federally funded. However, if scientists were using existing cell lines approved by the government, they could continue to receive funding. This ruling greatly slowed progress in understanding how stem cells can be used to treat diseases and traumas.

In 2007, breakthroughs in stem cell research have made it possible to create pluripotent stem cells (called induced pluripotent cells, or iPCs) using cells from adult skin and other tissues. This work bypasses the government restrictions on embryonic stem cells, and in the future, may allow tissues and organs to be grown from stem cells created from ordinary body cells collected from individuals. This might eliminate the need to use embryonic stem cells in research.

Some of the families in our case studies might benefit from stem cell research in the future. Harriet Abeline (see Chapter 11) might be able to use stem cells to stimulate her immune system to fight her cancer, and Julio Sanchez (see Chapter 13) may not need a kidney transplant if he can grow a new kidney from his own stem cells.

Some of the ethical questions that involve stem cells are as follows:

- Should we be using human embryos for this research?
- Will this research result in anything important?
- Who should control scientific experimentation?

What might happen if...

1. Politicians decided what scientists could work on?

2. Scientists decided what scientists could work on?

3. The general public decided what scientists could work on?

4. Stem cell research showed that it didn't do what was expected?

5. Stem cell research showed that it did everything that was expected?

How Drastic Should Treatment Be?

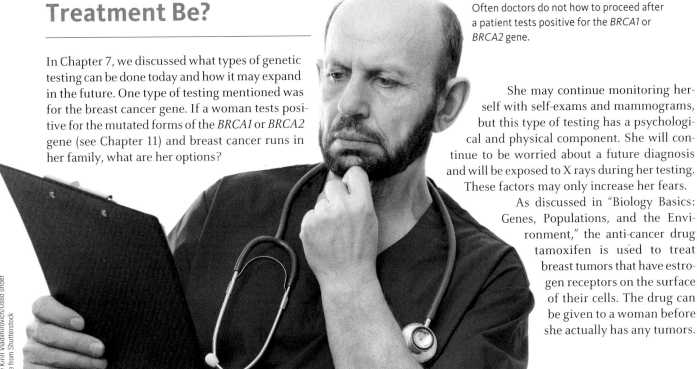

In Chapter 7, we discussed what types of genetic testing can be done today and how it may expand in the future. One type of testing mentioned was for the breast cancer gene. If a woman tests positive for the mutated forms of the *BRCA1* or *BRCA2* gene (see Chapter 11) and breast cancer runs in her family, what are her options?

■ **Figure 15.3 What's Next?**
Often doctors do not how to proceed after a patient tests positive for the *BRCA1* or *BRCA2* gene.

She may continue monitoring herself with self-exams and mammograms, but this type of testing has a psychological and physical component. She will continue to be worried about a future diagnosis and will be exposed to X rays during her testing. These factors may only increase her fears.

As discussed in "Biology Basics: Genes, Populations, and the Environment," the anti-cancer drug tamoxifen is used to treat breast tumors that have estrogen receptors on the surface of their cells. The drug can be given to a woman before she actually has any tumors.

Because the U.S. Food and Drug Administration (FDA) has approved this drug for prevention of breast cancer, women who test positive for *BRCA1* or *BRCA2* mutations can begin taking it. However, the drug has side effects, and some women may not want to take such a drug every day.

Some physicians have proposed that removing all breast tissue before cancerous cells can begin growing, a procedure called a **subcutaneous mastectomy,** may be an answer for women carrying the mutant *BRCA1* or *BRCA2* gene. But is this treatment too drastic? Some surgeons are not happy with the removal of healthy tissue because it *may* become cancerous. Isn't this true of all cells?

If Harriet Abeline (see Chapter 11) could have identified herself as *BRCA* positive, she might have been able to take some precautions such as a mastectomy.

Should an insurance company pay for this type of surgery? If the company is looking at the long-term costs, it may see that this surgery would save money in the long run. But in the short run, it is an expense that it may not have to incur.

What might happen if...

1. A woman who tested positive for a *BRCA1* mutation applied to her insurance company for a precancerous mastectomy?

2. A woman with a family history of breast cancer (but no test results) applied to her insurance company for a precancerous mastectomy?

3. A woman with no history of breast cancer and no genetic test applied to her insurance company for a precancerous mastectomy?

Get the Genetic Test

In Chapter 7, we learned about genetic testing and how it is changing the way doctors analyze patients' conditions. Recently, companies and labs that offer genetic tests have been advertising directly to consumers.

One example of this is the test for the breast cancer gene, *BRCA1* or *BRCA2*, as discussed above. Television ads have been popping up to inform the public that these tests exist and to encourage patients to ask their doctors about having these tests.

The first step in this type of advertising is to give a recognizable name to the test. In the case of the breast cancer gene, the test is called **BRACAnalysis®** (■ Figure 15.4). The ads for the test, as shown on television, and on the Internet are placing the decision whether to have such tests squarely in the hands of the individual.

Direct-to-consumer marketing is a recent development. Twenty years ago, prescription drugs were marketed only to physicians, and drug companies sent their sales representatives into doctors' offices to encourage doctors to prescribe their medicines.

■ **Figure 15.4 BRACAnalysis®**
This test is sent to your doctor or to your home when you order it from Myriad Genetics. An ad for the test often runs in major magazines.

More recently, television ads began targeting the actual consumers or buyers of the prescription drugs. The ads made drugs such as Prilosec and Ambien household names. Patients who didn't really understand what these drugs treated went to their doctors asking for them by name. After this form of advertising became widespread, more legislation was enacted to ensure that ads clearly identify side effects and include other information.

Many questions have been raised about advertisements for genetic tests because they target an entire population when only a few people carry the gene and do not disclose that both false positive and false negative results are possible.

What might happen if...

1. Every woman demanded a genetic test for breast cancer?

2. Hundreds of tests for different genes were developed?

3. Doctors had trouble explaining why patients didn't need these tests?

The Dog Genome

Geneticists have recently identified a mutant allele of the **myostatin gene** in whippets, a type of racing dog (see ■ Figure 15.5). This mutant gene causes the muscle mass in affected dogs to double. A dog that carries one copy of the mutated myostatin gene can run much faster than a dog with two normal myostatin genes. A dog that carries two mutant copies of the gene becomes the overmuscled **"bully" whippet.**

If two fast whippets (each carrying one copy of the mutant gene) are bred, some of the offspring may be fast, but they can also produce "bully" puppies. This outcome is similar to the inheritance of recessive traits in humans. In the case of the whippets, the fast parents are heterozygotes.

Mutations in the myostatin gene have been found in more than a dozen breeds of cattle, where it is called "double muscling." This condition has attracted attention as commercially useful. In humans, a young boy with this mutation was identified several years ago. He may have a future in competitive athletics, where his genetics may offer him a definite advantage.

Research on the dog genome also involves studying more than one gene. A number of years ago a couple who wanted their dog cloned offered $2 million to anyone who

Stuart Isett Photography

■ **Figure 15.5 The "Bully" Whippet**
The dog on the right is the "bully" whippet with double the musculature of the normal whippet.

would work on the project. Scientists at Texas A&M University set up a lab to collect cell samples from cats and dogs around the country. In 2002, they succeeded in cloning a cat by transplanting DNA from Rainbow, a female three-colored (tortoiseshell or calico) cat into an egg whose nucleus had been surgically removed. This embryo was then implanted into Allie, a cat that served as a surrogate mother.

Cloning dogs however, has turned out to be more difficult. Finally, in 2005, a Korean scientist, Hwang Wook-Suk, announced the cloning of a dog he called Snuppy. Although other work by Dr. Hwang has been discredited, his dog has been confirmed as a clone. In 2007, a South Korean company associated with Dr. Hwang, RNL Bio, began advertising that it will clone dogs for a $150,000 fee, to be paid when a cloned dog is delivered to the customer. In 2008, a California company, also associated with Dr. Hwang announced plans to offer a series of online auctions for dog cloning, with the opening bid priced at $100,000.

On December 7, 2005, an international team led by researchers at the Broad Institute of MIT and Harvard University, announced the publication of the dog genome sequence. Dog lovers have been breeding dogs the natural way for many years. Through this process, many traits have been bred in and out of certain types of dogs. This includes "personality traits" such as gentleness and aggressiveness as well as physical traits such as size and color. Now with the dog genome sequence known, and clones of specific genes available, coupled with the ability to clone dogs, dog breeding may be changing.

What might happen if...

1. All dogs were clones?

2. You could pick exactly the way you wanted your dog or cat to look?

3. We could pick the traits of our pets with much more accuracy than breeding allows?

Knockout Mice Win Big Prize

The Nobel Prize for medicine in 2007 was awarded to three scientists whose work led to the creation of the **knockout mouse.** The ability to isolate stem cells from mouse embryos made it possible to target certain genes in mice and then turn them off, using *knockout technology.* In experiments from many labs, more than 10,000 genes in the mouse have been knocked out using this technology. This is almost half of the genes in the mouse genome. One example was discussed in Chapter 12, where the behavior of the mice was changed when one gene was knocked out.

With a gene switched off by knocking it out, scientists can discover the action of the normal gene by observing changes in the mice. When behavior or traits change in the mouse, we know the effect of the gene that was turned off. Studying how the phenotypes of knockout mice develop gives researchers an insight into how conditions progress. This technology has proved invaluable in developing and testing new drug therapies for human diseases. Gene targeting has already produced more than 500 different mouse models of human disorders, including cardiovascular and neurodegenerative diseases, diabetes, and cancer.

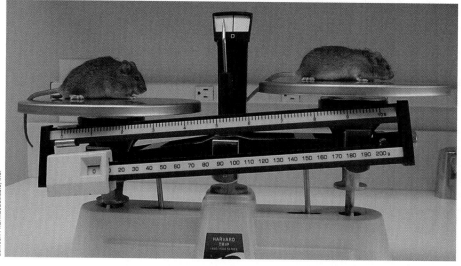

Lexicon Pharmaceuticals, Inc.

■ **Figure 15.6 Knockout Mice**
A knockout mouse (left) that is a model of obesity, compared with a normal mouse (right). When a certain gene is turned off, the mouse gains quite a bit of weight.

15.3 What Is in Store for the Future?

Some of the people in our case studies were concerned about the impact of genetic technology on future generations. Looking closely at human genetics reveals an ever-expanding outlook for this science. To emphasize the importance of informed decisions about both the present and future genetic technology available to us, this book brings to mind important genetic questions about you, your family, and society.

Even more amazing is how the Human Genome Project and future scientific discoveries may be applied to our near and distant future. When genetically engineered food becomes even more widespread on grocery store shelves, will we accept it? When everyone's unborn fetus has its genome sequenced and analyzed for all genetic traits and diseases, will insurance pay for it? When a woman wants to have a baby but she has had a hysterectomy, will she be able to? The answer to all of these questions is probably yes. All of these things and much more will change our genetic future.

Here are just a few of the possible future scenarios in human genetics.

Artificial Uterus Developed

In 1999, Yoshinori Kuwabara and his colleagues at Juntendo University in Tokyo began an experiment in which they constructed an **artificial uterus** made from a clear acrylic tank. The tank was filled with eight quarts of amniotic fluid kept at body temperature. As shown in ■ Figure 15.7, the umbilical cord of a goat fetus (with its artery and vein) was threaded into two heart–lung machines to supply oxygen and food for the fetus and to clean the blood of waste products.

The experiment began with a goat embryo that was transferred from its mother's uterus and lived for four months of its five-month gestational period in the artificial uterus. At "birth" the goat weighed six pounds, blinked its eyes, and kicked its limbs around.

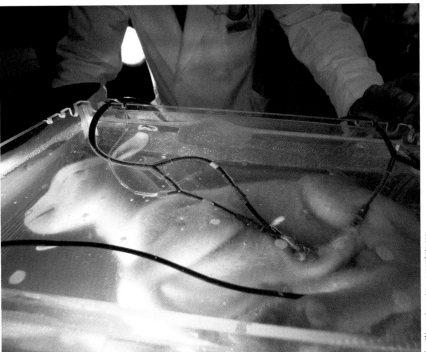

■ Figure 15.7 The Artificial Uterus
This artificial uterus, developed in Japan, is shown with a goat fetus near the time of birth.

Tom Wagner/www.tomwagnerphoto.com

Because goats are mammals, it is clear that with only minor modifications this technology could be adapted to growing humans in artificial wombs.

Some people think that this type of assisted reproduction is unethical. If this technology became available, these machines might be in laboratories and parents could come to visit their babies and watch them grow.

In legal cases dealing with parental rights, courts usually consider motherhood to be determined either biologically or legally. If a child is adopted, legal papers determine who the parents are. Genetics usually determines a child's biological parents. DNA testing for paternity is often used (see Chapter 8) to help the court make this determination. In some states, the woman who gives birth to a child is automatically considered the mother, even though she may be a surrogate hired by a couple to bear the child. If the artificial uterus becomes available, motherhood will be difficult to determine without DNA samples from the egg and sperm donors and the fetus.

What might happen if...

1. The artificial uterus were available for human babies?

2. Large numbers of women began choosing to use an artificial uterus to carry their babies? Would this affect physicians? How?

3. A private company developed this technology and wanted to make up the money it spent in research and development?

ASSIGNMENTS

1. Develop a marketing plan for the artificial uterus.

2. List four types of people who might use this technology.

A Mother at Any Age

In Chapter 2, Brian and Laura questioned how to address their infertility. But what might happen if a woman decided later in life to have a baby? Might this result in society deciding that women of a certain age should have to follow nature's plan and not have children?

Scientists have been studying human eggs for years to find out why older eggs are difficult to fertilize. If an older woman wants to conceive a child, she will have some problems. If she is undergoing or has completed menopause, her eggs are not released on a regular basis and she has trouble conceiving. However, the eggs can be removed in a surgical procedure (see Chapter 2) and analyzed for their ability to be fertilized.

Many older women have become mothers using donor eggs if their eggs are not viable, but many want to have their own genetic children. How would this be possible?

Gregory Bull/AP Photos

■ **Figure 15.8 An Older Mother**
Aleta St. James of New York gave birth to twins just before her 57th birthday.

Recently it has been found that the problem with older eggs is in the cytoplasm, which may cause failure of the embryo to divide by mitosis. Using a technique called **nuclear transfer**, Drs. James Grifo and John Zhang of New York University Medical Center removed the nucleus from an older woman's egg and transferred it to a younger woman's egg that had its nucleus removed. They taught this technique to a group of Chinese doctors who used it to initiate a pregnancy in China in 1995. Even though the twins conceived by this method later died, there were no chromosomal problems, indicating that with further refinement, this method may become a way for older women to have genetically related children.

In the United States, a similar version of this technique has been successful, leading to a live birth. At St. Barnabas Medical Center in Livingston, New Jersey, Drs. Richard Scott and Jacques Cohen removed cytoplasm from a young woman's eggs and injected it into the eggs of older women. One live birth has resulted from this process, and another birth was expected sometime in the late spring of 2008.

Using this same technique, some labs in Britain transferred a human nucleus into the egg of another mammal to study early stages of human embryo formation. The development of this technique raises the possibility that a resulting child might carry the cytoplasm of another species.

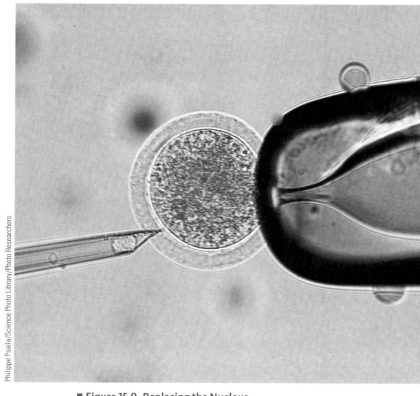

■ **Figure 15.9 Replacing the Nucleus**
In this photomicrograph, a new nucleus is being placed into an egg cell.

What might happen if...

1. Legislatures decided that a law was needed to control the age at which a woman could become a mother?

2. Women could freeze their own eggs at an early age to use later?

3. A woman with no eggs, no access to sperm, and no uterus wanted a child?

QUESTION

Some people call these embryos "three-parent embryos." Why?

DNA of your baby

The discussion of genetic screening in Chapter 7 described how Al and Victoria's baby was tested after birth for a number of genetic conditions. Some, such as PKU, the genetic disorder their baby had, are treatable; others are not. Many parents wonder whether this type of screening may increase. As companies create more and more tests for genetic disorders, such tests are likely to be used in state testing programs or by parents who want them done on their newborn.

As we become better and better at analyzing the human genome, some people have suggested that we take DNA samples from every newborn and create a database similar to the one used by law enforcement. Both positive and negative opinions exist about this type of sampling. The chart below shows some of the issues related to this type of database.

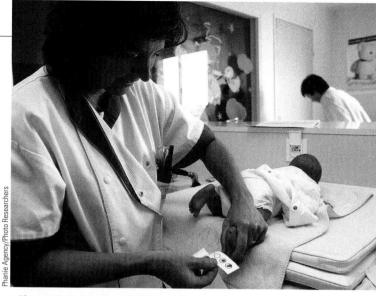

■ **Figure 15.10 Newborn DNA Sampling**
A blood sample is being taken from the newborn in order to collect DNA. This method may be expanded to collect DNA from every newborn.

Question	Pro	Con
Who will have access to this database?	Scientists, parents, doctors, and medical professionals would be able to do large population studies.	Lawyers, police, insurance companies, and possible employers will have access as well.
Will parents have the right to opt out of having their child's DNA in the database?	This would allow parents to make sure their child's DNA was not available to everyone.	If people were allowed to opt out, the database would be incomplete and possibly useless.
Can this database give us useful information about the genetics of our children?	When all or most human genes are identified and sequenced, we will know everything about someone's genome.	If we knew about the presence of only a few conditions, it would have little use.
Is this the future?	?	?

What might happen if...

1. This type of DNA database existed?

2. Laws were passed that mandated this type of sampling?

3. You were asked to vote on such a law?

Our closest relative

The Human Genome Project has revealed that to a large extent, the evolution of the human species is written in its genome. Fossil evidence indicates that our species, **Homo sapiens,** originated about 200,000 years ago in Africa. From here, humans migrated to all parts of the world and drove related species such as the **Neanderthals (*Homo neanderthalensis*)** into extinction.

Neanderthals lived in Europe and western Asia from about 300,000 to 30,000 years ago. For about 30,000 years, Neanderthals lived side by side with *Homo sapiens* in several regions of the Middle East. This evidence raises several important questions:

1. Are Neanderthals direct ancestors of humans?
2. Did Neanderthals and *Homo sapiens* interbreed, so that descendants of the Neanderthals are alive today, or did Neanderthals die off and become extinct?
3. How does the human genome compare with that of the Neanderthals?

These and other questions are being answered by researchers who are extracting and analyzing DNA from Neanderthal bones. The bones of Neanderthals were found in Belgium and Europe as long ago as 1857. In 1997, scientists compared mitochondrial DNA sequences from our species with those from Neanderthal skeletons recovered in Germany and Russia and concluded that Neanderthals are a distant relative of modern humans.

Although results from studies using mitochondrial DNA support the conclusion that Neanderthals and modern humans are only distantly related and there is little or no Neanderthal contribution to the human genome, these studies did not examine DNA from genes carried in human nuclei.

Recently, two research groups have successfully sequenced nuclear DNA recovered from Neanderthal remains. Sequence analysis indicates that although Neanderthals and modern humans have genomes that are more than 99.5% identical, Neanderthals are not direct ancestors of humans.

These findings do not rule out the possibility that Neanderthals and *Homo sapiens* may have interbred, but the results indicate that Neanderthals did not make major con-

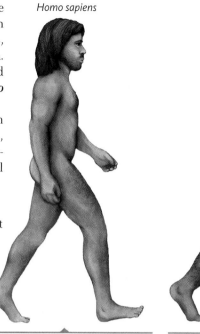

Homo sapiens   Neanderthal Man

Homo sapiens, our most recent ancestors, strongly resembled us physically as long ago as 50,000 years.

Recently, DNA evidence has ruled out the idea that the Neanderthals were our direct ancestors.

■ **Figure 15.11 Human Ancestors**
Fossil evidence has shown that *Homo sapiens* and Neanderthals were not as closely related as originally thought.

tributions to the human genome. As a species, Neanderthals are extinct, but it is possible that a small number of their genes may survive as part of our species' genome.

If the entire Neanderthal genome were available (which may be possible in the next few years), researchers could answer many more questions. To reach this goal, a number of groups are now working to isolate, sequence, and analyze the Neanderthal genome, and they hope to have most of the sequence completed sometime in 2009. Comparing the human genome with that of Neanderthals and primates such as chimpanzees will allow researchers to identify the genes that help make us uniquely human.

What might happen if...

1. Researchers found that Neanderthals were more closely related to humans than was originally thought? Should they still be classified as a separate species?

2. Researchers found that Neanderthals were not related to humans at all?

3. Scientists gave up on sequencing the Neanderthal fossils?

The Ultimate Question

Many people feel that the ultimate question that scientists and ethicists may have to deal with in the future is whether humans should be cloned.

Dolly the sheep was cloned in the late 1990s. Since then other animals have been cloned, including cats, dogs, monkeys, and cows. The method used was called *nuclear transfer,* is also referred to as **reproductive cloning** (see ■ Figures 15.12 and 15.13 on the next page).

To clone an organism using reproductive cloning, a body cell is removed from the "parent" organism and the diploid nucleus is transferred to an egg cell that has had its nucleus removed. The egg cell is then stimulated to activate cell division. An embryo is formed and transferred to the uterus of a surrogate mother.

In the quest to clone Dolly, the failure rate was very high; many embryos were transferred into surrogates and only one sheep was created. Although mice have been cloned with a much higher success rate using a method related to the one used to produce Dolly, cloning of animals by nuclear transfer has not been done on a large scale.

Most of the furor surrounding reproductive cloning involves scientists who have talked about cloning human embryos. On a smaller scale, human embryos have been created in the laboratory, usually for isolating stem cells (see Section 15.2). But full reproductive cloning of humans probably has not been done. Some groups have claimed to have a cloned baby, but this has not been shown to be true.

To see what might happen at some point in the future, let's look at the following case.

PLEASE CLONE MY DAUGHTER

Robert Denker made his money in real estate. He began with one apartment building, and now he owned enough land that he was the richest man in Charlotte, North Carolina. But being rich didn't keep him from tragedy. One day his five-year-old daughter, Lucy, was hit by a drunk driver as she was crossing the street. By the time the ambulance arrived, she had suffered serious brain damage. In the emergency room, she was put on a respirator and her heart was beating normally, but doctors told Mr. Denker that she probably would never regain consciousness. He was devastated.

In Charlotte, a group of scientists had recently cloned a cow from one that was a high milk producer. They rented Mr. Denker's land for their herd, and he had been following their work. As a businessman, he saw the potential for the process. Now he had another idea.

Dr. Meding-Smith was the head scientist on the project. Mr. Denker called her a week after the accident. He told her he was willing to spend every cent he had to bring his daughter back. He wanted Dr. Meding-Smith to clone his daughter from one of her body's cells.

What might happen if...

1. What should Dr. Meding-Smith do?

2. Give three reasons why she should not clone Lucy.

3. Give three reasons why she should clone Lucy.

4. If Mr. Denker offered Dr. Meding-Smith not only money, but also future funding for her research, would this make a difference?

5. Dr. Meding-Smith would be the first to clone a human being; she could have the most modern laboratory and everything she ever wanted. Should this play a part in her decision?

6. Should scientists do everything they *can* do? Give four reasons for your answer.

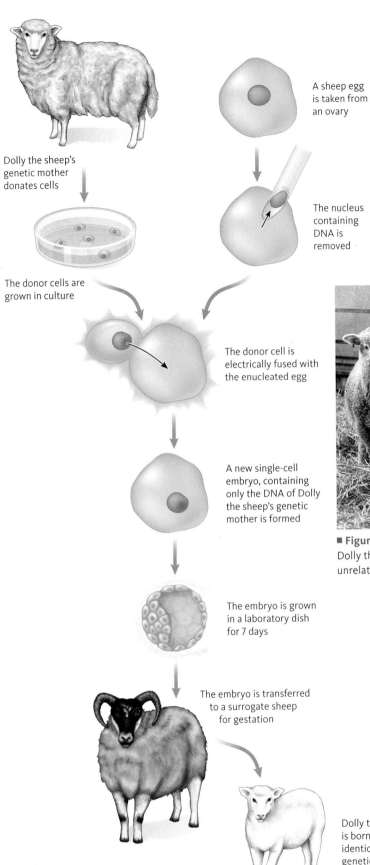

Dolly the sheep's genetic mother donates cells

The donor cells are grown in culture

A sheep egg is taken from an ovary

The nucleus containing DNA is removed

The donor cell is electrically fused with the enucleated egg

A new single-cell embryo, containing only the DNA of Dolly the sheep's genetic mother is formed

The embryo is grown in a laboratory dish for 7 days

The embryo is transferred to a surrogate sheep for gestation

Dolly the sheep is born and is identical to its genetic mother

■ **Figure 15.12 Nuclear Transfer**
Nuclear transfer was used to create Dolly the sheep. This process could be used to clone any mammal.

John Chadwick/AP Photos

■ **Figure 15.13 Dolly the Sheep**
Dolly the sheep and her baby. Dolly died of a sheep virus that was unrelated to the cloning process.

1. If analysis of the dog genome was used to define all breeds, what effect might this have on dog shows?

2. Research drugs are being developed to treat patients based on their genetic makeup. List three and tell their uses.

3. In the past, scientists tried to better the human race using eugenics. Is it wrong to encourage genetically healthy families to have more children? Why or why not?

4. Think back through the chapters of this book. Are any of the new techniques used today forms of eugenics? Which ones and why?

5. Research the knockout mouse. List four mice created from this technology and discuss how they are used.

6. When we can manipulate genes using recombinant DNA and other techniques, we may be able to change eggs and sperm and pick the genes our children will have. How do you see this being done scientifically? Will there be people who want this? Who and why?

Large Research Assignments

1. Create a reading notebook of articles on any or all of the topics in our book from newspapers and magazines and on the Internet.

2. Set up a blog of opinions on the topics that have been covered throughout the semester, and share it with either the public or members of the class.

3. Pick a topic covered this semester and discuss it with someone you know. Have the person write a short paragraph explaining his or her opinions.

4. Take a survey on any of the controversial topics we covered this semester and report the results.

5. Do research on the newest work being done in gene therapy, the human genome, gene testing and marketing, or another topic from the scientific journals available online and through the book's Web site or in your library. Write a short report on your findings.

6. Create a video on that addresses one of our questions or cases. Place it on YouTube and collect comments.

7. Research the marketing strategy used to advertise Gardasil, a vaccine for human papillomavirus.

KEY TERMS

adult stem cells (p. 236)
artificial uterus (p. 241)
BRACAnalysis (p. 238)
"bully" whippet (p. 239)
cell line (p. 236)

embryonic stem cells (p. 236)
Homo neanderthalensis (p. 245)
Homo sapiens (p. 245)
knockout mouse (p. 240)
myostatin gene (p. 239)

Neanderthals (p. 245)
negative eugenics (p. 235)
nuclear transfer (p. 243)
pluripotent stem cells
   (p. 236)

positive eugenics (p. 235)
reproductive cloning (p. 246)
subcutaneous mastectomy
   (p. 238)

Throughout the book, we have suggested that you begin a blog with members of your class or others. Now is the time to look back at what you and others have written and see how these writings can relate to the past, present, and future of genetics.

Don't just think about the genetics of humans or the genetics of your family, but try to remember what we covered throughout the book and relate it to the future. The cases we've asked you to look at are a good starting point. Make a list of them, pick the most intriguing, and write about them in your blog. Ask others to comment.

How do you think the future of genetics will turn out?

Do you think you will have some personal connection to genetics or genetic testing?

Even if you didn't start a blog, now is the time to talk to people you know (your friends, your parents, or your grandparents) about what you learned from this book. You'll be surprised at how interested they will be.

Do this via e-mail, phone, or blogging.

Appendix: Want to Learn More?

By now you are probably interested in reading more about some of the situations involving genetics and social issues that we have presented in the 15 chapters of *Human Genetics and Society*. Here is your opportunity to delve deeply into a selected number of topics related to those situations. This section includes real life cases, articles written by experts giving their opinions, and scenarios that involve interactions among genetics, law, medicine, and ethics.

The readings in this section will often present viewpoints you might not have thought of or even agree with. As part of your education, it is important to be able to read, understand, and analyze other people's positions on an issue and evaluate them using your knowledge of genetics and ethics as a framework.

I. ASSISTED REPRODUCTION

In Chapter 2, we discussed surrogacy. Go back and review the two types of surrogacy that are currently used. The difference between the Baby M case and the Calvert case (discussed below) is in the type of surrogacy. This difference played an important role in the court's decision in each of these cases. Who gets the baby in each situation, and why?

CASE PRESENTATION
Baby M and Mary Beth Whitehead: Surrogate Pregnancy in Court

On March 30, 1986, Elizabeth Stern, a professor of pediatrics, and her husband William accepted from Mary Beth Whitehead a baby who had been born four days earlier. The child's biological mother was Whitehead, but she had been engaged by the Sterns as a surrogate mother. Even so, it was not until almost exactly a year later that the Sterns were able to claim legal custody of the child.

The Sterns, working through the Infertility Center of New York, had first met with Whitehead and her husband Richard in January of 1985. Whitehead, who already had a son and a daughter, had indicated her willingness to become a surrogate mother by signing up at the Infertility Center. "What brought her there was empathy with childless couples who were infertile," her attorney later stated. Her own sister had been unable to conceive.

According to court testimony, the Sterns considered Mrs. Whitehead a "perfect person" to bear a child for them. Mr. Stern said it was "compelling" for him to have children, for he had no relatives "anywhere in the world." He and his wife planned to have children, but they put off attempts to conceive until his wife completed her medical residency in 1981. In 1979, however, she was diagnosed as having an eye condition indicating she probably had multiple sclerosis. When she learned the symptoms of the disease might be worsened by pregnancy and that she might become temporarily or even permanently paralyzed, the Sterns "decided the risk wasn't worth it." It was this decision that led them to the Infertility Center and to Mary Beth Whitehead.

The Sterns agreed to pay Whitehead $10,000 to be artificially inseminated with Mr. Stern's sperm and to bear a child. Whitehead would then turn the child over to the Sterns, and Elizabeth Stern would be allowed to adopt the child legally. The agreement was drawn up by a lawyer specializing in surrogacy. Mr. Stern later testified that Whitehead seemed perfectly pleased with the agreement and expressed no interest in keeping the baby she was to bear. "She said she would not come to our doorstep," he said. "All she wanted from us was a photograph each year and a little letter on what transpired that year."

Birth and Strife

The baby was born on March 27, 1986. According to Elizabeth Stern, the first indication Whitehead might not keep the agreement was her statement to the Sterns in the hospital two days after the baby's birth.

"She said she didn't know if 'I can go through with it,'" Dr. Stern testified. Although Whitehead did turn the baby over to the Sterns on March 30, she called a few hours later.

"She said she didn't know if she could live any more," Elizabeth Stern said. She called again the next morning and asked to see the baby, and she and her sister arrived at the Sterns' house before noon.

According to Elizabeth Stern, Whitehead told her she "woke up screaming in the middle of the night" because the baby was gone, her husband was threatening to leave her, and she had "considered taking a bottle of Valium." Stern quoted Whitehead as saying, "I just want her for a week, and I'll be out of your lives forever." The Sterns allowed Mrs. Whitehead to take the baby home with her.

Whitehead then refused to return the baby and took the infant with her to her parents' home in Florida. The Sterns obtained a court order, and on July 31 the child was seized from Whitehead. The Sterns were granted temporary custody. Then Mr. Stern, as the father of the child, and Mrs. Whitehead, as the mother, each sought permanent custody from the Superior Court of the State of New Jersey.

Trial

The seven-week trial attracted national attention, for the legal issues were without precedent. Whitehead was the first to challenge the legal legitimacy of a surrogate agreement in a U.S. court. She argued the agreement was "against public policy" and violated New Jersey prohibitions against selling babies. In contrast, Mr. Stern was the first to seek a legal decision to uphold the "specific performance" of the terms of a surrogate contract. In particular, he argued Whitehead should be ordered to uphold her agreement and to surrender her parental rights and permit his wife to become the baby's legal mother. In addition to the contractual issues, the judge had to deal with the "best interest" of the child as required by New Jersey child custody law. In addition to being a vague concept, the "best interest" standard had never been applied in a surrogacy case.

On March 31, 1987, Judge Harvey R. Sorkow announced his decision. He upheld the legality of the surrogate-mother agreement between the Sterns and Whitehead and dismissed all arguments that the contract violated public policy or prohibitions against selling babies.

Immediately after he read his decision, Judge Sorkow summoned Elizabeth Stern into his chambers and allowed her to sign documents permitting her to adopt the baby she and her husband called Melissa.

The court decision effectively stripped Mary Beth Whitehead of all parental rights concerning this same baby, the one she called Sara.

Appeal

The Baby M story did not stop with Judge Sorkow's decision. Whitehead's attorney appealed the ruling to the New Jersey Supreme Court, and on February 3, 1988, the seven members of the court, in a unanimous decision, reversed Judge Sorkow's ruling on the surrogacy agreement.

The court held that the agreement violated the state's adoption laws, because it involved a payment for a child. "This is the sale of a child, or at the very least, the sale of a mother's right to her child," Chief Justice Wilentz wrote. The agreement "takes the child from the mother regardless of her wishes and her maternal fitness ... ; and it accomplishes all of its goals through the use of money."

The court ruled that surrogacy agreements might be acceptable if they involved no payment and if a surrogate mother voluntarily surrendered her parental rights. In the present case, though, the court regarded paying for surrogacy as "illegal, perhaps criminal, and potentially degrading to women."

The court let stand the award of custody to the Sterns, because "their household and their personalities promise a much more likely foundation for Melissa to grow and thrive." Mary Beth Whitehead, having divorced her husband three months earlier, was romantically involved with a man named Dean Gould and was pregnant at the time of the court decision.

Despite awarding custody to the Sterns, the court set aside the adoption agreement signed by Elizabeth Stern. Whitehead remained a legal parent of Baby M, and the court ordered a lower court hearing to consider visitation rights for the mother.

The immediate future of the child known to the court and to the public as Baby M was settled. Neither the Sterns nor Mary Beth Whitehead had won exactly what they had sought, but neither had they lost all.

CASE PRESENTATION
The Calvert Case: A Gestational Surrogate Changes Her Mind

Disease forced Crispina Calvert of Orange County, California, to have a hysterectomy, but only her uterus was removed by surgery, not her ovaries. She and her husband, Mark, wanted a child of their own, but without a uterus Crispina would not be able to bear it. For a fee of $10,000 they arranged with Anna Johnson to act as a surrogate.

Unlike the more common form of surrogate pregnancy, Johnson would have no genetic investment in the child. The ovum that would be fertilized would not be hers. Mary Beth Whitehead, the surrogate in the controversial Baby M case, had received artificial insemination. Thus, she made as much genetic contribution to the child as did the biological father.

Johnson, however, would be the gestational surrogate. In an *in vitro* fertilization process, ova were extracted from Crispina Calvert and mixed with sperm from Mark. An embryo was implanted in Anna Johnson's uterus, and a fetus began to develop.

Johnson's pregnancy proceeded in a normal course, but in her seventh month she announced she had changed her mind about giving up the child. She filed suit against the Calverts to seek custody of the unborn child. "Just because you donate a sperm and an egg doesn't make you a parent," said Johnson's attorney. "Anna is not a machine, an incubator."

"That child is biologically Chris and Mark's," said the Calverts' lawyer. "That contract is valid."

Critics of genetic surrogate pregnancy are equally critical of gestational surrogate pregnancy. Both methods, some claim, exploit women, particularly poor women. Further, in gestational pregnancy the surrogate is the one who must run the risks and suffer the discomforts and dangers of pregnancy. She has a certain biological claim to be the mother, because it was her body that produced the child according to the genetic information supplied by the implanted embryo.

Defenders of surrogate pregnancy respond to the first criticism by denying surrogates are exploited. They enter freely into a contract to serve as a surrogate for pay, just as anyone might agree to perform any other service for pay. Pregnancy has hazards and leaves its marks on the body, but so do many other paid occupations. As far as gestational surrogacy is concerned, defenders say, since the surrogate makes no genetic contribution to the child, in no reasonable way can she be regarded as the child's parent.

The Ethics Committee of the American Fertility Society has endorsed a policy opposing surrogate pregnancy "for non-medical reasons. "The apparent aim of the policy is to permit the use of gestational surrogate pregnancy in cases like that of Mrs. Calvert, while condemning it when its motivation is mere convenience or an unwillingness to be pregnant. When a woman is fertile but, because of diabetes, uncontrollable hypertension, or some other life-threatening disorder, is unable to bear the burden of pregnancy, then gestational surrogacy would be a legitimate medical option.

Birth and Resolution

The child carried by Anna Johnson, a boy, was born on September 19, and for a while, under a court order, Johnson and the Calverts shared visitation rights. Then, in October 1990, a California Superior Court denied Johnson the parental right she had sought. Justice R. N. Parslow awarded complete custody of the child to the Calverts and terminated Johnson's visitation rights.

"I decline to split the child emotionally between two mothers," the judge said. He said Johnson had nurtured and fed the fetus in the way a foster parent might take care of a child, but she was still a "genetic stranger" to the boy and could not claim parenthood because of surrogacy.

Justice Parslow found the contract between the Calverts and Johnson to be valid, and he expressed doubt about Johnson's contention that she had "bonded" with the fetus she was carrying. "There is substantial evidence in the record that Anna Johnson never bonded with the child till she filed her lawsuit, if then," he said. While the trial was in progress, Johnson had been accused of planning to sue the Calverts from the beginning to attempt to make the case famous so she could make money from book and movie rights.

"I see no problem with someone getting paid for her pain and suffering," Parslow said. "There is nothing wrong with getting paid for nine months of what I understand is a lot of misery and a lot of bad days. They are not selling a baby; they are selling pain and suffering."

The Calverts were overjoyed by the decision.

READING
Surrogate Motherhood as Prenatal Adoption

Bonnie Steinbock

Bonnie Steinbock reviews the Baby M case and maintains that the court decision was inconsistent in considering the best interest of the child. The aim of legislation, she claims, should be to minimize potential harms and prevent cases like that of Baby M from happening again. This can be so only if surrogacy is not intrinsically wrong.

This leads Steinbock to examine three lines of argument and attempt to show that neither paternalism nor such considerations as threats of exploitation, loss of dignity, or harm to the child are adequate to show that surrogacy is inherently objectionable. In Steinbock's view, regulating surrogacy—and protecting liberty—is preferable to prohibiting it.

The recent case of "Baby M" has brought surrogate motherhood to the forefront of American attention. Ultimately, whether we permit or prohibit surrogacy depends on what we take to be good reasons for preventing people from acting as they wish. A growing number of people want to be, or hire, surrogates; are there legitimate reasons to prevent them? Apart from its intrinsic interest, the issue of surrogate motherhood provides us with an opportunity to examine different justifications for limiting individual freedom.

In the first section, I examine the Baby M case, and the lessons it offers. In the second section, I examine claims that surrogacy is ethically unacceptable because it is exploitive, inconsistent with human dignity, or harmful to the children born of such arrangements. I conclude that these reasons justify restrictions on surrogate contracts, rather than an outright ban.

Baby M

Mary Beth Whitehead, a married mother of two, agreed to be inseminated with the sperm of William Stern, and to give up the child to him for a fee of $10,000. The baby (whom Mrs. Whitehead named Sara, and the Sterns named Melissa) was born on March 27, 1986. Three days later, Mrs. Whitehead took her home from the hospital, and turned her over to the Sterns.

Then Mrs. Whitehead changed her mind. She went to the Sterns' home, distraught, and pleaded to have the baby temporarily. Afraid that she would kill herself, the Sterns agreed. The next week, Mrs. Whitehead informed the Sterns that she had decided to keep the child, and threatened to leave the country if court action was taken.

At that point, the situation deteriorated into a cross between the Keystone Kops and Nazi storm troopers. Accompanied by five policemen, the Sterns went to the Whitehead residence armed with a court order giving them temporary custody of the child. Mrs. Whitehead managed to slip the baby out of a window to her husband, and the following morning the Whiteheads fled with the child to Florida, where Mrs. Whitehead's parents lived. During the next three months, the Whiteheads lived in roughly twenty different hotels, motels, and homes to avoid apprehension. From time to time, Mrs. Whitehead telephoned Mr. Stern to discuss the matter: He taped these conversations on advice of counsel. Mrs. Whitehead threatened to kill herself, to kill the child, and falsely to accuse Mr. Stern of sexually molesting her older daughter.

At the end of July 1986, while Mrs. Whitehead was hospitalized with a kidney infection, Florida police raided her mother's home, knocking her down, and seized the child. Baby M was placed in the custody of Mr. Stern, and the Whiteheads returned to New Jersey, where they attempted to regain custody. After a long and emotional court battle, Judge Harvey R. Sorkow ruled on March 31, 1987, that the surrogacy contract was valid, and that specific performance was justified in the best interests of the child. Immediately after reading his decision, he called the Sterns into his chambers so that Mr. Stern's wife, Dr. Elizabeth Stern, could legally adopt the child.

This outcome was unexpected and unprecedented. Most commentators had thought that a court would be unlikely to order a reluctant surrogate to give up an infant merely on the basis of a contract. Indeed, if Mrs. Whitehead had never surrendered the child to the Sterns, but had simply taken her home and kept her there, the outcome undoubtedly would have been different. It is also likely that Mrs. Whitehead's failure to obey the initial custody order angered Judge Sorkow, and affected his decision.

The decision was appealed to the New Jersey Supreme Court, which issued its decision on February 3, 1988. Writing for a unanimous court, Chief Justice Wilentz reversed the lower court's ruling that the surrogacy contract was valid. The court held that a surrogacy contract which provides money for the surrogate mother, and which includes her irrevocable agreement to surrender her child at birth, is invalid and unenforceable. Since the contract was invalid, Mrs. Whitehead did not relinquish, nor were there any other grounds for terminating, her parental rights. Therefore, the adoption of Baby M by Mrs. Stern was improperly granted, and Mrs. Whitehead remains the child's legal mother.

The Court further held that the issue of custody is determined solely by the child's best interests, and it agreed with the lower court that it was in Melissa's best interests to remain with the Sterns. However, Mrs. Whitehead, as Baby M's legal as well as natural mother, is entitled to have her own interest in visitation considered. The determination of what kind of visitation rights should be granted to her, and under what conditions, was remanded to the trial court.

The distressing details of this case have led many people to reject surrogacy altogether. Do we really want police officers wrenching infants from their mothers' arms, and prolonged custody battles when surrogates find they are unable to surrender their children, as agreed? Advocates of surrogacy say that to reject the practice wholesale, because of one unfortunate instance, is an example of a "hard case" making bad policy. Opponents reply that it is entirely reasonable to focus on the worst potential outcomes when deciding public policy. Everyone can agree on at least one thing: This particular case seems to have been mismanaged from start to finish, and could serve as a manual of how not to arrange a surrogate birth.

First, it is now clear that Mary Beth Whitehead was not a suitable candidate for surrogate motherhood. Her ambivalence about giving up the child was recognized early on, although this information was not passed on to the Sterns.[1] Second, she had contact with the baby after birth, which is usually avoided in "successful" cases. Typically, the adoptive mother is actively involved in the pregnancy, often serving as the pregnant woman's coach in labor. At birth, the baby is given to the adoptive, not the biological, mother. The joy of the adoptive parents in holding their child serves both to promote their bonding, and to lessen the pain of separation of the biological mother.

At Mrs. Whitehead's request, no one at the hospital was aware of the surrogacy arrangement. She and her husband appeared as the proud parents of "Sara Elizabeth Whitehead," the name on her birth certificate. Mrs. Whitehead held her baby, nursed her, and took her home from the hospital—just as she would have done in a normal pregnancy and birth. Not surprisingly, she thought of Sara as her child, and she fought with every weapon at her disposal, honorable and dishonorable, to prevent her being taken away. She can hardly be blamed for doing so.[2]

Why did Dr. Stern, who supposedly had a very good relation with Mrs. Whitehead before the birth, not act as her labor coach? One possibility is that Mrs. Whitehead, ambivalent about giving up her baby, did not want Dr. Stern involved. At her request, the Sterns' visits to the hospital to see the newborn baby were unobtrusive. It is also possible that Dr. Stern was ambivalent about having a child. The original idea of hiring a surrogate was not hers, but her husband's. It was Mr. Stern who felt a "compelling" need to have a child related to him by blood, having lost all his relatives to the Nazis.

Furthermore, Dr. Stern was not infertile, as was stated in the surrogacy agreement. Rather, in 1979 she was diagnosed by two eye specialists as suffering from optic neuritis, which meant that she "probably" had multiple sclerosis. (This was confirmed by all four experts who testified.) Normal conception was ruled out by the Sterns in late 1982, when a medical colleague told Dr. Stern that his wife, a victim of multiple sclerosis, had suffered a temporary paralysis during pregnancy. "We decided the risk wasn't worth it," Mr. Stern said.[3]

Mrs. Whitehead's lawyer, Harold J. Cassidy, dismissed the suggestion that Dr. Stern's "mildest case" of multiple sclerosis determined their decision to seek a surrogate. He noted that she was not even treated for multiple sclerosis until after the Baby M dispute had started. "It's almost as though it's an afterthought," he said.[4]

Judge Sorkow deemed the decision to avoid conception "medically reasonable and understandable." The Supreme Court did not go so far, noting that "her anxiety appears to have exceeded the actual risk, which current medical authorities assess as minimal."[5] Nonetheless the court acknowledged that her anxiety, including fears that pregnancy might precipitate blindness and paraplegia, was "quite real." Certainly, even a woman who wants a child very much, may reasonably wish to avoid becoming blind and paralyzed as a result of pregnancy. Yet is it believable that a woman who really wanted a child would decide against pregnancy *solely* on the basis of *someone else's* medical experience? Would she not consult at least one specialist on her *own* medical condition before deciding it wasn't worth the risk? The conclusion that she was at best ambivalent about bearing a child seems irresistible.

This possibility conjures up many people's worst fears about surrogacy: That prosperous women, who do not want to interrupt their careers, will use poor and educationally disadvantaged women to bear their children. I will return shortly to the question of whether this is exploitive. The issue here is psychological: What kind of mother is Dr. Stern likely to be? If she is unwilling to undergo pregnancy, with its discomforts, inconveniences, and risks, will she be willing to make the considerable sacrifices which good parenting requires? Mrs. Whitehead's ability to be a good mother was repeatedly questioned during the trial. She was portrayed as immature, untruthful, hysterical, overly identified with her children, and prone to smothering their independence. Even if all this is true—and I think that Mrs. Whitehead's inadequacies were exaggerated—Dr. Stern may not be such a prize either. The choice for Baby M may have been between a highly strung, emotional, over-involved mother, and a remote, detached, even cold one.

The assessment of Mrs. Whitehead's ability to be a good mother was biased by the middle-class prejudices of the judge and mental health officials who testified.

Mrs. Whitehead left school at 15, and is not conversant with the latest theories on child rearing: She made the egregious error of giving Sara teddy bears to play with, instead of the more "age-appropriate," expert-approved pans and spoons. She proved to be a total failure at patty-cake. If this is evidence of parental inadequacy, we're all in danger of losing our children.

The Supreme Court felt that Mrs. Whitehead was "rather harshly judged" and acknowledged the possibility

that the trial court was wrong in its initial award of custody. Nevertheless, it affirmed Judge Sorkow's decision to allow the Sterns to retain custody, as being in Melissa's best interests. George Annas disagrees with the "best interests" approach. He points out that Judge Sorkow awarded temporary custody of Baby M to the Sterns in May 1986 without giving the Whiteheads notice or an opportunity to obtain legal representation. That was a serious wrong and injustice to the Whiteheads. To allow the Sterns to keep the child compounds the original unfairness: "... justice requires that reasonable consideration be given to returning Baby M to the permanent custody of the Whiteheads."[6]

But a child is not a possession, to be returned to the rightful owner. It is not fairness to all parties that should determine a child's fate, but what is best for her. As Chief Justice Wilentz rightly stated, "The child's interests comes first: We will not punish it for judicial errors, assuming any were made."[7]

Subsequent events have substantiated the claim that giving custody to the Sterns was in Melissa's best interests. After losing custody, Mrs. Whitehead, whose husband had undergone a vasectomy, became pregnant by another man. She divorced her husband and married Dean R. Gould last November. These developments indicate that the Whiteheads were not able to offer a stable home, although the argument can be made that their marriage might have survived, but for the strains introduced by the court battle, and the loss of Baby M. But even if Judge Sorkow had no reason to prefer the Sterns to the Whiteheads back in May 1986, he was still right to give the Sterns custody in March 1987. To take her away then, at nearly eighteen months of age, from the only parents she had ever known, would have been disruptive, cruel, and unfair to her.

Annas's preference for a just solution is premised partly on his belief that there is no "best interest" solution to this "tragic custody case." I take it that he means that however custody is resolved, Baby M is the loser. Either way, she will be deprived of one parent. However, a best interests solution is not a perfect solution. It is simply the solution which is on balance best for the child, given the realities of the situation. Applying this standard, Judge Sorkow was right to give the Sterns custody, and the Supreme Court was right to uphold the decision.

The best interests argument is based on the assumption that Mr. Stern has at least a *prima facie* claim to Baby M. We certainly would not consider allowing a stranger who kidnapped a baby, and managed to elude the police for a year, to retain custody on the grounds that he was providing a good home to a child who had known no other parent. However, the Baby M case is not analogous. First, Mr. Stern is Baby M's biological father and, as such, has at least some claim to raise her, which no non-parental kidnapper has. Second, Mary Beth Whitehead agreed to give him their baby. Unlike the miller's daughter in *Rumpelstiltskin,* the fairy tale to which the Baby M case is sometimes compared, she was not forced into the agreement. Because both Mary Beth Whitehead and Mr. Stern have *prima facie* claims to

Baby M, the decision as to who should raise her should be based on her present best interests. Therefore we must, regretfully, tolerate the injustice to Mrs. Whitehead, and try to avoid such problems in the future.

It is unfortunate that the Court did not decide the issue of visitation on the same basis as custody. By declaring Mrs. Whitehead Gould the legal mother, and maintaining that she is entitled to visitation, the Court has prolonged the fight over Baby M. It is hard to see how this can be in her best interests. This is no ordinary divorce case, where the child has a relation with both parents which it is desirable to maintain. As Mr. Stern said at the start of the court hearing to determine visitation, "Melissa has a right to grow and be happy and not be torn between two parents."[8]

The Court's decision was well-meaning but internally inconsistent. Out of concern for the best interests of the child, it granted the Sterns custody. At the same time, by holding Mrs. Whitehead Gould to be the legal mother, with visitation rights, it precluded precisely what is most in Melissa's interest, a resolution of the situation. Further, the decision leaves open the distressing possibility that a Baby M situation could happen again. Legislative efforts should be directed toward ensuring that this worse-case scenario never occurs.

Should Surrogacy Be Prohibited?

On June 27, 1988, Michigan became the first state to outlaw commercial contracts for women to bear children for others. Yet making a practice illegal does not necessarily make it go away: Witness black market adoption. The legitimate concerns which support a ban on surrogacy might be better served by careful regulation. However, some practices, such as slavery, are ethically unacceptable, regardless of how carefully regulated they are. Let us consider the arguments that surrogacy is intrinsically unacceptable.

Paternalistic Arguments

These arguments against surrogacy take the form of protecting a potential surrogate from a choice she may later regret. As an argument for banning surrogacy, as opposed to providing safeguards to ensure that contracts are freely and knowledgeably undertaken, this is a form of paternalism.

At one time, the characterization of a prohibition as paternalistic was a sufficient reason to reject it. The pendulum has swung back, and many people are willing to accept at least some paternalistic restrictions on freedom. Gerald Dworkin points out that even Mill made one exception to his otherwise absolute rejection of paternalism: He thought that no one should be allowed to sell himself into slavery, because to do so would be to destroy his future autonomy.

This provides a narrow principle to justify some paternalistic interventions. To preserve freedom in the long run, we give up the freedom to make certain choices, those which have results which are "farreaching, potentially dangerous and irreversible."[9] An example would be a ban on the sale of crack. Virtually everyone who uses crack becomes addicted and, once addicted, a slave to its use. We reasonably and willingly give up our freedom to buy the drug, to protect our ability to make free decisions in the future.

Can a Dworkinian argument be made to rule out surrogacy agreements? Admittedly, the decision to give up a child is permanent, and may have disastrous effects on the surrogate mother. However, many decisions may have long-term, disastrous effects (e.g., postponing childbirth for a career, having an abortion, giving a child up for adoption). Clearly we do not want the state to make decisions for us in all these matters. Dworkin's argument is rightly restricted to paternalistic interferences which protect the individual's autonomy or ability to make decisions in the future. Surrogacy does not involve giving up one's autonomy, which distinguishes it from both the crack and selling-oneself-into-slavery examples. Respect for individual freedom requires us to permit people to make choices which they may later regret.

Moral Objections

Four main moral objections to surrogacy were outlined in the Warnock Report.[10]

1. It is inconsistent with human dignity that a woman should use her uterus for financial profit.
2. To deliberately become pregnant with the intention of giving up the child distorts the relationship between mother and child.
3. Surrogacy is degrading because it amounts to child-selling.
4. Since there are some risks attached to pregnancy, no woman ought to be asked to undertake pregnancy for another in order to earn money.

We must all agree that a practice which exploits people or violates human dignity is immoral. However, it is not clear that surrogacy is guilty on either count.

Exploitation. The mere fact that pregnancy is *risky* does not make surrogate agreements exploitive, and therefore morally wrong. People often do risky things for money; why should the line be drawn at undergoing pregnancy? The usual response is to compare surrogacy and kidney-selling. The selling of organs is prohibited because of the potential for coercion and exploitation. But why should kidney-selling be viewed as intrinsically coercive? A possible explanation is that no one would do it, unless driven by poverty. The choice is both forced and dangerous, and hence coercive.

The situation is quite different in the case of the race car driver or stuntman. We do not think that they are *forced* to perform risky activities for money: They freely choose to do so. Unlike selling one's kidneys, these are activities which we can understand (intellectually, anyway) someone choosing to do. Movie stuntmen, for example, often enjoy their work, and derive satisfaction from doing it well. Of course they "do it for the money," in the sense that they would not do it without compensation; few people are willing to work "for free." The element of coercion is missing, however, because they enjoy the job, despite the risks, and could do something else if they chose.

The same is apparently true of most surrogates. "They choose the surrogate role primarily because the fee provides a better economic opportunity than alternative occupations, but also because they enjoy being pregnant and the respect and attention that it draws."[11] Some may derive a feeling of self-worth from an act they regard as highly altruistic: providing a couple with a child they could not otherwise have. If these motives are present, it is far from clear that the surrogate is being exploited. Indeed, it seems objectionably paternalistic to insist that she is.

Human Dignity. It may be argued that even if womb-leasing is not necessarily exploitive, it should still be rejected as inconsistent with human dignity. But why? As John Harris points out, hair, blood and other tissue is often donated or sold; what is so special about the uterus?[12]

Human dignity is more plausibly invoked in the strongest argument against surrogacy, namely, that it is the sale of a child. Children are not property, nor can they be bought or sold. It could be argued that surrogacy is wrong because it is analogous to slavery, and so is inconsistent with human dignity.

However, there are important differences between slavery and a surrogate agreement. The child born of a surrogate is not treated cruelly or deprived of freedom or resold; none of the things which make slavery so awful are part of surrogacy. Still, it may be thought that simply putting a market value on a child is wrong. Human life has intrinsic value; it is literally priceless. Arrangements which ignore this violate our deepest notions of the value of human life. It is profoundly disturbing to hear the boyfriend of a surrogate say, quite candidly in a television documentary on surrogacy, "We're in it for the money."

Judge Sorkow accepted the premise that producing a child for money denigrates human dignity, but he denied that this happens in a surrogate agreement. Mrs. Whitehead was not paid for the surrender of the child to the father: She was paid for her willingness to be impregnated and carry Mr. Stern's child to term. The child, once born, is his biological child. "He cannot purchase what is already his."

This is misleading, and not merely because Baby M is as much Mrs. Whitehead's child as Mr. Stern's. It is misleading because it glosses over the fact that the surrender of the child was part—indeed, the whole point—of the agreement. If the surrogate were paid merely for being willing to be impregnated and carrying the child to term, then she would fulfill the contract upon giving birth. She could take the money *and* the child. Mr. Stern did not agree to pay Mrs. Whitehead merely to *have* his child, but to provide him with a child. The New Jersey Supreme Court held that this violated New Jersey's laws prohibiting the payment or acceptance of money in connection with adoption.

One way to remove the taint of baby-selling would be to limit payment to medical expenses associated with the birth or incurred by the surrogate during pregnancy (as is allowed in many jurisdictions, including New Jersey, in ordinary adoptions).

Surrogacy could be seen, not as baby-selling, but as a form of adoption. Nowhere did the Supreme Court find any legal prohibition against surrogacy when there is no payment, and when the surrogate has the right to change her mind and keep the child. However, this solution effectively prohibits surrogacy, since few women would become surrogates solely for self-fulfillment or reasons of altruism.

The question, then, is whether we can reconcile paying the surrogate, beyond her medical expenses, with the idea of surrogacy as prenatal adoption. We can do this by separating the terms of the agreement, which include surrendering the infant at birth to the biological father, from the justification for payment. The payment should be seen as compensation for the risks, sacrifice, and discomfort the surrogate undergoes during pregnancy. This means that if, through no fault on the part of the surrogate, the baby is stillborn, she should still be paid in full, since she has kept her part of the bargain. (By contrast, in the Stern–Whitehead agreement, Mrs. Whitehead was to receive only $1,000 for a stillbirth.) If, on the other hand, the surrogate changes her mind and decides to keep the child, she would break the agreement, and would not be entitled to any fee, or compensation for expenses incurred during pregnancy.

The Right of Privacy

Most commentators who invoke the right of privacy do so in support of surrogacy. However, George Annas makes the novel argument that the right to rear a child you have borne is also a privacy right, which cannot be prospectively waived. He says:

> [Judge Sorkow] grudgingly concedes that [Mrs. Whitehead] could not prospectively give up her right to have an abortion during pregnancy. . . This would be an intolerable restriction on her liberty and under *Roe v. Wade*, the state has no constitutional authority to enforce a contract that prohibits her from terminating her pregnancy.
>
> But why isn't the same logic applicable to the right to rear a child you have given birth to? Her constitutional rights to rear the child she has given birth to are even stronger since they involve even more intimately, and over a lifetime, her privacy rights to reproduce and rear a child in a family setting.[13]

Absent a compelling state interest (such as protecting a child from unfit parents), it certainly would be an intolerable invasion of privacy for the state to take children from their parents. But Baby M has two parents, both of whom now want her. It is not clear why only people who can give birth (i.e., women) should enjoy the right to rear their children.

Moreover, we do allow women to give their children up for adoption after birth. The state enforces those agreements, even if the natural mother, after the prescribed waiting period, changes her mind. Why should the right to rear a child be unwaivable before, but not after birth? Why should the state have the constitutional authority to uphold postnatal, but not prenatal, adoption agreements? It is not clear why birth should affect the waivability of this right, or have the constitutional significance which Annas attributes to it.

Nevertheless, there are sound moral and policy, if not constitutional, reasons to provide a postnatal waiting period in surrogate agreements. As the Baby M case makes painfully clear, the surrogate may underestimate the bond created by gestation, and the emotional trauma caused by relinquishing the baby. Compassion requires that we acknowledge these feelings, and not deprive a woman of the baby she has carried because, before conception, she underestimated the strength of her feelings for it. Providing a waiting period, as in ordinary postnatal adoptions, will help protect women from making irrevocable mistakes, without banning the practice.

Some may object that this gives too little protection to the prospective adoptive parents. They cannot be sure that the baby is theirs until the waiting period is over. While this is hard on them, a similar burden is placed on other adoptive parents. If the absence of a guarantee serves to discourage people from entering surrogacy agreements, that is not necessarily a bad thing, given all the risks inherent in such contracts. In addition, this requirement would make stricter screening and counseling of surrogates essential, a desirable side effect.

Harm to Others

Paternalistic and moral objections to surrogacy do not seem to justify an outright ban. What about the effect on the offspring of such contracts? We do not yet have solid data on the effects of being a "surrogate child." Any claim that surrogacy creates psychological problems in the children is purely speculative. But what if we did discover that such children have deep feelings of worthlessness from learning that their natural mothers deliberately created them with the intention of giving them away? Might we ban surrogacy as posing an unacceptable risk of psychological harm to the resulting children?

Feelings of worthlessness are harmful. They can prevent people from living happy, fulfilling lives. However, a surrogate child, even one whose life is miserable because of these feelings, cannot claim to have been harmed by the surrogate agreement. Without the agreement, the child would never have existed. Unless she is willing to say that her life is not worth living because of these feelings, that she would be better off never having been born, she cannot claim to have been harmed by being born of a surrogate mother.

Children can be *wronged* by being brought into existence, even if they are not, strictly speaking, harmed. They are wronged if they are deprived of the minimally decent existence to which all citizens are entitled. We owe it to our children to see that they are not born with such serious impairments that their most basic interests will be doomed in advance. If being born to a surrogate is a handicap of this magnitude, comparable to being born blind or deaf or severely mentally retarded, then surrogacy can be seen as wronging the offspring. This would be a strong reason against permitting such contracts. However, it does not seem likely. Probably the problems arising from surrogacy will be like those faced by adopted children and children whose parents divorce. Such problems are not trivial, but neither are they so serious that the child's very existence can be seen as wrongful.

If surrogate children are neither harmed nor wronged by surrogacy, it may seem that the argument for banning surrogacy on grounds of its harmfulness to the offspring evaporates. After all, if the children themselves have no cause for complaint, how can anyone else claim to reject it on their behalf? Yet it seems extremely counter-intuitive to suggest that the risk of emotional damage to the children born of such arrangements is not even relevant to our deliberations. It seems quite reasonable and proper—even morally obligatory—for policymakers to think about the possible detrimental effects of new reproductive technologies, and to reject those likely to create physically or emotionally damaged people. The explanation for this must involve the idea that it is wrong to bring people into the world in a harmful condition, even if they are not, strictly speaking, harmed by having been brought into existence. Should evidence emerge that surrogacy produces children with serious psychological problems, that would be a strong reason for banning the practice.

There is some evidence on the effect of surrogacy on the other children of the surrogate mother. One woman reported that her daughter, now 17, who was 11 at the time of the surrogate birth, "… is still having problems with what I did, and as a result she is still angry with me." She explains, "Nobody told me that a child could bond with a baby while you're still pregnant. I didn't realize then that all the times she listened to his heartbeat and felt his legs kick that she was becoming attached to him."[14]

A less sentimental explanation is possible. It seems likely that her daughter, seeing one child given away, was fearful that the same might be done to her. We can expect anxiety and resentment on the part of children whose mothers give away a brother or sister. The psychological harm to these children is clearly relevant to a determination of whether surrogacy is contrary to public policy. At the same time, it should be remembered that many things, including divorce, remarriage, and even moving to a new neighborhood, create anxiety and resentment in children. We should not use the effect on children as an excuse for banning a practice we find bizarre or offensive.

Conclusion

There are many reasons to be extremely cautious of surrogacy. I cannot imagine becoming a surrogate, nor would I advise anyone else to enter into a contract so fraught with peril. But the fact that a practice is risky, foolish, or even morally distasteful is not sufficient reason to outlaw it. It would be better for the state to regulate the practice, and minimize the potential for harm, without infringing on the liberty of citizens.

Notes

1. Had the Sterns been informed of the psychologist's concerns as to Mrs. Whitehead's suitability to be a surrogate, they might have ended the arrangement, costing the Infertility Center its fee. As Chief Justice Wilentz said, "It is apparent that the profit motive got the better of the Infertility Center." In the matter of Baby M, Supreme Court of New Jersey, A–39, at 45.

2. "[W]e think it is expecting something well beyond normal human capabilities to suggest that this mother should have parted with her newly born infant without a struggle.... We ... cannot conceive of any other case where a perfectly fit mother was expected to surrender her newly born infant, perhaps forever, and was then told she was a bad mother because she did not." *Id. at 79.*

3. Father recalls surrogate was "perfect." *New York Times*, January 6, 1987, B2.

4. *Id.*

5. In the matter of Baby M, *supra* note 1, at 8.

6. Annas, G. J.: Baby M: babies (and justice) for sale. *Hastings Center Report* 17 (3): 15, 1987.

7. In the matter of Baby M, *supra* note 1, at 75.

8. Anger and Anguish at Baby M Visitation Hearing, *New York Times*, March 29, 1988, 17.

9. Dworkin, G.: Paternalism. In Wasserstrom, R. A., ed.: *Morality and the Law.* Belmont, Calif., Wadsworth, 1971; reprinted in Feinberg, J., Gross, H., eds., *Philosophy of Law,* 3rd ed. Wadsworth, 1986, p. 265.

10. Warnock, M., chair: *Report of the committee of inquiry into human fertilisation and embryology.* London: Her Majesty's Stationery Office, 1984.

11. Robertson, J. A.: Surrogate mothers: not so novel after all. *Hastings Center Report* 13 (5): 29,1983. Citing Parker, P.: Surrogate mother's motivations: initial findings. *American Journal of Psychiatry* (140): 1, 1983.

12. Harris, J.: *The Value of Life.* London: Routledge & Kegan Paul, 1985, 144.

13. Annas, *supra* note 6.

14. Baby M case stirs feelings of surrogate mothers. *New York Times*, March 2, 1987, B1.

READING
Is Women's Labor a Commodity?

Elizabeth S. Anderson

Elizabeth Anderson argues that commercial surrogacy should not be allowed. The practice of paying women to be surrogate mothers involves a "commodification" of both children and women. It treats women and their children as things to be used, instead of as persons deserving respect. Hence, surrogacy contracts should be unenforceable, and those who arrange them should be subject to criminal penalties.

Anderson holds that the introduction of market values and norms into a situation previously based on respect, consideration, and unconditional love has the effect of harming children and degrading and exploiting women. The values of the market contribute to a tendency to view children as property. When this happens, they are no longer valued unconditionally (as is the case with parental love), but are valued only because they possess characteristics with a market value.

Market values require that surrogate mothers repress whatever parental love they may feel for their children. Hence, the feelings of women are manipulated, degraded, and denied legitimacy. Further, women are exploited by having the personal feelings that incline them to become surrogates turned into something that can be marketed as part of a commercial enterprise.

In the past few years the practice of commercial surrogate motherhood has gained notoriety as a method for acquiring children. A commercial surrogate mother is anyone who is paid money to bear a child for other people and terminate her parental rights, so that the others may raise the child as exclusively their own. The growth of commercial surrogacy has raised with new urgency a class of concerns regarding the proper scope of the market. Some critics have objected to commercial surrogacy on the ground that it improperly treats children and women's reproductive capacities as commodities.[1] The prospect of reducing children to consumer durables and women to baby factories surely inspires revulsion. But are there good reasons behind the revulsion? And is this an accurate description of what commercial surrogacy implies? This article offers a theory about what things are properly regarded as commodities which supports the claim that commercial surrogacy constitutes an unconscionable commodification of children and of women's reproductive capacities.

What Is a Commodity?

The modern market can be characterized in terms of the legal and social norms by which it governs the production, exchange, and enjoyment of commodities. To say that something is properly regarded as a commodity is to claim that the norms of the market are appropriate for regulating its production, exchange, and enjoyment. To the extent that moral principles or ethical ideals preclude the application of market norms to a good, we may say that the good is not a (proper) commodity.

Why should we object to the application of a market norm to the production or distribution of a good? One reason may be that to produce or distribute the good in accordance with the norm is to *fail to value it in an appropriate way.* Consider, for example, a standard Kantian argument against slavery, or the commodification of persons. Slaves are treated in accordance with the market norm that owners may use commodities to satisfy their own interests without regard for the interests of the commodities themselves. To treat a person without regard for her interests is to fail to respect her. But slaves are persons who may not be merely used in this fashion, since as rational beings they possess a dignity which commands respect. In Kantian theory, the problem with slavery is that it treats beings worthy of *respect* as if they were worthy merely of *use.* "Respect" and "use" in this context denote what we may call different *modes of valuation*....

These considerations support a general account of the sorts of things which are appropriately regarded as commodities. Commodities are those things which are properly treated in accordance with the norms of the modern market. We can question the application of market norms to the production, distribution, and enjoyment of a good by appealing to ethical ideals which support arguments that the good should be valued in some other way than use. Arguments of the latter sort claim that to allow certain market norms to govern our treatment of a thing expresses a mode of valuation not worthy of it. If the thing is to be valued appropriately, its production, exchange, and enjoyment must be removed from market norms and embedded in a different set of social relationships.

The Case of Commercial Surrogacy

Let us now consider the practice of commercial surrogate motherhood in the light of this theory of commodities. Surrogate motherhood as a commercial enterprise is based upon contracts involving three parties: the intended father, the broker, and the surrogate mother. The intended father agrees to pay a lawyer to find a suitable surrogate mother and make the requisite medical and legal arrangements for the conception and birth of the child, and for the transfer of legal custody to himself.[2] The surrogate mother agrees to become impregnated with the intended father's sperm, to carry the resulting child to term, and to relinquish her parental rights to it, transferring custody to the father in return for a fee and medical expenses. Both she and her husband (if she has one) agree not to form a parent–child bond with her child and to do everything necessary to effect the transfer of the child to the intended father. At current market prices, the lawyer arranging the contract can expect to gross $15,000 from the contract, while the surrogate mother can expect a $10,000 fee.[3]

The practice of commercial surrogacy has been defended on four main grounds. First, given the shortage of children available for adoption and the difficulty of qualifying as adoptive parents, it may represent the only hope for some people to be able to raise a family. Commercial surrogacy should be accepted as an effective means for realizing this highly significant good. Second, two fundamental human rights support commercial surrogacy: the right to procreate and freedom of contract. Fully informed autonomous adults should have the right to make whatever arrangements they wish for the use of their bodies and the reproduction of children, so long as the children themselves are not harmed. Third, the labor of the surrogate mother is said to be a labor of love. Her altruistic acts should be permitted and encouraged.[4] Finally, it is argued that commercial surrogacy is no different in its ethical implications from many already accepted practices which separate genetic, gestational, and social parenting, such as artificial insemination by donor, adoption, wet-nursing, and day care. Consistency demands that society accept this new practice as well.[5]

In opposition to these claims, I shall argue that commercial surrogacy does raise new ethical issues, since it represents an invasion of the market into a new sphere of conduct, that of specifically women's labor—that is, the labor of carrying children to term in pregnancy. When women's labor is treated as a commodity, the women who perform it are degraded. Furthermore, commercial surrogacy degrades children by reducing their status to that of commodities. Let us consider each of the goods of concern in surrogate motherhood—the child, and women's reproductive labor—to see how the commercialization of parenthood affects people's regard for them.

Children as Commodities

The most fundamental calling of parents to their children is to love them. Children are to be loved and cherished by their parents, not to be used or manipulated by them for merely personal advantage. Parental love can be understood as a passionate, unconditional commitment to nurture one's child, providing it with the care, affection, and guidance it needs to develop its capacities to maturity. This understanding of the way parents should value their children informs our interpretation of parental rights over their children. Parents' rights over their children are trusts, which they must always exercise for the sake of the child. This is not to deny that parents have their own aspirations in raising children. But the child's interests beyond subsistence are not definable independently of the flourishing of the family, which is the object of specifically parental aspirations. The proper exercise of parental rights includes those acts which promote their shared life as a family, which realize the shared interests of the parents and the child.

The norms of parental love carry implications for the ways other people should treat the relationship between parents and their children. If children are to be loved by their parents, then others should not attempt to compromise the integrity of parental love or work to suppress the emotions supporting the bond between parents and their children. If the rights to children should be understood as trusts, then if those rights are lost or relinquished, the duty of those in charge of transferring custody to others is to consult the best interests of the child.

Commercial surrogacy substitutes market norms for some of the norms of parental love. Most importantly, it requires us to understand parental rights no longer as trusts but as things more like property rights—that is, rights of use and disposal over the things owned. For in this practice the natural mother deliberately conceives a child with the intention of giving it up for material advantage. Her renunciation of parental responsibilities is not done for the child's sake, nor for the sake of fulfilling an interest she shares with the child, but typically for her own sake (and possibly, if "altruism" is a motive, for the intended parents' sakes). She and the couple who pay her to give up her parental rights over her child thus treat her rights as a kind of property right. They thereby treat the child itself as a kind of commodity, which may be properly bought and sold.

Commercial surrogacy insinuates the norms of commerce into the parental relationship in other ways. Whereas parental love is not supposed to be conditioned upon the child having particular characteristics, consumer demand is properly responsive to the characteristics of commodities. So the surrogate industry provides opportunities to adoptive couples to specify the height, I.Q., race, and other attributes of the surrogate mother, in the expectation that these traits will be passed on to the child.[6] Since no industry assigns agents to look after the "interests" of its commodities, no one represents the child's interests in the surrogate industry. The surrogate agency promotes the adoptive parents' interests and not the child's interests where matters of custody are concerned. Finally, as the agent of the adoptive parents, the broker has the task of policing the surrogate (natural) mother's relationship to her child, using persuasion, money, and the threat of a lawsuit to weaken and destroy whatever parental love she may develop for her child.[7]

All of these substitutions of market norms for parental norms represent ways of treating children as commodities which are degrading to them. Degradation occurs when something is treated in accordance with a lower mode of valuation than is proper to it. We value things not just "more" or "less," but in qualitatively higher and lower ways. To love or respect someone is to value her in a higher way than one would if one merely used her. Children are properly loved by their parents and respected by others. Since children are valued as mere use-objects by the mother and the surrogate agency when they are sold to others, and by the adoptive parents when they seek to conform the child's genetic makeup to their own wishes, commercial surrogacy degrades children insofar as it treats them as commodities.[8]

One might argue that since the child is most likely to enter a loving home, no harm comes to it from permitting the natural mother to treat it as property. So the purchase and sale of infants is unobjectionable, at least from the point of view of children's interests.[9] But the sale of an infant has an expressive significance which this argument fails to recognize. By engaging in the transfer of children by sale, all of the parties to the surrogate contract express a set of attitudes toward children which undermine the norms of parental love. They all agree in treating the ties between a natural mother and her children as properly loosened by a monetary incentive. Would it be any wonder if a child born of a surrogacy agreement feared resale by parents who have such an attitude? And a child who knew how anxious her parents were that she have the "right" genetic makeup might fear that her parents' love was contingent upon her expression of these characteristics.[10]

The unsold children of surrogate mothers are also harmed by commercial surrogacy. The children of some surrogate mothers have reported their fears that they may be sold like their half-brother or half-sister, and express a sense of loss at being deprived of a sibling.[11] Furthermore, the widespread acceptance of commercial surrogacy would psychologically threaten all children. For it would change the way children are valued by people (parents and surrogate brokers)—from being loved by their parents and respected by others, to being sometimes used as objects of commercial profit-making.[12]

Proponents of commercial surrogacy have denied that the surrogate industry engages in the sale of children. For it is impossible to sell to someone what is already his own, and the child is already the father's own natural offspring. The payment to the surrogate mother is not for her child, but for her services in carrying it to term.[13] The claim that the parties to the surrogate contract treat children as commodities, however, is based on the way they treat the *mother's* rights over her child. It is irrelevant that the natural father also has some rights over the child; what he pays for is exclusive rights to it. He would not pay her for the "service" of carrying the child to term if she refused to relinquish her parental rights to it. That the mother regards only her labor and not her child as requiring compensation is also irrelevant. No one would argue that the baker does not treat his bread as property just because he sees the income from its sale as compensation for his labor and expenses and not for the bread itself, which he doesn't care to keep.[14]

Defenders of commercial surrogacy have also claimed that it does not differ substantially from other already accepted parental practices. In the institutions of adoption and artificial insemination by donor (AID), it is claimed, we already grant parents the right to dispose of their children.[15] But these practices differ in significant respects from commercial surrogacy. The purpose of adoption is to provide a means for placing children in families when their parents cannot or will not discharge their parental responsibilities. It is not a sphere for the existence of a supposed parental right to dispose of one's children for profit. Even AID does not sanction the sale of fully formed human beings. The semen donor sells only a product of his body, not his child, and does not initiate the act of conception.

Two developments might seem to undermine the claim that commercial surrogacy constitutes a degrading commerce in children. The first is technological: the prospect of transplanting a human embryo into the womb of a genetically unrelated woman. If commercial surrogacy used women only as gestational mothers and not as genetic mothers, and if it was thought that only genetic and not gestational parents could properly claim that a child was "theirs," then the child born of a surrogate mother would not be hers to sell in the first place. The second is a legal development: the establishment of the proposed "consent-intent" definition of parenthood.[16] This would declare the legal parents of a child to be whoever consented to a procedure which leads to its birth, with the intent of assuming parental responsibilities for it. This rule would define away the problem of commerce in children by depriving the surrogate mother of any legal claim to her child at all, even if it was hers both genetically and gestationally.[17]

There are good reasons, however, not to undermine the place of genetic and gestational ties in these ways. Consider first the place of genetic ties. By upholding a system of involuntary (genetic) ties of obligation among people, even when the adults among them prefer to divide their rights and obligations in other ways, we help to secure children's interests in having an assured place in the world, which is more firm than the wills of their parents. Unlike the consent–intent rule, the principle of respecting genetic ties does not make the obligation to care for those whom one has created (intentionally or not) contingent upon an arbitrary desire to do so. It thus provides children with a set of pre-existing social sanctions which give them a more secure place in the world. The genetic principle also places the children in a far wider network of associations and obligations than the consent–intent rule sanctions. It supports the roles of grandparents and other relatives in the nurturing of children, and provides children with a possible focus of stability and an additional source of claims to care if their parents cannot sustain a well-functioning household.

In the next section I will defend the claims of gestational ties to children. To deny these claims, as commercial surrogacy does, is to deny the significance of reproductive labor to the mother who undergoes it and thereby to dehumanize and degrade the mother herself. Commercial surrogacy would be a corrupt practice even if it did not involve commerce in children.

Women's Labor as a Commodity

Commercial surrogacy attempts to transform what is specifically women's labor—the work of bringing forth children into the world—into a commodity. It does so by replacing the parental norms which usually govern the practice of gestating children with the economic norms which govern ordinary production processes. The application of commercial norms to women's labor reduces the surrogate mothers from persons worthy of respect and consideration to objects of mere use.

Respect and consideration are two distinct modes of valuation whose norms are violated by the practices of the surrogate industry. To respect a person is to treat her in accordance with principles she rationally accepts—principles consistent with the protection of her autonomy and her rational interests. To treat a person with consideration is to respond with sensitivity to her and to her emotional relations with others, refraining from manipulating or denigrating these for one's own purposes....

The application of economic norms to the sphere of women's labor violates women's claims to respect and consideration in three ways. First, by requiring the surrogate mother to repress whatever parental love she feels for the child, these norms convert women's labor into a form of alienated labor. Second, by manipulating and denying legitimacy to the surrogate mother's evolving perspective on her

own pregnancy, the norms of the market degrade her. Third, by taking advantage of the surrogate mother's noncommercial motivations without offering anything but what the norms of commerce demand in return, these norms leave her open to exploitation. The fact that these problems arise in the attempt to commercialize the labor of bearing children shows that women's labor is not properly regarded as a commodity.

The key to understanding these problems is the normal role of the emotions in noncommercialized pregnancies. Pregnancy is not simply a biological process but also a social practice. Many social expectations and considerations surround women's gestational labor, marking it off as an occasion for the parents to prepare themselves to welcome a new life into their family. For example, obstetricians use ultrasound not simply for diagnostic purposes but also to encourage maternal bonding with the fetus.[18] We can all recognize that it is good, although by no means inevitable, for loving bonds to be established between the mother and her child during this period.

In contrast with these practices, the surrogate industry follows the putting-out system of manufacturing. It provides some of the raw materials of production (the father's sperm) to the surrogate mother, who then engages in production of the child. Although her labor is subject to periodic supervision by her doctors and by the surrogate agency, the agency does not have physical control over the product of her labor as firms using the factory system do. Hence, as in all putting-out systems, the surrogate industry faces the problem of extracting the final product from the mother. This problem is exacerbated by the fact that the social norms surrounding pregnancy are designed to encourage parental love for the child. The surrogate industry addresses this problem by requiring the mother to engage in a form of emotional labor.[19] In the surrogate contract, she agrees not to form or to attempt to form a parent–child relationship with her offspring.[20] Her labor is alienated, because she must divert it from the end which the social practices of pregnancy rightly promote—an emotional bond with her child. The surrogate contract thus replaces a norm of parenthood, that during pregnancy one create a loving attachment to one's child, with a norm of commercial production, that the producer shall not form any special emotional ties to her product....

Commercial surrogacy is also a degrading practice. The surrogate mother, like all persons, has an independent evaluative perspective on her activities and relationships. The realization of her dignity demands that the other parties to the contract acknowledge rather than evade the claims which her independent perspective makes upon them. But the surrogate industry has an interest in suppressing, manipulating, and trivializing her perspective, for there is an ever-present danger that she will see her involvement in her pregnancy from the perspective of a parent rather than from the perspective of a contract laborer.

How does this suppression and trivialization take place? The commercial promoters of surrogacy commonly describe the surrogate mothers as inanimate objects: mere "hatcheries," "plumbing," or "rented property"—things without emotions which could make claims on others.[21] They also refuse to acknowledge any responsibility for the consequences of the mother's emotional labor. Should she suffer psychologically from being forced to give up her child, the father is not liable to pay for therapy after her pregnancy, although he is liable for all other medical expenses following her pregnancy.[22]

The treatment and interpretation of surrogate mothers' grief raises the deepest problems of degradation. Most surrogate mothers experience grief upon giving up their children—in 10 percent of cases, seriously enough to require therapy.[23] Their grief is not compensated by the $10,000 fee they receive. Grief is not an intelligible response to a successful deal, but rather reflects the subject's judgment that she has suffered a grave and personal loss. Since not all cases of grief resolve themselves into cases of regret, it may be that some surrogate mothers do not regard their grief, in retrospect, as reflecting an authentic judgment on their part. But in the circumstances of emotional manipulation which pervade the surrogate industry, it is difficult to determine which interpretation of her grief more truly reflects the perspective of the surrogate mother. By insinuating a trivializing interpretation of her emotional responses to the prospect of losing her child, the surrogate agency may be able to manipulate her into accepting her fate without too much fuss, and may even succeed in substituting its interpretation of her emotions for her own. Since she has already signed a contract to perform emotional labor—to express or repress emotions which are dictated by the interests of the surrogate industry—this might not be a difficult task.[24] A considerate treatment of the mothers' grief, on the other hand, would take the evaluative basis of their grief seriously.

Some defenders of commercial surrogacy demand that the provision for terminating the surrogate mother's parental rights in her child be legally enforceable, so that peace of mind for the adoptive parents can be secured.[25] But the surrogate industry makes no corresponding provision for securing the peace of mind of the surrogate. She is expected to assume the risk of a transformation of her ethical and emotional perspective on herself and her child with the same impersonal detachment with which a futures trader assumes the risk of a fluctuation in the price of pork bellies. By applying the market norms of enforcing contracts to the surrogate mother's case, commercial surrogacy treats a moral transformation as if it were merely an economic change.[26]

The manipulation of the surrogate mother's emotions which is inherent in the surrogate parenting contract also leaves women open to grave forms of exploitation. A kind of exploitation occurs when one party to a transaction is oriented toward the exchange of "gift" values, while the other party operates in accordance with the norms of the market exchange of commodities. Gift values, which include love, gratitude, and appreciation of others, cannot be bought or obtained through piecemeal calculations of individual advantage. Their exchange requires a repudiation of a self-interested attitude, a willingness to give gifts to others without demanding some specific equivalent good in return each time one gives. The surrogate mother often operates according to the norms of gift relationships. The surrogate agency, on the other hand, follows market norms. Its job is to get the best deal for its clients and itself, while leaving the surrogate mother to look after her own interests as best as she can. The situation puts the surrogate agencies in a position to manipulate the surrogate mothers' emotions to gain favorable terms for themselves. For example, agencies screen prospective surrogate mothers for submissiveness, and emphasize to them the importance of the motives of generosity and love. When applicants question some of the terms of the contract, the broker sometimes intimidates them by questioning their character and morality: if they were really generous and loving they would not be so solicitous about their own interests.[27] ...

Many surrogate mothers see pregnancy as a way to feel "adequate," "appreciated," or "special." In other words, these women feel inadequate, unappreciated, or unadmired when they are not pregnant.[28] Lacking the power to achieve some worthwhile status in their own right, they must subordinate themselves to others' definitions of their proper place (as baby factories) in order to get from them the appreciation they need to attain a sense of self-worth. But the sense of selfworth one can attain under such circumstances is precarious and ultimately self-defeating. For example, those who seek gratitude on the part of the adoptive parents and some opportunity to share the joys of seeing their children grow discover all too often that the adoptive parents want nothing to do with them.[29] For while the surrogate mother sees in the arrangement some basis for establishing the personal ties she needs to sustain her emotionally, the adoptive couple sees it as an impersonal commercial contract, one of whose main advantages to them is that all ties between them and the surrogate are ended once the terms of the contract are fulfilled.[30] To them, her presence is a threat to marital unity and a competing object for the child's affections.

These considerations should lead us to question the model of altruism which is held up to women by the surrogacy industry. It is a strange form of altruism which demands such radical self-effacement, alienation from those whom one benefits, and the subordination of one's body, health, and emotional life to the independently defined interests of others.[31]

The primary distortions which arise from treating women's labor as a commodity—the surrogate mother's alienation from loved ones, her degradation, and her exploitation—stem from a common source.

This is the failure to acknowledge and treat appropriately the surrogate mother's emotional engagement with her labor. Her labor is alienated, because she must suppress her emotional ties with her own child, and may be manipulated into reinterpreting these ties in a trivializing way. She is degraded, because her independent ethical perspective is denied, or demoted to the status of a cash sum. She is exploited, because her emotional needs and vulnerabilities are not treated as characteristics which call for consideration, but as factors which may be manipulated to encourage her to make a grave self-sacrifice to the broker's and adoptive couple's advantage. These considerations provide strong grounds for sustaining the claims of women's labor to its "product," the child. The attempt to redefine parenthood so as to strip women of parental claims to the children they bear does violence to their emotional engagement with the project of bringing children into the world.

Commercial Surrogacy, Freedom, and the Law

In the light of these ethical objections to commercial surrogacy, what position should the law take on the practice? At the very least, surrogate contracts should not be enforceable. Surrogate mothers should not be forced to relinquish their children if they have formed emotional bonds with them. Any other treatment of women's ties to the children they bear is degrading.

But I think these arguments support the stronger conclusion that commercial surrogate contracts should be illegal, and that surrogate agencies who arrange such contracts should be subject to criminal penalties. Commercial surrogacy constitutes a degrading and harmful traffic in children, violates the dignity of women, and subjects both children and women to a serious risk of exploitation....

If commercial surrogate contracts were prohibited, this would be no cause for infertile couples to lose hope for raising a family. The option of adoption is still available, and every attempt should be made to open up opportunities for adoption to couples who do not meet standard requirements—for example, because of age. While there is a shortage of healthy white infants available for adoption, there is no shortage of children of other races, mixed-race children, and older and handicapped children who desperately need to be adopted. Leaders of the surrogate industry have proclaimed that commercial surrogacy may replace adoption as the method of choice for infertile couples who wish to raise families. But we should be wary of the racist and eugenic motivations which make some people rally to the surrogate industry at the expense of children who already exist and need homes.

The case of commercial surrogacy raises deep questions about the proper scope of the market in modern industrial societies. I have argued that there are principled grounds for rejecting the substitution of market norms for parental norms to govern the ways women bring children into the world. Such substitutions express ways of valuing mothers and children which reflect an inferior conception of human flourishing. When market norms are applied to the ways we allocate and understand parental rights and responsibilities, children are reduced from subjects of love to objects of use. When market norms are applied to the ways we treat and understand women's reproductive labor, women are reduced from subjects of respect and consideration to objects of use. If we are to retain the capacity to value children and women in ways consistent with a rich conception of human flourishing, we must resist the encroachment of the market upon the sphere of reproductive labor.

Women's labor is *not* a commodity.

Notes

The author thanks David Anderson, Steven Darwall, Ezekiel Emanuel, Daniel Hausman, Don Herzog, Robert Nozick, Richard Pildes, John Rawls, Michael Sandel, Thomas Scanlon, and Howard Wial for helpful comments and criticisms.

1. See, for example, Gena Corea, *The Mother Machine* (New York: Harper and Row, 1985), pp. 216, 219; Angela Holder, "Surrogate Motherhood: Babies for Fun and Profit," *Case and Comment* 90 (1985): 3–11; and Margaret Jane Radin, "Market Inalienability," *Harvard Law Review* 100 (June 1987): 1849–1937.

2. State laws against selling babies prevent the intended father's wife (if he has one) from being a party to the contract.

3. See Katie Marie Brophy, "A Surrogate Mother Contract to Bear a Child," *Journal of Family Law* 20 (1981–82): 263–91, and Noel Keane, "The Surrogate Parenting Contract," *Adelphia Law Journal 2 (1983):* 45–53, for examples and explanations of surrogate parenting contracts.

4. Mary Warnock, *A Question of Life* (Oxford: Blackwell, 1985), p. 45. This book reprints the Warnock Report on Human Fertilization and Embryology, which was commissioned by the British government for the purpose of recommending legislation concerning surrogacy and other issues. Although the Warnock Report mentions the promotion of altruism as one defense of surrogacy, it strongly condemns the practice overall.

5. John Robertson, "Surrogate Mothers: Not So Novel After All," *Hastings Center Report*. October 1983, pp. 28–34; John Harris, *The Value of Life* (Boston: Routledge and Kegan Paul, 1985).

6. See "No Other Hope for Having a Child," *Time,* 19 January 1987, pp. 50–51. Radin argues that women's traits are also commodified in this practice. See "Market Inalienability," pp. 1932–35.

7. Here I discuss the surrogate industry as it actually exists today. I will consider possible modifications of commercial surrogacy in the final section below.

8. Robert Nozick has objected that my claims about parental love appear to be culture-bound. Do not parents in the Third World, who rely on children to provide for the family subsistence, regard their children as economic goods? In promoting the livelihood of their families, however, such children need not be treated in accordance with market norms—that is, as commodities. In particular, such children usually remain a part of their families and hence can still be loved by their parents. But insofar as children are treated according to the norms of modern capitalist markets, this treatment is deplorable wherever it takes place.

9. See Elizabeth Landes and Richard Posner, "The Economics of the Baby Shortage," *Journal of Legal Studies* 7 (1978): 323–48, and Richard Posner, "The Regulation of the Market in Adoptions," *Boston University Law Review* 67 (1987): 59–72.

10. Of course, where children are concerned it is irrelevant whether these fears are reasonable. One of the greatest fears of children is separation from their parents. Adopted children are already known to suffer from separation anxiety more acutely than children who remain with their natural mothers, for they feel that the original mother did not love them. In adoption, the fact that the child would be even worse off if the mother did not give it up justifies her severing ties and can help to rationalize this event to the child. But in the case of commercial surrogacy, the severing of ties is done not for the child's sake, but for the parents' sakes. In the adoption case there are explanations for the mother's action which may quell

the child's doubts about being loved which are unavailable in the case of surrogacy.

11. Kay Longcope, "Surrogacy: Two Professionals on Each Side of Issue Give Their Argument for Prohibition and Regulation," *Boston Globe,* 23 March 1987, pp. 18–19; and Iver Peterson, "Baby M Case: Surrogate Mothers Vent Feelings, *New York Times,* 2 March 1987, pp. B1, B4.

12. Herbert Krimmel, "The Case Against Surrogate Parenting," *Hastings Center Report,* October 1983, pp. 35–37.

13. Judge Sorkow made this argument in ruling on the famous case of Baby M. See *In Re Baby M,* 217 N.J. Super 313. Reprinted in *Family Law Reporter* 13 (1987): 2001–30. Chief Justice Wilentz of the New Jersey Supreme Court overruled Sorkow's judgment. See *In the Matter of Baby M,* 109 N.J. 396, 537 A. 2d 1227 (1988).

14. Sallyann Payton has observed that the law does not permit the sale of parental rights, only their relinquishment or forced termination by the state, and these acts are subject to court review for the sake of the child's best interests. But this legal technicality does not change the moral implications of the analogy with baby-selling. The mother is still paid to do what she can to relinquish her parental rights and to transfer custody of the child to the father. Whether or not the courts occasionally prevent this from happening, the actions of the parties express a commercial orientation to children which is degrading and harmful to them. The New Jersey Supreme Court ruled that surrogacy contracts are void precisely because they assign custody without regard to the child's best interests. See *In the Matter of Baby M,* p. 1246.

15. Robertson, "Surrogate Mothers: Not So Novel After All, p. 32; Harris, *The Value of Life,* pp. 144–45.

16. See Philip Parker, "Surrogate Motherhood: The Interaction of Litigation, Legislation and Psychiatry," *International Journal of Law and Psychiatry* 5 (1982): 341–54.

17. The consent–intent rule would not, however, change the fact that commercial surrogacy replaces parental norms with market norms. For the rule itself embodies the market norm which acknowledges only voluntary, contractual relations among people as having moral force. Whereas familial love invites children into a network of unwilled relationships broader than those they have with their parents, the willed contract creates an exclusive elationship between the parents and the child only.

18. I am indebted to Dr. Ezekiel Emanuel for this point.

19. One engages in emotional labor when one is paid to express or repress certain emotions. On the concept of emotional labor and its consequences for workers, see Arlie Hochschild, *The Managed Heart* (Berkeley and Los Angeles: University of California Press, 1983).

20. Noel Keane and Dennis Breo, *The Surrogate Mother* (New York: Everest House, 1981), p. 291; Brophy, "A Surrogate Mother Contract," p. 267. The surrogate's husband is also required to agree to this clause of the contract.

21. Corea, *The Mother Machine,* p. 222.

22. Keane and Breo, *The Surrogate Mother,* p. 292.

23. Kay Longcope, "Standing Up for Mary Beth," *Boston Globe,* 5 March 1987, p. 83; Daniel Goleman, "Motivations of Surrogate Mothers," *New York Times,* 20 January 1987, p. C1; Robertson, "Surrogate Mothers: Not So Novel After All," pp. 30, 34 n. 8. Neither the surrogate mothers themselves nor psychiatrists have been able to predict which women will experience such grief.

24. See Hochschild, *The Managed Heart,* for an important empirical study of the dynamics of commercialized emotional labor.

25. Keane and Breo, *The Surrogate Mother,* pp. 236–37.

26. For one account of how a surrogate mother who came to regret her decision viewed her own moral transformation, see Elizabeth Kane: *Birth Mother: The Story of America's First Legal Surrogate Mother* (San Diego: Harcourt Brace Jovanovich, 1988). I argue below that the implications of commodifying women's labor are not significantly changed even if the contract is unenforceable.

27. Susan Ince, "Inside the Surrogate Industry," in *Test-Tube Women,* ed. Rita Ardith, Ranate Duelli Klein, and Shelley Minden (Boston: Pandora Press, 1984), p. 110.

28. The surrogate broker Noel Keane is remarkably open about reporting the desperate emotional insecurities which shape the lives of so many surrogate mothers, while displaying little sensitivity to the implications of his taking advantage of these motivations to make his business a financial success. See especially Keane and Breo, *The Surrogate Mother,* pp. 247ff.

29. See, for example, the story of the surrogate mother Nancy Barrass in Arlene Fleming, "Our Fascination with Baby M," *New York Times Magazine,* 29 March 1987, p. 38.

30. For evidence of these disparate perspectives, see Peterson, "Baby M Case: Surrogate Mothers Vent Feelings," p. B4.

31. The surrogate mother is required to obey all doctor's orders made in the interests of the child's health. (See Brophy, "A Surrogate Mother Contract"; Keane, "The Surrogate Parenting Contract"; and Ince, "Inside the Surrogate Industry.") These orders could include forcing her to give up her job, travel plans, and recreational activities. The doctor could confine her to bed, and order her to submit to surgery and take drugs. One can hardly exercise an autonomous choice over one's health if one could be held in breach of contract and liable for $35,000 damages for making a decision contrary to the wishes of one's doctor.

SCENARIO

In January 1985 the British High Court took custody of a five-day-old girl, the first child known to be born in Britain to a woman paid to be a surrogate mother.

An American couple, known only as "Mr. and Mrs. A," were reported to have paid about $7500 to a twenty-eight-year-old woman who allowed herself to be artificially inseminated with sperm from Mr. A. The woman, Kim Cotton, was prevented from turning the child over to Mr. and Mrs. A by a court order issued because of the uncertainty over the legal status of a surrogate mother.

The court permitted "interested parties, including the natural father" to apply for custody of the child. Mr. A applied, and Judge Sir John Latey ruled that the couple could take the baby girl out of the country because they could offer her the chance of "a very good upbringing."

1. Are there moral reasons that might have made the court hesitate before turning over the child to her biological father? For example, could it be persuasively argued that Kim Cotton was in effect selling her baby to Mr. and Mrs. A?

2. Kim Cotton agreed to be a surrogate mother for the sake of the money. Is surrogate pregnancy a practice that tends to exploit the poor? Or is it a legitimate way to earn money by providing a needed service?

3. Is serving as a surrogate mother essentially the same as prostitution? If it is not, then what are the relevant differences?

II. TREATMENT OF CRITICALLY ILL NEWBORNS

As discussed in Chapter 3, there are many decisions to be made when a fetus is diagnosed as having a chromosomal abnormality. In addition to the types of decisions Ms. Lawrence had to make, there are others to be made after an affected child is born. How do we decide what happens to these children?

CASE PRESENTATION
Baby Owens: Down Syndrome and Duodenal Atresia

On a chilly December evening in 1976, Dr. Joan Owens pushed through the plateglass doors of Midwestern Medical Center and walked over to the admitting desk. Dr. Owens was a physician in private practice and regularly visited Midwestern to attend to her patients.

But this night was different. Dr. Owens was coming to the hospital to be admitted as a patient. She was pregnant, and shortly after 9:00 she began having periodic uterine contractions. Dr. Owens recognized them as the beginnings of labor pains. She was sure of this not only because of her medical knowledge but also because the pains followed the same pattern they had before her other three children were born.

While her husband, Phillip, parked the car, Dr. Owens went through the formalities of admission. She was not particularly worried, for the birth of her other children had been quite normal and uneventful. But the pains were coming more frequently now, and she was relieved when she completed the admission process and was taken to her room. Phillip came with her, bringing her small blue suitcase of personal belongings.

At 11:30 that evening, Dr. Owens gave birth to a 4.5-pound baby girl. The plastic bracelet fastened around her wrist identified her as Baby Owens.

Bad News

Dr. Owens was groggy from exhaustion and from the medication she had received. But when the baby was shown to her, she saw at once that it was not normal. The baby's head was misshapen and the skin around her eyes strangely formed.

Dr. Owens recognized that her daughter had Down syndrome.

"Clarence," she called to her obstetrician. "Is the baby mongoloid?"

"We'll talk about it after your recovery,"Dr. Clarence Ziner said.

"Tell me now," said Dr. Owens. "Examine it!"

Dr. Ziner made a hasty examination of the child. He had already seen that Dr. Owens was right and was doing no more than making doubly certain. A more careful examination would have to be made later.

When Dr. Ziner confirmed Joan Owens's suspicion, she did not hesitate to say what she was thinking. "Get rid of it," she told Dr. Ziner. "I don't want a mongoloid child."

Dr. Ziner tried to be soothing. "Just sleep for a while now," he told her. "We'll talk about it later."

Four hours later, a little after 5:00 in the morning and before it was fully light, Joan Owens woke up. Phillip was with her, and he had more bad news to tell. A more detailed examination had shown that the child's small intestine had failed to develop properly and was closed off in one place—the condition known as duodenal atresia. It could be corrected by a relatively simple surgical procedure, but until surgery was performed the child could not be fed. Phillip had refused to consent to the operation until he had talked to his wife. Joan Owens had not changed her mind: She did not want the child.

"It wouldn't be fair to the other children to raise them with a mongoloid," she told Phillip. "It would take all of our time, and we wouldn't be able to give David, Sean, and Melinda the love and attention they need."

"I'm willing to do whatever you think best," Phillip said. "But what can we do?"

"Let the child die," Joan said. "If we don't consent to the surgery, the baby will die soon. And that's what we have to let happen."

Phillip put in a call for Dr. Ziner, and, when he arrived in Joan's room, they told him of their decision. He was not pleased with it.

"The surgery has very low risk," he said. "The baby's life can almost certainly be saved. We can't tell how retarded she'll be, but most DS children get along quite well with help from their families. The whole family will grow to love her."

"I know," Joan said. "And I don't want that to happen. I don't want us to center our lives around a defective child. Phillip and I and our other children will be forced to lose out on many of life's pleasures and possibilities."

"We've made up our minds," Phillip said. "We don't want the surgery."

"I'm not sure the matter is as simple as that," Dr. Ziner said. "I'm not sure we can legally just let the baby die. I'll have to talk to the director and the hospital attorney."

Applying for a Court Order

At 6:00 in the morning, Dr. Ziner called Dr. Felix Entraglo, the director of Midwestern Medical Center, and Isaac Putnam, the head of the center's legal staff. They agreed to meet at 9:00 to talk over the problem presented to them by the Owenses.

They met for two hours. It was Putnam's opinion that the hospital would not be legally liable if Baby Owens were allowed to die because her parents refused to give consent for necessary surgery.

"What about getting a court order requiring surgery?" Dr. Entraglo asked. "That's the sort of thing we do when an infant requires a blood transfusion or immunization and his parents' religious beliefs make them refuse consent."

"This case is not exactly parallel," said Mr. Putnam. "Here we're talking about getting a court to force parents to allow surgery to save the life of a defective infant. The infant will still be defective after the surgery, and I think a court would be reluctant to make a family undergo significant emotional and financial hardships when the parents have seriously deliberated about the matter and decided against surgery."

"But doesn't the child have some claim in this situation?" Dr. Ziner asked.

"That's not clear," said Mr. Putnam. "In general, we assume that parents will act for the sake of their child's welfare, and when they are reluctant to do so we look to the courts to act for the child's welfare. But in a situation like this ... who can say? Is the Owens baby really a person in any legal or moral sense?"

"I think I can understand why a court would be hesitant to order surgery," said Dr. Entraglo. "What sort of life would it be for a family when they had been pressured into accepting a child they didn't want? It would turn a family into a cauldron of guilt and resentment mixed in with love and concern. In this case, the lives of five normal people would be profoundly altered for the worse."

"So we just stand by and let the baby die?" asked Dr. Ziner.

"I'm afraid so," Dr. Entraglo said.

The Final Days

It took twelve days for Baby Owens to die. Her lips and throat were moistened with water to lessen her suffering, and in a small disused room set apart from the rooms of patients, she was allowed to starve to death.

Many nurses and physicians thought it was wrong that Baby Owens was forced to die such a lingering death. Some thought it was wrong for her to have to die at all, but such a protracted death seemed needlessly cruel. Yet they were cautioned by Dr. Entraglo that anything done to shorten the baby's life would probably constitute a criminal action. Thus, fear of being charged with a crime kept the staff from administering any medication to Baby Owens.

The burden of caring for the dying baby fell on the nurses in the obstetrics ward. The physicians avoided the child entirely, and it was the nurses who had to see to it that she received her water and was turned in her bed. This was the source of much resentment among the nursing staff, and a few nurses refused to have anything to do with the dying child. Most kept their ministrations to an absolute minimum.

But one nurse, Sara Ann Moberly, was determined to make Baby Owens's last days as comfortable as possible. She held the baby, rocked her, and talked soothingly to her when she cried. Doing all for the baby that she could do soothed Sara Ann as well. But even Sara Ann was glad when Baby Owens died. "It was a relief to me," she said. "I almost couldn't bear the frustration of just sitting there day after day and doing nothing that could really help her."

READING
Life-and-Death Decisions
in the Midst of Uncertainty

Robert F. Weir

Robert Weir argues for a position midway between those of Robertson and Engelhardt. He agrees with Robertson that decisions about extremely premature or impaired infants ought not be based on considerations of economic, social, or emotional costs, but he also agrees with Engelhardt that infants are not persons in the full sense. Accordingly, in some cases we may reasonably decide that it is not in the best interest of the infant to be treated.

In Weir's view, neonates not suffering from severe neurological impairments are "potential persons," and as such, they possess basic human rights, including the right not to be killed. However, all infants, including those lacking the potential to become persons in the full sense, are entitled to have their "best interest" considered, and Weir provides eight criteria for determining the best interest of an infant. In accordance with these criteria, an infant's interest may sometimes be served best by withholding treatment and allowing the child to die. Physicians and parents, Weir maintains, ought to make treatment decisions based exclusively on the benefits and burdens of treatment to the infant.

A Neonatal Intensive Care Unit (NICU) is characterized by premature and disabled patients with life-threatening conditions, highly trained medical and nursing specialists, state-of-the-art medical technology, an endless stream of medical consultants, parents grappling with frightening possibilities, and numerous decisions that have to be made in the midst of impenetrable uncertainty. Whether made while looking down at an imperiled baby, in consultation with the baby's parents, or in a conference room near the NICU, many of these decisions are crucial because a baby will continue to live or will die as a consequence of the decisions....

Do Neonates Count as Persons?

... To the extent that there is consensus among philosophers on the concept of personhood, that consensus focuses on the intrinsic rather than the extrinsic qualities of persons. Most philosophers agree on at least the core properties or traits of personhood, if not on all of their applications. Joel Feinberg, in his discussion of "commonsense personhood," puts forth the consensus view of personhood as being the possession of three necessary and jointly sufficient properties: consciousness, self-awareness, and at least minimum rationality. Such properties, for him and many others, represent "person-making characteristics."

The possession of personhood, therefore, has to do with neurological development and, at least among human beings, the absence of profound neurological dysfunction or impairment. The answer to the question of whether neonates are to be counted as persons depends on three interrelated factors:

1. How much neurological development is required for personhood;
2. How much neurological impairment is necessary to rule out personhood; and
3. Whether any significance is to be placed on the principle of potentiality as it applies to personhood.

In my judgment, there are three basic positions regarding the personhood of neonates (and other human beings whose personhood may be questioned), and the positions are distinguishable largely because of their handling of the factors of neurological development, neurological impairment, and potential personhood. The first position holds that *all neonates,* whether normal or neurologically impaired, *are nonpersons....*

The second position stands at the other end of a philosophical and political spectrum, and represents a very common view of neonates held by many physicians, nurses, and other people as well....

The third position stands between the other positions, differing from the first position's insufficient claims and the second position's excessive claims regarding the personhood of human newborns. This position, which holds that *most neonates are potential persons,* can be compared with the alternative views on the basis of its four claims:

1. Personhood is a moral category attaching to beings (of any species) with certain characteristics, principally cognitive capacities;

2. Neonates lack the intrinsic qualities that make a human into a person, as do fetuses;

3. Having the potential to become a person through the normal course of development does count, and neonates without severe neurological impairment (and fetuses having exhibited brain activity) have this potential; and

4. All *potential persons* have a *prima facie* claim to the moral benefits of personhood, including the right not to be killed, because they will subsequently acquire an *actual* person's moral and legal right to life.

The last of these positions, in my view, is the correct way of describing the ontological status of neonates. This position is preferable to the neonates-are-not-persons view of some philosophers, because it grants more than a species value to human newborns—and avoids the major weakness of having to allow, in principle, for the indiscriminate termination of an indeterminate number of neonatal lives, whether these lives are cognitively impaired, physically disabled, or normal. The third position is also preferable to the neonates-are-actual-persons view (especially as put forth by several prolife groups), because it takes the philosophical and psychological concept of personhood seriously—and avoids the major weakness of having to say, in principle, that a baby has no more claim to the moral benefits of personhood than an early human embryo does.

What Is the Best Ethical Option for Making Decisions to Initiate, Continue, or Abate Life-Sustaining Treatment?

... The options range from very conservative to very liberal, and they differ from one another regarding the substantive and, to a lesser extent, the procedural aspects of making life-and-death decisions for nonautonomous young patients.

The most conservative of these ethical options is the ethical perspective that was enacted into public policy by the Reagan administration through the "Baby Doe" regulations and subsequent Child Abuse regulations. Incensed that one "Baby Doe" (the 1982 Bloomington, Indiana, case) had died who could have lived with surgical intervention and concerned that other disabled infants were unnecessarily being allowed to die in other hospitals, the leaders of the Reagan administration went to great lengths to advocate the ethical perspective held by Surgeon General C. Everett Koop, and many of the administration's prolife supporters.

This ethical position holds that there is one and only one acceptable moral reason for not sustaining an infant's life, namely, the medical futility in a very limited number of cases of trying to do so. According to this ethical perspective, decisions not to sustain a severely disabled infant's life are acceptable only when such an infant is irretrievably dying (or, for some persons holding this position, an infant whose condition is some form of permanent unconsciousness). Therefore, the only cases in which such decisions by physicians or parents are justifiable are those unusual cases in which there is actually no moral decision to make: God, nature, fate, the roll of the genetic "dice," or some force beyond our control prevents medical efforts at sustaining life from working.

The most liberal option is a position that carries significant weight in some philosophical circles, but not, as we have already discussed, among physicians and others who are more oriented toward a practical, empirically based view of reality. This position, in sharp contrast to the first position, is based on the ontological status of the young lives at risk in critical care units rather than on the severity of their medical conditions. Instead of calling for life-sustaining treatment to be administered to all neonates or young children who are not dying (or permanently unconscious), the philosophers holding this position (e.g., Michael Tooley, Mary Anne Warren, and Peter Singer) argue that physicians and parents are obligated only to provide life-sustaining treatment for neonates and young children who count as persons. The catch is, as we have seen, that according to this perspective no neonates meet the criteria for personhood, and no moral weight is placed on the potential they may have to become persons later in the course of their development. An unresolved problem for these philosophers—and one of the reasons that this position will never become public policy—is that of defining the "magic moment" beyond the neonatal period when young children do meet the criteria for personhood and are thus protected from having their lives arbitrarily terminated.

The third position is the first of three positions that reside closer to the middle of the philosophical spectrum than either of the views just discussed. The physicians, philosophers, and other individuals who hold this view do not believe that all nondying neonates should be given life-sustaining treatment, nor do they believe that the lives of neonates can be terminated morally on the basis of a definitional point about personhood. Rather, they are convinced that the most important aspect of decisions not to sustain some infants' lives is the procedural question of who should make these difficult decisions. The correct answer to that question, according to the advocates of this position, is that the appropriate decision maker is the parent or parents of the neonate or young child whose life is threatened by his or her medical condition, even though the current federal regulations do not permit this kind of parental discretion. Since the parents of a disabled infant are the ones who stand to gain or lose the most, depending on what happens to the infant, it is they—instead of the physicians, an ethics committee, or anybody else—who should have the right to make the life-or-death decision in all cases over which there is some disagreement about whether a disabled infant should continue to live with or die in the absence of life-sustaining treatment.

Advocates of a fourth ethical position are convinced that quality-of-life judgments are unavoidable in cases of severe neurological or physiological malformation, in spite of what the federal regulations say to the contrary. All of the responsible parties in cases of serious neonatal abnormalities are morally obligated—and should be legally permitted—to raise important questions about the most likely future ahead of these children if their lives are to be prolonged with medical treatment. Of fundamental importance in such cases is not only the question of whether a given child can be salvaged with the abnormalities he or she has, but also what kind of life he or she is most likely to have with those abnormalities. The most important abnormalities to consider are neurological in nature. If a neurological disorder is sufficiently serious that pediatric neurologists and neonatologists project a life with severe disabilities for the child, virtually all persons holding a quality-of-life position would find the abatement of life-sustaining treatment in such a case to be morally justifiable.

The fifth position is held by individuals who are convinced that life-sustaining treatment should be provided to normal and disabled neonates whenever such treatment is in their best interests, and that life-sustaining treatment should be abated in the care of severely premature or severely disabled neonates (and other young children) whenever such treatment is judged not to be in their best interests.

Persons holding this position tend to be in agreement with quality-of-life advocates whose projections of a given child's future focus *entirely* on that child's likely abilities and disabilities, not on the child's impact on anybody else or ability to attain somebody else's minimal standard of acceptability for personal human life. By contrast, persons having a best-interests position disagree with quality-of-life advocates who tend to compare mentally and physically abnormal children with normal children, emphasize the problems that disabled children cause for their families and society, and try to protect families and society from having to deal with disabled children who cannot meet some arbitrary standard of acceptability. Like all advocates of the quality-of-life position, proponents of the best-interests

position hold a view that is more liberal than the current federal regulations.

The best-interests position, in my judgment, is the preferable ethical perspective to take in regard to difficult decisions about initiating, continuing, or abating life-sustaining treatment with any patients having life-threatening medical conditions. Neonatal and other young pediatric patients are no exception. They, like other nonautonomous patients, should receive life-sustaining treatment whenever the decision makers are convinced that the treatments available provide a balance of benefit to burden for the child. Such decisions should focus on the child's medical condition, concern suffering and irremediable handicap rather than projected social worth, and involve comparative judgments about the continuation of the child's injurious existence as opposed to the child's nonexistence.

What Does "Best Interests" Mean When Patients Are Neonates?

Even though widely supported in theory, the best-interests position is not without problems. Some of the advocates of the position admit that the concept of the patient's best interests is inherently vague, especially when the patient is a neonate. Nevertheless, they argue that the concept is helpful in decision making about life-sustaining treatment for neonates, because it focuses the decision-making process on precisely the human lives that ought to be the primary focus of concern.

Some of the critics of the best-interests position, at least as it applies to neonates, think that the conceptual foundation on which it stands is fundamentally flawed. Martin Benjamin argues that neonates simply do not yet possess the cognitive awareness, much less the specific wants and purposes, that are necessary for ascribing to them an interest in continued life. Howard Brody is convinced that any attempt to apply the concept of best interests to infants is bound to fail, because the concept is either incoherent or inadequate as a guide for tough clinical decisions. He argues that, even if infants can intelligibly be said to have interests, such interests would be unknowable by adult decision makers....

... One's "interests" consist of relationships, activities, and things in which one has a stake and on which one places value. To have interests (as opposed to sensations or instincts) normally requires as necessary conditions that one be conscious, aware of oneself, and able cognitively to have wants and purposes. In other words, to have interests normally requires that one be a person.

Yet, as Joel Feinberg points out in a discussion of fetal interests, it is plausible to ascribe *future* interests to a "prepersonal fetus." Even though a fetus "presumably has no actual interests," it can correctly be said to have future interests on the assumption that it will at some future point in its normal development (at birth or subsequent to birth) become a person and, thus, the possessor of actual interests. In a similar manner, the law recognizes that fetuses can have "contingent rights," such as the right to property, that will become actual rights the moment the fetus becomes a baby. Any contingent right of a fetus is instantly voided if the fetus dies before birth.

The same kind of reasoning about interests can be used in analyzing the interests that are ascribable to neonates, even by philosophers who claim that all neonates are non-persons. For even if neonates as nonpersons cannot correctly be said to be the possessors of actual interests, they can be said to have future interests (assuming that they will at some future point become persons) that can be interfered with or damaged by decisions or actions by adults long before these developing human lives become persons. For example, a neonate with myelomeningocele could reasonably be said to have a future interest in physical mobility, but come to realize later in life that a decision by physicians or parents during the neonatal period not to have the lesion surgically corrected had preempted that future interest from being actualized.

An alternative conceptual framework for discussing the interests of neonates was presented earlier, namely the philosophical view that neonates without severe neurological impairments are to be regarded as potential persons. In this framework, an analysis of the interests of neonates does not involve the ascription of future interests to them because they are thought likely to become persons at some "magic moment" in the future, but ascribes future interests to them because they have the potential to become the possessors of interests through the normal course of their development.

The point is a fundamental one. Just as potentiality is an important aspect of the concept of personhood, so potentiality is an important feature of a philosophical understanding of interests (but not of legal rights). Interests change from time to time after one becomes a person, with some interests intensifying over time, others waning, and others appearing as though newly born. For that reason, a discussion of the future interests of any given neonate becomes problematic if one can only project the actual interests that child will have when he or she meets the criteria for personhood at some future point in time. By contrast, the principle of potentiality, as it applies both to the possession of personhood and the possession of interests, permits one reasonably to ascribe to any given neonate the most general and basic kinds of interests that most individuals tend to have as they develop from young children to older children and on through the various phases of personal life....

When applied to neonates, the concept of "best interests" can obviously not refer to the specific wants and purposes any given neonate may have in continued life, much less to the specific wants and purposes that the neonate may have later in life. However, the concept of "best interests" can be used to capture the most fundamental future interest that persons have when they are patients, namely, an interest in not being harmed on balance during the course of medical treatment. For most patients in most clinical situations, this vital interest in not being harmed on balance means that they prefer continued life to death—unless intractable pain and

other suffering have made continued life more harmful than the prospect of death. To ascribe this general and basic interest to neonates is to claim that all neonates lacking severe neurological impairment can reasonably be said to have this future interest in not being harmed, an interest that will become actualized as they become persons during the normal course of their development.

... [T]he toughest aspect of using the concept of best interests in decision making in NICUs is determining the factors that should be considered in any given case. How can physicians and parents assess the beneficial and detrimental aspects of medical treatment in a case? How can they decide if life-sustaining treatment is in a neonate's best interests or is contrary to those interests? My suggestion is to regard the patient's-best-interests standard as having eight variables. In neonatal (and other young pediatric) cases, the variables are as follows:

1. Severity of the patient's medical condition;
2. Availability of curative or corrective treatment;
3. Achievability of important medical goals;
4. Presence of serious neurological impairments;
5. Extent of the infant's suffering;
6. Multiplicity of other serious medical problems;
7. Life expectancy of the infant; and
8. Proportionality of treatment-related benefits and burdens to the infant.

The last of these variables is, in many respects, a summation of the preceding variables. For decision makers in such cases, a consideration of the benefits of the treatment (both short-term and long-term) to the patient is the "bottom line" for determining whether life-sustaining treatment or the abatement of life-sustaining treatment is in a particular neonate's best interests. In making this assessment, decision makers arrive at a subjective judgment that includes objective factors, but is not finally reducible to quantifiable information. For to decide in rare clinical situations that-treatment is, on balance, harmful to the infant rather than beneficial is to make a moral judgment....

Should Life-Sustaining Treatment for Neonates and Other Young Children Ever Be Abated for Economic Reasons?

Neonatologists and other pediatric specialists place considerable importance on providing good patient care. In terms of the patient's-best-interests position, this emphasis on the medical needs and interests of individual patients is the morally preferable perspective for pediatricians and other physicians to have. According to this view, the needs and interests of each patient related to continued life correctly outweigh any competing interests of parents, siblings, or society. Simply put, no neonate or other young, nonautonomous patient should die merely because their medical and hospital care is expensive, even when the physicians and parents in a given case know that the family's income and insurance cannot cover the costs involved in the patient's care.

In the last few years, however, a number of factors have combined to create uncertainty about this basic moral premise for the provision of medical care, especially as it applies to cases of extremely premature or severely disabled neonates. Physicians, hospital administrators, and other concerned persons often question the importance that should be placed on the economic aspects of sustaining the lives of some neonates and other young children, especially when these lives predictably will be characterized by severe mental and physical disabilities. Case discussions in NICUs, PICUs, and specialized chronic care units for young children increasingly have comments and questions by staff physicians, residents, nurses, social workers, and ethicists regarding the costs of the ongoing treatment and who will have to pay for those costs....

... [A] factor contributing to uncertainty about the role of economics in neonatal cases pertains to the escalating costs of the care needed by extremely premature or severely disabled newborns. This uncertainty is brought about not only by an awareness of the escalating costs of providing care for these babies, but also by the realization that efforts to provide comparative cost figures for neonatal care have proven less than satisfactory, that the application of diagnosis-related group (DRG) categories has not worked well in NICUs, and that the cost-effectiveness of neonatal care for low-birthweight neonates is still questionable....

Studies published in recent years all document the increasing cost of providing care for disabled neonates and young children in chronic care units. For example, one study from Canada (using 1978 Canadian dollars) found that the costs of intensive care for infants weighing less than 1000 grams averaged $102,500 per survivor. A study from Australia (using 1984 Australian dollars) determined that the total direct cost for level-III, high-dependency care in one hospital was $690 per day. Studies in the United States, varying greatly in methodology, have found the total cost for selected survivors of neonatal intensive care in a Boston hospital to range from $14,600 to $40,700, for long-term survivors in a Washington, DC, pediatric hospital to be $182,500 for a year, and for extremely low-birthweight survivors of NICUs in six medical centers to range from $72,110 to $524,110, with a mean cost of care per infant of $158,800 for 137 days.

... [Another] factor has been the increased recognition that the financial pressures created by expensive neonatal and pediatric treatment can greatly damage and sometimes destroy families. For example, a 1988 Minnesota followup study of disabled infants and their families had a number of disturbing findings: the proportion of families with young children but lacking health insurance is increasing, 16% of the families in the study pay the entire cost of their health insurance, middle-income families have not qualified for state financial assistance, several of the families have filed for bankruptcy, and at least one family still owes a hospital and physicians over $300,000 for the care of their young child. The report concludes: "Families should not

have to lose their homes, mortgage their future, or neglect other children's needs to pay for the care of a chronically ill or disabled child."

A related, but different factor has to do with the long-term costs of providing medical, nursing, and surgical care for severely disabled children who remain in hospitals for months and years. Sometimes called "boarder babies," these children have complicated, chronic medical conditions, are usually dependent on mechanical ventilation and other technological assistance for survival, and frequently come from low-income, single-parent families that simply cannot afford (in terms of money and time) to have the child at home.

If no other institutional home can be arranged (usually because of the cost and technology involved), and if foster parents are not a realistic option, such children may reside for several years in a specialized chronic care unit in the hospital in which they were born. When that happens, the children become living symbols of a "second generation" type of problem brought about by the successes of neonatal intensive care: They are survivors of the NICU, but remain captives of medical technology in an institution that nobody would choose to call home....

Given the uncertainty generated by these variables, what should be done? For two of the ethical positions described earlier, the answer is reasonably simple: revise or ignore the federal regulations, abate life-sustaining treatment more quickly on the basis of (1) parental discretion or (2) projected quality of life for the neonates involved (including the impact of a neonate's later life on others), and thus cut down on the costs in NICUs, to families, and to institutions.

To go that way, for that reason, would be a mistake. The economic aspects of neonatal intensive care would become a dominant factor in decision making by parents and physicians, and many premature and disabled neonates would have their lives cut short to save money. To establish a policy that would encourage parents to make life-and-death decisions in individual cases as a money-saving strategy for themselves (or for physicians to do the same to save money for their hospitals) is not the best policy for addressing the very real problem of escalating costs for neonatal intensive care, especially if that policy is to be guided by the ethical principles of beneficence, nonmaleficence, and justice.

There is, in my judgment, a better alternative. That alternative is a combination of:

1. Continued use of the patient's-best-interests standard in clinical settings, including increased emphasis on the eight variables that comprise the standard;
2. The establishment of a national policy, based on sound clinical evidence, that would restrict the use of neonatal intensive care in terms of infants' birthweights; and
3. The establishment of a national health insurance program that would pay for the catastrophic health-care expenses generated by providing care for extremely premature severely disabled newborns.

The results of this combined approach would be threefold. A more consistent application of the best-interests standard would result in an increased number of decisions, as difficult as they are, by parents and physicians to discontinue life-sustaining treatment in individual cases. Such decisions would not be made to save money, but would be based on an honest conclusion that the treatment available, although capable of sustaining a neonate's life, is contrary to the infant's best interests.

In addition, the establishment of a national policy that would limit life-sustaining treatment to neonates over a certain birthweight (e.g., 600 grams) would not only cut down on the enormously high costs of caring for extremely low-birthweight infants, but could also be defended, depending on the rationale and details of the policy, as meeting the requirements of justice. Such a policy would surely not solve all of the problems of uncertainty in NICUs, but could provide a measure of greater certainty, if based on a consensus among neonatologists, in establishing a minimum weight limit for neonates who would be given life-sustaining treatment.

Finally, by establishing a national insurance program, the federal government would help pay for the enormous costs that are involved in neonatal intensive care and specialized chronic-care units for young children. For the federal government to mandate that virtually all neonates, unless dying or permanently unconscious, be kept alive, and then to make no serious effort to help parents and institutions pay for that expensive care is unjust. In the absence of such a program, parents and physicians will continue to be faced with the task of making life-and-death decisions for newborns in the midst of great uncertainty—including whether the family will be destroyed financially by costs of the medical care.

SCENARIO

Irene Towers had been a nurse for almost twelve years; for the last three of those years she had worked in the Neonatal Unit of Halifax County Hospital. It was a job she loved. Even when the infants were ill or required special medical or surgical treatment, she found the job of caring for them immensely rewarding. She knew that without her efforts many of the babies would simply die.

Irene Towers was on duty the night that conjoined twins were born to Corrine Couchers and brought at once to the Neonatal Unit. Even Irene, with all her experience, was distressed to see them. The twin boys were joined at their midsections in a way that made it impossible to separate them surgically. Because of the position of the single liver and the kidneys, not even one twin could be saved at the expense of the life of the other. Moreover, both children were severely deformed, with incompletely developed arms and legs and misshapen heads. As best as the neurologist could determine, both suffered severe brain damage.

The father of the children was Dr. Harold Couchers. Dr. Couchers, a slightly built man in his early thirties, was a specialist in internal medicine with a private practice.

Irene felt sorry for him the night the children were born. When he went into the room with the obstetrician to examine his sons, he had already been told what to expect. He showed no signs of grief as he stood over the slat-sided crib, but the corners of his mouth were drawn tight, and his face was almost unnaturally empty of expression. Most strange for a physician, Irene thought, he merely looked at the children and did not touch them. She was sure that in some obscure way he must be blaming himself for what had happened to them.

Later that evening, Irene saw Dr. Couchers sitting in the small conference room at the end of the hall with Dr. Cara Rosen, Corrine Couchers's obstetrician. They were talking earnestly and quietly when Irene passed the open door. Then, while she was looking over the assignment sheet at the nursing station, the two of them walked up. Dr. Rosen took a chart from the rack behind the desk and made a notation. After returning the chart, she shook hands with Dr. Couchers, and he left.

It was not until the end of her shift that Irene read the chart; Dr. Rosen's note said that the twin boys were to be given neither food nor water. At first Irene couldn't believe the order. But when she asked her supervisor, she was told that the supervisor had telephoned Dr. Rosen and that the obstetrician had confirmed the order.

Irene said nothing to the supervisor or to anyone else, but she made her own decision. She believed it was wrong to let the children die, particularly in such a horrible way. They deserved every chance to fight for their lives, and she was going to help them the way she had helped hundreds of other babies in the unit.

For the next week and a half, Irene saw to it that the children were given water and fed the standard infant formula. She did it all herself, on her own initiative. Although some of the other nurses on the floor saw what she was doing, none of them said anything to her. One even smiled and nodded to her when she saw Irene feeding the children.

Apparently someone else also disapproved of the order to let the twins die. Thirteen days after their birth, an investigator from the state Family Welfare Agency appeared in the neonatal ward. The rumor was that his visit had been prompted by an anonymous telephone call.

Late in the afternoon of the day of that visit, the deformed twins were made temporary wards of the Agency, and the orders on the chart were changed—the twins were now to be given food and water. On the next day, the county prosecutor's office announced publicly that it would conduct an investigation of the situation and decide whether criminal charges should be brought against Dr. Couchers or members of the hospital staff.

Irene was sure that she had done the right thing. Nevertheless, she was glad to be relieved of the responsibility.

1. Is there a morally relevant distinction between not treating (and allowing to die) and not providing such minimal needs as food and water (and allowing to die)?
2. Does any line of reasoning support the action taken by Irene Towers?
3. Did Irene Towers exceed the limits of her responsibility, or did she act in a morally heroic way?

III. RESEARCH ETHICS

The discussion of biotechnology in Chapter 6 led us to consider its uses in gene therapy in Chapter 9. Review these topics to see that the use of this technology may lead to a cure for genetic diseases. But how do we test these potential cures?

CASE PRESENTATION
Jesse Gelsinger: The First Gene-Therapy Death

When Jesse Gelsinger was three months short of his third birthday, he was watching cartoons on TV when he fell asleep. Except it was a sleep from which his parents were unable to arouse him. Panicked, they rushed him to a local hospital.

When Jesse was examined, he responded to stimuli but didn't awaken. The physicians classified him as being in a level-one coma. Laboratory tests showed he had a high level of ammonia in his blood, but it was only after several days and additional blood assays that Jesse's physicians arrived at a diagnosis of ornithine transcarbamylase deficiency—OTC.

OTC is a rare genetic disorder in which the enzyme ornithine transcarbamylase, one of the five involved in the urea cycle, is either missing or in short supply. The enzymes in the cycle break down the ammonia that is a by-product of protein metabolism.

A deficiency of OTC means the body cannot get rid of the ammonia, and it gradually accumulates in the blood. When the ammonia reaches a crucial level, it causes coma, brain damage, and eventually death. The disease results from a mutation on the X chromosome; thus females are carriers of the gene, which they pass on to their sons. The disorder occurs in 1 of every 40,000 births. Infants with the mutation usually become comatose and die within seventy-two hours of birth. Half die within a month of birth, and half of those who remain die before age five.

Although OTC is a genetic disease, no one else in Jesse's immediate family or ancestry had ever been diagnosed with the disease. His disease was probably the result of a spontaneous mutation. He was a genetic mosaic, which meant his body contained a mixture of normal and mutated cells.

For this reason, Jesse had a comparatively mild form of OTC. His body produced enough of the enzyme that he could remain in stable health if he stuck to a low-protein diet and took his medications. These included substances, like sodium benzoate, that chemically bind to ammonia and make it easier for the body to excrete it.

At age ten, after an episode of consuming too much protein, Jesse once again fell into a coma and was hospitalized. But five days later, he was back home with no apparent neurological damage. During his teens, Jesse's condition was monitored by semiannual visits to a metabolic clinic in his hometown of Tucson, Arizona.

In 1998 Jesse, now seventeen, and his father, Paul Gelsinger, heard from Dr. Randy Heidenreich, a doctor at the clinic, about a clinical trial at the University of Pennsylvania. Researchers at the Institute for Human Gene Therapy, Heidenreich told the Gelsingers, were trying to use gene therapy to supply the gene for OTC. Their success would not be a cure for the disease, but it would be a treatment that might be able to bring babies out of comas and prevent their having brain damage.

The Gelsingers were interested, but Jesse was still a year short of being old enough to participate. In April 1999, during another visit to the clinic, they again talked to Dr. Heidenreich about the trial, and Paul mentioned that the family would be taking a trip to New Jersey in June. They would be able to make a side trip to Philadelphia and talk to the investigators.

Dr. Heidenreich contacted an investigator at the Institute and mentioned the Gelsingers' interest in the research, and Paul received a letter from him in April. Jesse would be interviewed and tested at the university hospital on June 22 to determine whether he met the criteria for becoming a research participant.

A bioethicist at the university, Arthur Caplan, had advised the researchers that it would be morally wrong to use infants born with OTC as participants in the gene-therapy trial. Because they could not be expected to live, Caplan reasoned, their parents would be desperate to find a way to save their child's life. Hence, driven by desperation, their consent would not be free. The appropriate participants would be women who were carriers of the gene or men in stable health with only a mild form of the disease. Jesse would celebrate his eighteenth birthday the day the family flew to the East Coast, and his age would then make him eligible to become a participant.

On June 22, 1999, Jesse and Paul Gelsinger met with Dr. Steven Raper for forty-five minutes to review the consent forms and discuss the procedure for which Jesse might volunteer, if he qualified. Dr. Raper, a surgeon, would be the one performing the gene-therapy procedure.

According to Paul Gelsinger's recollections, Raper explained that Jesse would be sedated and two catheters inserted: one in the artery leading to his liver, the second in the vein leaving it. A weakened strain of adenovirus (the virus causing colds), genetically modified to include the

OTC gene, would be injected into the hepatic artery. Blood would then be taken from the vein to monitor whether the viral particles were being taken up by the liver cells.

To reduce the risk of a blood clot's breaking loose from the infusion site, Jesse would have to remain in bed for eight hours after the procedure. Most likely, he would soon develop flu-like symptoms lasting for a few days. He might develop hepatitis, an inflammation of the liver. The consent form mentioned that if hepatitis progressed, Jesse might need a liver transplant. The consent form also mentioned that death was a possible outcome.

Paul Gelsinger saw this as such a remote possibility that he was more concerned about the needle biopsy of the liver to be performed a week after the procedure. The risk of death from the biopsy was given as 1 in 10,000. Paul urged Jesse to read the consent document carefully and to make sure he understood it. Paul thought the odds looked very good.

Dr. Raper explained that Jesse couldn't expect to derive any personal medical benefit from participating in the clinical trial. Even if the genes became incorporated into his cells and produced OTC, the effect would only be transitory. His immune system would attack the viral particles and destroy them within a month to six weeks.

Jesse, at the end of the information session, agreed to undergo tests to determine how well the OTC he produced got rid of ammonia in his blood—a measure of OTC efficiency. Samples of his blood were taken; then he drank a small amount of radioactively tagged ammonia. Later, samples of his blood and urine were taken to see how much of the ingested ammonia had been eliminated. The results showed his body's efficiency was only 6 percent of a normal performance.

A month later, the Gelsingers received a letter from Dr. Mark Bratshaw, the pediatrician at the Institute who proposed the clinical trial. Bratshaw confirmed the 6 percent efficiency figure from additional test results and expressed his wish to have Jesse take part in the study. A week later, Bratshaw called Jesse and talked to him. Jesse had already expressed to his father a wish to participate, but he told Bratshaw to talk to his father.

Bratshaw told Paul about the results of their animal studies. The treatment had worked well in mice, preventing the death of those given a lethal injection of ammonia. Also, the most recent patient treated had shown a 50 percent increase in her ability to excrete ammonia. Paul Gelsinger later recalled saying, "Wow! This really works. So, with Jesse at 6 percent efficiency, you may be able to show exactly how well this works."

Bratshaw said their real hope was to find a treatment for newborns lacking any OTC efficiency and with little chance of survival. Also, another twenty-five liver disorders could potentially be treated with the same gene-therapy technique. The promise, then, was that hundreds of thousands, if not millions, of lives might be saved. Bratshaw and Paul never talked about the dangers to Jesse of becoming a subject in the clinical trial.

Paul discussed participation with Jesse. They both agreed that it was the right thing to do. Jesse would be helping babies stay alive and, perhaps in the long run, he might even be helping himself.

Approval

The clinical trial was supported by a National Institutes of Health grant awarded to Dr. James Wilson, the head of the Institute, and Mark Bratshaw. Their protocol had been reviewed by the federal Recombinant-DNA Advisory Committee (RAC) and the FDA. The animal studies Bratshaw had mentioned to Paul included twenty studies on mice to show the efficacy of the proposed technique. Wilson and his group had also conducted studies on monkeys and baboons to demonstrate the safety of the procedure.

Three of the treated monkeys had died of severe liver inflammation and a blood-clotting disorder when they had been given a stronger strain of adenovirus at a dose twenty times that proposed in the human trial. Both of the scientists assigned by the RAC to review the proposal thought the trial was too dangerous to include stable, asymptomatic volunteers. But Wilson and Bratshaw, using Caplan's argument, convinced the panel that using subjects capable of giving consent was morally preferable to using OTC newborns.

The initial protocol called for the modified viruses to be injected into the right lobe of the liver. The thinking was that if the treatment caused damage, the right lobe could be removed and the left lobe spared. But the RAC objected to injecting the viruses into the liver and the investigators agreed to change the protocol. The decision was later reversed by the FDA, on the grounds that wherever the viruses were injected, they would end up in the liver. The RAC was in the process of being reorganized and, in effect, taken out of the approval loop for proposals; it never received notice of the change. The investigators continued to operate under the modified protocol.

Protocol

The study was a Phase I clinical trial. According to its protocol, eighteen patients were to receive an infusion of the genetically modified adenovirus. The aim of the study was to determine "the maximum tolerated dose." The investigators wanted to determine the point at which the transferred gene would be producing OTC in the maximum amount compatible with side effects that could be tolerated.

The eighteen patients were divided into six groups of three. Each successive group was to receive a slightly higher dose than the preceding one. The idea behind this common procedure is to protect the safety of the study participants. By increasing doses slightly, the hope is to spot the potential for serious side effects in time to avoid causing harm to the participants.

Preparation

On Thursday, September 9, Jesse Gelsinger, carrying one suitcase of clothes and another of videos, caught a plane for Philadelphia. He checked into the hospital alone. His father, a self-employed handyman, stayed in Tucson to work. Paul planned to arrive on the 18th to be present for what he considered the most dangerous part of the trial—the liver biopsy.

"You're my hero," Paul told Jesse. He looked him in the eye, then gave him a big hug.

The level of ammonia in Jesse's blood was tested on Friday and Sunday. Sunday night he called his father, worried. His ammonia level was high, and his doctors had put him on IV medication to lower it. Paul reassured his son, reminding him that the doctors at the Institute knew more about OTC than anybody else in the world.

Tragedy

On the morning of Monday, September 13, Gelsinger became the eighteenth patient treated. He was transported from his room to the hospital's interventional radiology suite, where a catheter was snaked through an artery in his groin to the hepatic artery. A second catheter was placed in the vein exiting the liver.

Dr. Raper then slowly injected thirty milliliters of the genetically altered virus into Jesse's hepatic artery. This was the highest dose given to any participant. Patient 17, however, had received the same size dose from a different lot of the virus and had done well. The procedure was completed around noon, and Jesse was returned to his room.

That evening Gelsinger, as expected, began to develop flu-like symptoms. He was feeling ill and feverish when he talked to his father and his stepmother, Mickie, that evening. "I love you, Dad," Jesse told his father. They all said what turned out to be their last goodbyes.

During the night, Jesse's fever soared to 104.5 degrees. A nurse called Dr. Raper at home, and, when he arrived at the hospital around 6:15 that morning, the whites of Jesse's eyes had a yellowish tinge. This was a sign of jaundice, not something the doctors had encountered with the other trial participants. Laboratory findings revealed that Jesse's bilirubin, the product of red blood cell destruction, was four times the normal level.

Raper called Dr. Bratshaw, who was in Washington, to tell him their patient had taken a serious turn. Bratshaw said he would catch the train and arrive in Philadelphia in two hours. Raper also called Paul Gelsinger to explain the situation.

The jaundice was worrying to Jesse's physicians. Either his liver was not functioning adequately or his blood was not clotting properly and his red blood cells were breaking down faster than his liver could process them. Such a breakdown was life threatening for someone with OTC, because the destroyed cells released protein the body would have to metabolize. Jesse was showing the same problem as the monkeys that had been given the stronger strain of the virus.

Tuesday afternoon Paul received a call from Dr. Bratshaw. Jesse's blood-ammonia level had soared to 250 micromoles per deciliter, with 35 being a normal measure. He had slipped into a coma and was on dialysis to try to clear the ammonia from his blood. Paul said he would catch a plane and be at the hospital the next morning.

By the time Paul arrived at eight o'clock on Wednesday and met Bratshaw and Raper, Jesse had additional problems. Dialysis had brought his ammonia level down to 70 from its peak of 393, but he was definitely having a blood-clotting problem. Also, although placed on a ventilator, he continued to breathe for himself, causing hyperventilation. This increased the pH of his blood, which increased the level of ammonia circulating to his brain. Paul gave his permission for the doctors to give Jesse medications that would paralyze his breathing muscles and allow the machine to take over completely.

By Wednesday afternoon, Jesse's breathing was under control. His blood pH had fallen back to normal, and the clotting disorder was improving. Bratshaw returned to Washington. Paul began to relax, and at 5:30 he went out to dinner with his brother and his wife. But he returned to the hospital to find Jesse had been moved to a different intensive care ward, and as he watched the monitors, he saw the oxygen content of Jesse's blood was dropping. A nurse asked him to wait outside.

At 10:30 that evening, a doctor told Paul that Jesse's lungs were failing. Even by putting him on pure oxygen, they were unable to get an adequate amount of oxygen into his blood. The doctors had also talked with a liver transplant team and learned that Jesse was not a good candidate for a transplant.

Raper, very worried, discussed Jesse's problems with Bratshaw and Wilson, and the three of them decided to put Jesse on extracorporeal membrane oxygenation—ECMO. The machine would remove carbon dioxide from Jesse's blood and supply it with the needed oxygen. The procedure was far from standard, however. Only half of the 1000 people placed on ECMO had lived, but Paul was informed that Jesse had only a 10 percent chance of surviving without ECMO.

"If we could just buy his lungs a day or two," Raper later told a reporter, "maybe he would go ahead and heal up."

Jesse was not hooked up to the ECMO unit until five o'clock Thursday morning. Bratshaw attempted to return from Washington, but he was trapped in an Amtrak train outside Baltimore. Hurricane Floyd was headed toward the East Coast; Jesse's stepmother arrived from Tucson just before the airport closed.

The ECMO appeared to be working. But Paul was told that Jesse's lungs were so severely damaged that, if he survived, it would take a long time for him to recover.

When Paul finally saw his son at mid-morning, Jesse was still comatose and bloated beyond recognition. Only the tattoo on his right calf and a scar on his elbow assured Paul that the person in the bed was Jesse.

That evening, unable to sleep, Paul walked the half-mile from his hotel to the hospital to check on Jesse. His son was no better, and Paul noticed that the urine-collecting bag attached to Jesse's bed contained blood. He realized that this meant Jesse's kidneys were shutting down. "He was sliding into multiple-organ-system failure," Raper later recalled.

The next morning, Friday, September 17, Raper and Bratshaw met with Paul and Mickie to give them the bad news that Paul had already predicted. Jesse had suffered irreversible brain damage, and the doctors wanted Paul's permission to turn off the ventilator. At Paul's request, he and Mickie were left alone for a few minutes. He then told the doctors he wanted to bring in his family and have a brief service for Jesse.

Paul and Mickie, seven of Paul's fifteen siblings and their spouses, and about ten staff members crowded into Jesse's room. Paul leaned over Jesse, then turned and told the crowd, "Jesse was a hero." The chaplain said a prayer; then Paul gave a signal. Someone flipped one switch to turn off the ventilator, and flipped a second to turn off the ECMO unit.

Dr. Raper watched the heart monitor. When the line went flat, he put his stethoscope against Jesse's chest. At 2:30 P.M. Raper officially pronounced him dead. "Goodbye, Jesse," he said. "We'll figure this out."

Gathering Storm

Dr. James Wilson, the head of the Institute, immediately reported Jesse's death to the FDA. Paul Gelsinger, sad as he was, didn't blame Jesse's physicians for what had happened. Indeed, he supported them in the face of an initial round of criticism. "These guys didn't do anything wrong," he told reporters.

Then journalists began to bring to light information that raised questions about whether Jesse and his father had been adequately informed about the risks of the trial that claimed Jesse's life. Also, it raised questions about a conflict of interest that might have led researchers to minimize the risks. The FDA initiated an investigation, and the University of Pennsylvania conducted an internal inquiry.

Paul Gelsinger decided to attend the December 1999 RAC that discussed his son's death. He learned for the first time at that meeting, according to his account, that gene therapy had never been shown to work in humans. He had been misled, not necessarily deliberately, by the researcher's accounts of success in animals. As Paul listened to criticisms of the clinical trial, his faith in the researchers waned and was replaced by anger and a feeling of betrayal.

Other information fed his anger. When a month earlier he had asked James Wilson, "What is your financial position in this?" Wilson's reply, as Paul recalled, was that he was an unpaid consultant to the biotech company Genovo that was partially funding the Institute. Then later Paul learned that both Wilson and the University of Pennsylvania were major stockholders in Genovo and that Wilson had sold his 30 percent share of the company for $13.5 million.

Wilson and the university, as Paul saw it, had good reason to recruit volunteers for the clinical trial and produce positive results. Thus, they might not have been as careful in warning the Gelsingers about the risks of the study as they should have been. Also, the bioethicist approving the trial was someone who held an appointment in the department headed by Wilson. This, in effect, made Wilson his superior and thus automatically raised a question about the independence of his judgment.

A year and a day after Jesse's death, the Gelsinger family filed a wrongful-death lawsuit against the people conducting the clinical trial and the University of Pennsylvania. The university settled the suit out of court. The terms of the settlement were not disclosed.

FDA Findings

An investigation by the FDA resulted in a report to Wilson and the University of Pennsylvania pointing to two flaws in the way the clinical trial was conducted. First, the investigators failed to follow their protocol and failed to report liver toxicity in four patients treated prior to Gelsinger. Second, the investigators failed to acknowledge the death of two rhesus monkeys injected with a high level of a similar vector.

Wilson's response was that he had sent the FDA the liver-toxicity information prior to the final approval of the protocol, although his report had been late. Further, the two monkeys that died were part of another study that used a different, stronger virus. In effect, then, Wilson was claiming that he and his colleagues had done nothing wrong and the FDA criticisms were unjustified.

Critics point out, apart from the question of how legitimate the criticisms were, that the FDA itself does not have enough power to oversee clinical trials properly. Most important, it is prohibited by law from distributing some so-called "adverse-event" reports. Difficulties encountered by patients in the fifty or so gene-therapy trials are often not made public, or even shared with investigators conducting similar trials, because drug-company sponsors regard information about adverse events as proprietary. This, critics say, puts participants in the position of having to take risks that they know nothing about. The law seems to favor protecting the investments of the pharmaceutical industry more than protecting human subjects.

Outcome

What caused the death of Jesse Gelsinger? Even after the autopsy, the answer isn't clear. The most suggestive finding was that Jesse had abnormal cells in his bone marrow. This may have been a preexisting condition, and it may account for why his immune system reacted in such an unpredicted way to the viral injection. He apparently died from an immunological response.

The FDA, after Jesse's death, shut down all gene therapy operations temporarily for review. The University of Pennsylvania, after its internal review, restricted the role of the Institute for Human Gene Therapy to conducting basic biological research. Unable to carry out clinical trials, the Institute was de facto put out of business. A year or so later, it ceased to exist.

Because of Jesse's death, the Office for the Protection of Human Research Subjects committed itself to a major effort to educate researchers in the requirements for protecting participants in clinical trials and to stress the importance of Institutional Review Boards in seeing to the safety of participants. Even so, adverse event reporting is still prohibited by law, when it can be deemed to constitute proprietary information. Critics continue to see this as incompatible with the idea behind informed consent.

READING: *Belmont Report*
National Commission for the Protection of Human Subjects

The 1974 National Research Act mandated that every institution receiving federal funding and conducting research involving human subjects establish an institutional review board (IRB) to oversee such research. The Act was prompted by public revelations about the United States Public Health Service–sponsored Tuskegee syphilis study, in which investigators enrolled patients without their consent and treated them in ways that were condemned at the 1947 Nuremberg trials of Nazi physicians and researchers.

The 1974 act also established the National Commission for the Protection of Human Subjects of Biomedical and Behavioral Research and charged it with identifying the ethical principles basic to human research and formulating guidelines to guarantee that they are followed in its conduct. The Commission met for four days of discussion at the Smithsonian Institution's Belmont Conference Center, and its deliberations were published as the Belmont Report *in the Federal Register in 1979. The report was accepted by the Secretary of what is now the Department of Health and Human Services as the department's policy statement on the use of human research subjects.*

The Belmont Report *is not a set of regulations, but a framework for identifying, discussing, and settling ethical matters, while leaving open the possibility that reasonable people may sometimes differ irreconcilably. The report distinguishes medical practice from research (see the Briefing Session in this chapter for a discussion) and identifies three principles as most relevant for evaluating the ethical legitimacy of research involving human subjects. The principles, which in our society are generally accepted as so uncontroversial as not to require argument, are respect for persons, beneficence, and justice.*

Ethical Principles and Guidelines for Research Involving ... Human Subjects

Boundaries Between Practice and Research

It is important to distinguish between biomedical and behavioral research, on the one hand, and the practice of accepted therapy on the other, in order to know what activities ought to undergo review for the protection of human subjects of research. The distinction between research and practice is blurred partly because both often occur together (as in research designed to evaluate a therapy) and partly because notable departures from standard practice are often called "experimental" when the terms "experimental" and "research" are not carefully defined.

For the most part, the term "practice" refers to interventions that are designed solely to enhance the well-being of an individual patient or client and that have a reasonable expectation of success. The purpose of medical or behavioral practice is to provide diagnosis, preventive treatment or therapy to particular individuals. By contrast, the term "research" designates an activity designed to test an hypothesis, permit conclusions to be drawn, and thereby to develop or contribute to generalizable knowledge (expressed, for example, in theories, principles, and statements of relationships). Research is usually described in a formal protocol that sets forth an objective and a set of procedures designed to reach that objective.

When a clinician departs in a significant way from standard or accepted practice, the innovation does not, in and of itself, constitute research. The fact that a procedure is "experimental," in the sense of new, untested or different, does not automatically place it in the category of research. Radically new procedures of this description should, however, be made the object of formal research at an early stage in order to determine whether they are safe and effective. Thus, it is the responsibility of medical practice committees, for example, to insist that a major innovation be incorporated into a formal research project.

Research and practice may be carried on together when research is designed to evaluate the safety and efficacy of a therapy. This need not cause any confusion regarding whether or not the activity requires review; the general rule is that if there is any element of research in an activity, that activity should undergo review for the protection of human subjects.

Basic Ethical Principles

The expression "basic ethical principles" refers to those general judgments that serve as a basic justification for the many particular ethical prescriptions and evaluations of human actions. Three basic principles, among those generally accepted in our cultural tradition, are particularly relevant to the ethics of research involving human subjects: the principles of respect of persons, beneficence and justice.

Respect for Persons. Respect for persons incorporates at least two ethical convictions: first, that individuals should be treated as autonomous agents, and second, that persons with diminished autonomy are entitled to protection. The principle of respect for persons thus divides into two separate moral requirements: the requirement to acknowledge autonomy and the requirement to protect those with diminished autonomy.

An autonomous person is an individual capable of deliberation about personal goals and of acting under the direction of such deliberation. To respect autonomy is to give weight to autonomous persons' considered opinions and choices while refraining from obstructing their actions unless they are clearly detrimental to others. To show lack of respect for an autonomous agent is to repudiate that person's considered judgments, to deny an individual the freedom to act on those considered judgments, or to withhold information necessary to make a considered judgment, when there are no compelling reasons to do so.

However, not every human being is capable of self-determination. The capacity for self-determination matures during an individual's life, and some individuals lose this capacity wholly or in part because of illness, mental disability, or circumstances that severely restrict liberty. Respect for the immature and the incapacitated may require protecting them as they mature or while they are incapacitated.

Some persons are in need of extensive protection, even to the point of excluding them from activities which may harm them; other persons require little protection beyond making sure they undertake activities freely and with awareness of possible adverse consequence. The extent of protection afforded should depend upon the risk of harm and the likelihood of benefit. The judgment that any individual lacks autonomy should be periodically reevaluated and will vary in different situations.

In most cases of research involving human subjects, respect for persons demands that subjects enter into the research voluntarily and with adequate information. In some situations, however, application of the principle is not obvious. The involvement of prisoners as subjects of research provides an instructive example. On the one hand, it would seem that the principle of respect for persons requires that prisoners not be deprived of the opportunity to volunteer for research. On the other hand, under prison conditions they may be subtly coerced or unduly influenced to engage in research activities for which they would not otherwise volunteer. Respect for persons would then dictate that prisoners be protected. Whether to allow prisoners to "volunteer" or to "protect" them presents a dilemma. Respecting persons, in most hard cases, is often a matter of balancing competing claims urged by the principle of respect itself.

Beneficence. Persons are treated in an ethical manner not only by respecting their decisions and protecting them from harm, but also by making efforts to secure their well-being. Such treatment falls under the principle of beneficence. The term "beneficence" is often understood to cover acts of kindness or charity that go beyond strict obligation. In this document, beneficence is understood in a stronger sense, as an obligation. Two general rules have been formulated as complementary expressions of beneficent actions in this sense: (1) do not harm and (2) maximize possible benefits and minimize possible harms.

The Hippocratic maxim "do no harm" has long been a fundamental principle of medical ethics. Claude Bernard extended it to the realm of research, saying that one should not injure one person regardless of the benefits that might come to others. However, even avoiding harm requires learning what is harmful; and, in the process of obtaining this information, persons may be exposed to risk of harm. Further, the Hippocratic Oath requires physicians to benefit their patients "according to their best judgment." Learning what will in fact benefit may require exposing persons to risk. The problem posed by these imperatives is to decide when it is justifiable to seek certain benefits despite the risks involved, and when the benefits should be foregone because of the risks.

The obligations of beneficence affect both individual investigators and society at large, because they extend both to particular research projects and to the entire enterprise of research. In the case of particular projects, investigators and members of their institutions are obliged to give forethought to the maximization of benefits and the reduction of risk that might occur from the research investigation. In the case of scientific research in general, members of the larger society are obliged to recognize the longer term benefits and risks that may result from the improvement of knowledge and from the development of novel medical, psychotherapeutic, and social procedures.

The principle of beneficence often occupies a well-defined justifying role in many areas of research involving human subjects. An example is found in research involving children. Effective ways of treating childhood diseases and fostering healthy development are benefits that serve to justify research involving children—even when individual research subjects are not direct beneficiaries. Research also makes it possible to avoid the harm that may result from the application of previously accepted routine practices that on closer investigation turn out to be dangerous. But the role of the principle of beneficence is not always so unambiguous. A difficult ethical problem remains, for example, about research that presents more than minimal risk without immediate prospect of direct benefit to the children involved. Some have argued that such research is inadmissible, while others have pointed out that this limit would rule out much research promising great benefit to children in the future. Here again, as with all hard cases, the different claims covered by the principle of beneficence may come into conflict and force difficult choices.

Justice. Who ought to receive the benefits of research and bear its burdens? This is a question of justice, in the sense of "fairness in distribution" or "what is deserved." An injustice occurs when some benefit to which a person is entitled is denied without good reason or when some burden is imposed unduly. Another way of conceiving the principle of justice is that equals ought to be treated equally. However, this statement requires explication. Who is equal and who is unequal? What considerations justify departure from equal distribution? Almost all commentators allow that distinctions based on experience, age, deprivation, competence, merit and position do sometimes constitute criteria justifying differential treatment for certain purposes. It is necessary, then, to explain in what respects people should be treated equally. There are several widely accepted formulations of just ways to distribute burdens and benefits. Each formulation mentions some relevant property on the basis of which burdens and benefits should be distributed. These formulations are (1) to each person an equal share, (2) to each person according to individual need, (3) to each person according to individual effort, (4) to each person according to societal contribution, and (5) to each person according to merit.

Questions of justice have long been associated with social practice such as punishment, taxation and political representation. Until recently these questions have not generally been associated with scientific research. However, they are foreshadowed even in the earliest reflections on the ethics of research involving human subjects. For example, during the 19th and early 20th centuries the burdens of serving as research subjects fell largely upon poor ward patients, while the benefits of improved medical care flowed primarily to private patients. Subsequently, the exploitation of unwilling prisoners as research subjects in Nazi concentration camps was condemned as a particularly flagrant injustice. In this country, in the 1940's, the Tuskegee syphilis study used disadvantaged, rural black men to study the untreated course of a disease that is by no means confined to that population. These subjects were deprived of demonstrably effective treatment in order not to interrupt the project, long after such treatment became generally available.

Against this historical background, it can be seen how conceptions of justice are relevant to research involving human subjects. For example, the selection of research subjects needs to be scrutinized in order to determine whether some classes (e.g., welfare patients, particular racial and ethnic minorities, or persons confined to institutions) are being systematically selected simply because of their easy availability, their compromised position, or their manipulability, rather than for reasons directly related to the problem being studied. Finally, whenever research supported by public funds leads to the development of therapeutic devices and procedures, justice demands both that these not provide advantages only to those who can afford them and that such research should not unduly involve persons from groups unlikely to be among the beneficiaries of subsequent *applications of the research.*

READING
Clinical Trials: Are They Ethical?

Eugene Passamani

Eugene Passamani argues that randomized clinical trials (RCTs) are the most reliable means of evaluating new therapies. Without RCTs, chance and bias may affect our conclusions.

Passamani rejects the argument that the physician–patient relationship demands that physicians recommend the "best" therapy for patients, no matter how poor the data on which the recommendation is based. He acknowledges that RCTs pose ethical problems for physician-researchers but believes the difficulties can be overcome by employing three procedural safeguards.

First, all participants must give their informed consent. They must be told about the goals of the research and its potential benefits and risks. Moreover, they must be informed about alternatives to their participation, and they must be permitted to withdraw from the trial at any time they choose. Second, for an RCT to be legitimate, a state of clinical equipoise must exist. Competent physicians must be genuinely uncertain about which of the alternative therapies in the trial is superior and content to allow their patients to be treated with any of them. Finally, the clinical trial must be designed as a critical test of the therapeutic alternatives. Properly carried out, Passamani holds, RCTs protect physicians and patients from therapies that are ineffective or toxic.

[A Type I error consists in deciding that therapy A is better than therapy B when, in fact, both are of equal worth (i.e., a true null hypothesis is rejected). A Type II error consists in deciding that the treatments are equally good when A is actually better than B (i.e., a false null hypothesis is accepted).— Ronald Munson.]

Biomedical research leads to better understanding of biology and ultimately to improved health. Physicians have for millenniums attempted to understand disease, to use this knowledge to cure or palliate, and to relieve attendant suffering. Improving strategies for prevention and treatment remains an ethical imperative for medicine. Until very recently, progress depended largely on a process of carefully observing groups of patients given a new and promising therapy; outcome was then compared with that previously observed in groups undergoing a standard treatment. Outcome in a series of case patients as compared with that in nonrandomized controls can be used to assess the treatment of disorders in which therapeutic effects are dramatic and the pathophysiologic features are relatively uncomplicated, such as vitamin deficiency or some infectious diseases. Observational methods are not very useful, however, in the detection of small treatment effects in disorders in which there is substantial variability in expected outcome and imperfect knowledge of complicated pathophysiologic features (many vascular disorders and most cancers, for example). The effect of a treatment cannot easily be extracted from variations in disease severity and the effects of concomitant treatments. Clinical trials have thus become a preferred means of evaluating an ever increasing flow of innovative diagnostic and therapeutic maneuvers. The randomized, double-blind clinical trial is a powerful technique because of the efficiency and credibility associated with treatment comparisons involving randomized concurrent controls.

The modern era of randomized trials began in the early 1950s with the evaluation of streptomycin in patients with tuberculosis.[1] Since that time trial techniques and methods have continuously been refined.[2] In addition, the ethical aspects of these experiments in patients have been actively discussed.[3–7]

In what follows I argue that randomized trials are in fact the most scientifically sound and ethically correct means of evaluating new therapies. There is potential conflict between the roles of physician and physician-scientist, and for this reason society has created mechanisms to ensure that the interests of individual patients are served should they elect to participate in a clinical trial.[6]

Clinical Research

The history of medicine is richly endowed with therapies that were widely used and then shown to be ineffective or frankly toxic. Relatively recent examples of such therapeutic maneuvers include gastric freezing for peptic ulcer disease, radiation therapy for acne, MER-29 (triparanol) for cholesterol reduction, and thalidomide for sedation in pregnant women. The 19th century was even more gruesome, with purging and bloodletting. The reasons for this march of folly are many and include, perhaps most importantly, the lack of complete understanding of human biology and pathophysiology, the use of observational methods coupled with the failure to appreciate substantial variability between patients in their response to illness and to therapy, and the shared desire of physicians and their patients for cure or palliation.

Chance or bias can result in the selection of patients for innovative treatment who are either the least diseased or the most severely affected. Depending on the case mix, a treatment that has no effect can appear to be effective or toxic when historical controls are used. With the improvement in diagnostic accuracy and the understanding of disease that has occurred with the passage of time, today's patients are identified earlier in the natural history of their disease. Recently selected case series therefore often have patients who are less ill and an outcome that is considerably better than that of past case series, even without changes in treatment.

Randomization tends to produce treatment and control groups that are evenly balanced in both known and unrecognized prognostic factors, which permits a more accurate estimate of treatment effect in groups of patients assigned to experimental and standard therapies. A number of independent randomized trials with congruent results are powerful evidence indeed.

A physician's daily practice includes an array of preventive, diagnostic, and therapeutic maneuvers, some of which have been established by a plausible biologic mechanism and substantial evidence from randomized clinical trials (e.g., the use of beta blockers, thrombolytic therapy, and aspirin in patients with myocardial infarction).[8] It is unlikely that our distant descendants in medicine will discover that we late-20th-century physicians were wrong in these matters. However, new therapeutic maneuvers that have not undergone rigorous assessment may well turn out to be ineffective or toxic. Every therapy adopted by common consent on the basis of observational studies and plausible mechanism, but without the benefit of randomized studies, may be categorized by future physicians as useless or worse. Physicians are aware of the fragility of the evidence supporting many common therapies, and this is why properly performed randomized clinical trials have profound effects on medical practice. The scientific importance of randomized, controlled trials is in safeguarding current and future patients from our therapeutic passions. Most physicians recognize this fact.

Like any human activity, experimentation involving patients can be performed in an unethical and even criminal fashion. Nazi war crimes led to substantial efforts to curb abuse, beginning with the Nuremberg Code and the Helsinki Declaration and culminating in the promulgation of clearly articulated regulations in the United States and elsewhere.[4–6] There are abuses more subtle than those of the Gestapo and the SS. Involving patients in experiments that are poorly conceived and poorly executed is unethical. Patients who participate in such research may incur risk without the hope of contributing to a body of knowledge that will benefit them or others in the future. The regulations governing human experimentation are very important, as is continuing discussion and debate to improve the scientific and ethical aspects of this effort.

Several general features must be part of properly designed trials. The first is informed consent, which involves explicitly informing a potential participant of the goals of the research, its potential benefits and risks, the alternatives to participating, and the right to withdraw from the trial at any time. Whether informed consent is required in all trials has been debated.[9] I believe that patients must always be aware that they are part of an experiment. Second, a state of clinical equipoise must exist. Clinical equipoise means that on the basis of the available data, a community of competent physicians would be content to have their patients pursue any of the treatment strategies being tested in a randomized trial, since none of them have been clearly established as preferable.[7] The chief purpose of a data-monitoring committee is to stop the trial if the accumulating data destroy the state of clinical equipoise—that is, indicate efficacy or suggest toxicity. Finally, the trial must be designed as a critical test of the therapeutic alternatives being assessed. The question must be clearly articulated, with carefully defined measures of outcome; with realistic estimates of sample size, including probable event rates in the control group and a postulated and plausible reduction in the event rates in the treatment group; with Type I and II errors specified; and with subgroup hypotheses clearly stated if appropriate. The trial must have a good chance of settling an open question.[2]

Ethical Dimensions of Properly Constituted Trials

Experimentation in the clinic by means of randomized, controlled clinical trials has been periodically attacked as violating the covenant between doctor and patient.[10–12] Critics have charged that physicians engaged in clinical trials sacrifice the interests of the patient they ask to participate to the good of all similarly affected patients in the future. The argument is that physicians have a personal obligation to use their best judgment and recommend the "best" therapy, no matter how tentative or inconclusive the data on which that judgment is based. Physicians must play their hunches. According to this argument, randomized clinical trials may be useful in seeking the truth, but carefully designed, legitimate trials are unethical and perhaps even criminal because they prevent individual physicians from playing their hunches about individual patients. Therefore, it is argued, physicians should not participate in such trials.

It is surely unethical for physicians to engage knowingly in an activity that will result in inferior therapy for their patients. It is also important that the community of physicians be clear in distinguishing between established therapies and those that are promising but unproved. It is this gulf between proved therapies and possibly effective therapies (all the rest) that defines the ethical and unethical uses of randomized clinical trials. Proved therapies involve a consensus of the competent medical community that the data in hand justify using a treatment in a given disorder. It is this consensus that defines an ethical boundary. The physician-investigator who asks a patient to participate in a randomized, controlled trial represents this competent medical community in asserting that the community is unpersuaded by existing data that an innovative treatment is superior to standard therapy. Arguments that a physician who believes that such a treatment *might be* useful commits an unethical act by randomizing patients are simply wrong. Given the history of promising but discarded therapies, hunches about potential effectiveness are not the ideal currency of the patient–doctor interchange.

Lest readers conclude that modern hunches are more accurate than older ones, I have selected an example from the current cardiovascular literature that reveals the problems inherent in relying on hunches to the exclusion of carefully done experiments.

The Cardiac Arrhythmia Suppression Trial

Sudden death occurs in approximately 300,000 persons in the United States each year and is thus a problem worthy of our best efforts. In the vast majority of cases the mechanism is ventricular fibrillation superimposed on a scarred or ischemic myocardium. It had been observed that the ventricular extrasystoles seen on the ambulatory electrocardiographic recordings of survivors of myocardial infarction were independently and reproducibly associated with an increased incidence of subsequent mortality.[13, 14] It had been established that a variety of antiarrhythmic drugs can suppress ventricular extrasystoles. Accordingly, physicians had the hunch that suppressing ventricular extrasystoles in the survivors of myocardial infarction would reduce the incidence of ventricular fibrillation and sudden death.

The Cardiac Arrhythmia Suppression Trial (CAST) investigators decided to test this hypothesis in a randomized, controlled trial. They sought survivors of myocardial infarction who had frequent extrasystoles on electrocardiographic recordings. The trial design included a run-in period during which one of three active drugs was administered and its effect on extrasystoles noted. Those in whom arrhythmias were suppressed were randomly assigned to active drug or placebo. The trial had to be stopped prematurely because of an unacceptable incidence of sudden death in the treatment group.[15] During an average follow-up of 10 months, 56 of 730 patients (7.7 percent) assigned to active drug and 22 of 725 patient (3.0 percent) assigned to placebo died. Clinical equipoise was destroyed by this striking effect. It is quite unlikely that observational (nonrandomized) methods would have detected this presumably toxic effect.

The CAST trial was a major advance in the treatment of patients with coronary disease and ventricular arrhythmia. It clearly revealed that the hunches of many physicians were incorrect. The trial's results are applicable not only to future patients with coronary disease and ventricular arrhythmia but also to the patients who participated in the study. By randomizing, investigators ensured that half the participants received the better therapy—in this case placebo—and, contrary to intuition, most of them ultimately received the better therapy after the trial ended prematurely and drugs were withdrawn.

To summarize, randomized clinical trials are an important element in the spectrum of biomedical research. Not all questions can or should be addressed by this technique; feasibility, cost, and the relative importance of the issues to be addressed are weighed by investigators before they elect to proceed. Properly carried out, with informed consent, clinical equipoise, and a design adequate to answer the question posed, randomized clinical trials protect physicians and their patients from therapies that are ineffective or toxic.

Physicians and their patients must be clear about the vast gulf separating promising and proved therapies. The only reliable way to make this distinction in the face of incomplete information about pathophysiology and treatment mechanism is to experiment, and this will increasingly involve randomized trials. The alternative—a retreat to older methods—is unacceptable.

Physicians regularly apply therapies tested in groups of patients to an individual patient. The likelihood of success in an individual patient depends on the degree of certainty evident in the group and the scientific strength of the methods used. We owe patients involved in the assessment of new therapies the best that science and ethics can deliver. Today, for most unproved treatments, that is a properly performed randomized clinical trial.

Notes

1. Streptomycin in Tuberculosis Trials Committee, Medical Research Council. Streptomycin treatment of pulmonary tuberculosis: a Medical Research Council investigation. *BMJ* 1948: 2: 769–82.

2. Friedman LM, Furberg CD, DeMets DL. *Fundamentals of clinical trials.* Boston: John Wright/PSG, 1981.

3. Beecher HK. Ethics and clinical research. *N Engl J Med* 1966: 274: 1354–60.

4. Appendix II (The Nuremberg Code). In: Beauchamp TL, Childress JF. *Principles of biomedical ethics.* New York: Oxford University Press, 1979: 287–9.

5. Appendix II (The World Medical Association Declaration of Helsinki). In: Beauchamp TL, Childress JF. *Principles of biomedical ethics.* New York: Oxford University Press, 1979: 289–93.

6. The National Commission for the Protection of Human Subjects of Biomedical and Behavioral Research. The Belmont report: ethical principles and guidelines for the protection of human subjects of research. Washington, D.C.: Government Printing Office, 1978. (DHEW publication no. (05) 78-0012.)

7. Freedman B. Equipoise and the ethics of clinical research. *N Engl J Med* 1987: 317: 141–5.

8. Yusuf S, Wittes J, Friedman L. Overview of results of randomized clinical trials in heart disease. I. Treatments following myocardial infarction. *JAMA* 1988: 260: 2088–93.

9. Brahams D. Randomized trials and informed consent. *Lancet* 1988: 1033–4.

10. Burkhardt R, Kienle G. Controlled clinical trials and medical ethics. *Lancet* 1978: 2: 1356–9.

11. Marquis D. Leaving therapy to chance. *Hastings Cent Rep* 1983: 13(4): 40–7.

12. Gifford F. The conflict between randomized clinical trials and the therapeutic obligation. *J Med Philos* 1986: 1: 347–66.

13. Ruberman W, Weinblatt E, Goldberg JD, Frank CW, Shapiro S. Ventricular premature beats and mortality after myocardial infarction. *N Engl J Med* 1977: 297: 750–7.

14. Lown B. Sudden cardiac death: the major challenge confronting contemporary cardiology. *Am J Cardiol* 1979: 43: 313–28.

15. The Cardiac Arrhythmia Suppression Trial (CAST) investigators. Preliminary report: effect of encainide and flecainide on mortality in a randomized trial of arrhythmia suppression after myocardial infarction. *N Engl J Med* 1989: 321: 406–12.

SCENARIO

"You realize," Dr. Thorne said, "that you may not be in the group that receives medication. You may be in the placebo group for at least part of the time."

"Right," Ms. Ross said. "You're just going to give me some medicine."

"And do you understand the aims of the research?"

"You want to help me get better," Ms. Ross suggested hesitantly.

"We hope you get better, of course. But that's not what we're trying to accomplish here. We're trying to find out if this medication will help other people in your condition if we can treat them earlier than we were able to treat you."

"You want to help people," Ms. Ross said.

"That's right. But you do understand that we may not be helping you in this experiment?"

"But you're going to try?"

"Not exactly. I mean, we aren't going to try to harm you. But we aren't necessarily going to be giving you the preferred treatment for your complaint either. Do you know the difference between research and therapy?"

"Research is when you're trying to find something out. You're searching around."

"That's right. And we're asking you to be part of a research effort. As I told you, there are some risks. Besides the possibility of not getting treatment that you need, the drug may produce limited hepatic portal damage. We're not sure how much."

"I think I understand," Ms. Ross said.

"I'm sure you do," said Dr. Thorne. "I understand that you are freely volunteering to participate in this research."

"Yes, sir. Mrs. Woolerd, she told me if I volunteered I'd get a letter put in my file and I could get early release."

"Mrs. Woolerd told you the review board would take your volunteering into account when they considered whether you should be put on work-release."

"Yes, sir. And I'm awfully anxious to get out of here. I've got two children staying with my aunt, and I need to get out of this place as quick as I can."

"I understand. We can't promise you release, of course. But your participation will look good on your record. Now I have some papers here I want you to sign."

1. Discuss some of the difficulties involved in explaining research procedures to nonexperts and determining whether they are aware of the nature and risks of their participation.

2. What reasons are there for believing that Ms. Ross does not understand what she is volunteering for?

3. Discuss the problems involved in securing free and voluntary consent from a person involuntarily confined to an institution (a prisoner, for example).

4. Is it possible to obtain genuine consent from patients in Phase I cancer trials even if they are not in prison?

IV. TRANSPLANTATION

The problems associated with organ transplantation were presented in Chapter 13. One of the most important questions is deciding how these organs are allocated. Who should receive an organ when many are waiting?

CASE PRESENTATION
Transplants for the Mentally Impaired

Sandra Jensen was born with a deformed heart, but it wasn't until she was thirty-five that it began to make her so sick that she needed a heart–lung transplant to extend her life. She was young and otherwise healthy, but transplant centers at both Stanford University and the University of California, San Diego rejected her as a candidate.

Sandra Jensen also had Down syndrome, and the transplanters doubted she had sufficient intelligence to care for herself after the surgery. She would have to follow the complicated routine of taking doses of dozens of medications daily that is the lot of every transplant recipient. If she failed to adhere to the postoperative requirements, she would die, and the organs that might have saved the life of one or two other people would be wasted.

William Bronston, a state rehabilitation administrator and a friend of Jensen, became her advocate. He pointed out that she had demonstrated a high level of intellectual functioning. She was a high school graduate who worked with people with Down syndrome, and she had lived on her own for several years. She spoke for the disabled in California and attended the Washington signing by George H. Bush of the Americans with Disabilities Act in 1990.

Thanks to strong lobbying by Bronston and the threat of adverse publicity, Stanford reversed its decision. On January 23, 1996, in a five-hour operation, Ms. Jensen became the first seriously mentally retarded person in the United States to receive a major organ transplant.

More than a year later, on May 4, 1997, after her health began deteriorating, Ms. Jensen entered Sutter General Hospital in San Francisco. She had been admitted to the hospital several times before because of her reaction to the immunosuppressive drug. But this time was the last, and she died there on May 25, 1997. "Every day was always precious and lived well by her," her friend William Bronston said.

Prompted by Ms. Jensen's struggle to be accepted for a transplant, the California Assembly passed a bill to prohibit transplant centers from discriminating against impaired people needing a transplant.

READING
The Prostitute, the Playboy, and the Poet: Rationing Schemes for Organ Transplantation

George J. Annas

George Annas takes a position on transplant selection that introduces a modification of the first-come, first-served principle. He reviews four approaches to rationing scarce medical resources—market, selection committee, lottery, and customary—and finds each has disadvantages so serious as to make them all unacceptable. An acceptable approach, he suggests, is one that combines efficiency, fairness, and a respect for the value of life. Because candidates should both want a transplant and be able to derive significant benefits from one, the first phase of selection should involve a screening process that is based exclusively on medical criteria that are objective and as free as possible of judgments about social worth.

Since selection might still have to be made from this pool of candidates, it might be done by social-worth criteria or by lottery. However, social-worth criteria seem arbitrary, and a lottery would be unfair to those who are in more immediate need of a transplant—ones who might die quickly without it. After reviewing the relevant considerations, a committee operating at this stage might allow those in immediate need of a transplant to be moved to the head of a waiting list. To those not in immediate need, organs would be distributed in a first-come, first-served fashion. Although absolute equality is not embodied in this process, the procedure is sufficiently flexible to recognize that some may have needs that are greater (more immediate) than others.

In the public debate about the availability of heart and liver transplants, the issue of rationing on a massive scale has been credibly raised for the first time in United States medical care. In an era of scarce resources, the eventual arrival of such a discussion was, of course, inevitable.[1] Unless we decide to ban heart and liver transplantation, or make them available to everyone, some rationing scheme must be used to choose among potential transplant candidates. The debate has existed throughout the history of medical ethics. Traditionally it has been stated as a choice between saving one of two patients, both of whom require the immediate assistance of the only available physician to survive.

National attention was focused on decisions regarding the rationing of kidney dialysis machines when they were first used on a limited basis in the late 1960s. As one commentator described the debate within the medical profession:

Shall machines or organs go to the sickest, or to the ones with most promise of recovery; on a first-come, first-served basis; to the most 'valuable' patient (based on wealth, education, position, what?); to the one with the most dependents; to women and children first; to those who can pay; to whom? Or should lots be cast, impersonally and uncritically?[2]

In Seattle, Washington, an anonymous screening committee was set up to pick who among competing candidates would receive the life-saving technology. One lay member of the screening committee is quoted as saying:

The choices were hard ... I remember voting against a young woman who was a known prostitute. I found I couldn't vote for her, rather than another candidate, a young wife and mother. I also voted against a young man who, until he learned he had renal failure, had been a ne'er do-well, a real playboy. He promised he would reform his character, go back to school, and so on, if only he were selected for treatment. But I felt I'd lived long enough to know that a person like that won't really do what he was promising at the time.[3]

When the biases and selection criteria of the committee were made public, there was a general negative reaction against this type of arbitrary device. Two experts reacted to the "numbing accounts of how close to the surface lie the prejudices and mindless cliches that pollute the committee's deliberations," by concluding that the committee was "measuring persons in accordance with its own middle-class values." The committee process, they noted, ruled out "creative nonconformists" and made the Pacific Northwest "no place for a Henry David Thoreau with bad kidneys."[4]

To avoid having to make such explicit, arbitrary, "social worth" determinations, the Congress, in 1972, enacted legislation that provided federal funds for virtually all kidney dialysis and kidney transplantation procedures in the United States.[5] This decision, however, simply served to postpone the time when identical decisions will have to be made about candidates for heart and liver transplantation in a society that does not provide sufficient financial and medical resources to provide all "suitable" candidates with the operations.

There are four major approaches to rationing scarce medical resources: the market approach; the selection committee approach; the lottery approach; and the "customary" approach.[1]

The Market Approach

The market approach would provide an organ to everyone who could pay for it with their own funds or private insurance. It puts a very high value on individual rights, and a very low value on equality and fairness. It has properly been criticized on a number of bases, including that the transplant technologies have been developed and are supported with public funds, that medical resources used for transplantation will not be available for higher priority care, and that financial success alone is an insufficient justification for demanding a medical procedure. Most telling is its complete lack of concern for fairness and equity.[6]

A "bake sale" or charity approach that requires the less financially fortunate to make public appeals for funding is demeaning to the individuals involved, and to society as a whole. Rationing by financial ability says we do not believe in equality, but believe that a price can and should be placed on human life and that it should be paid by the individual whose life is at stake. Neither belief is tolerable in a society in which income is inequitably distributed.

The Committee Selection Process

The Seattle Selection Committee is a model of the committee process. Ethics Committees set up in some hospitals to decide whether or not certain handicapped newborn infants should be given medical care may represent another.[7] These committees have developed because it was seen as unworkable or unwise to explicitly set forth the criteria on which selection decisions would be made. But only two results are possible, as Professor Guido Calabresi has pointed out: either a pattern of decision-making will develop or it will not. If a pattern does develop (e.g., in Seattle, the imposition of middle-class values), then it can be articulated and those decision "rules" codified and used directly, without resort

to the committee. If a pattern does not develop, the committee is vulnerable to the charge that it is acting arbitrarily, or dishonestly, and therefore cannot be permitted to continue to make such important decisions.[1]

In the end, public designation of a committee to make selection decisions on vague criteria will fail because it too closely involves the state and all members of society in explicitly preferring specific individuals over others, and in devaluing the interests those others have in living. It thus directly undermines, as surely as the market system does, society's view of equality and the value of human life.

The Lottery Approach

The lottery approach is the ultimate equalizer which puts equality ahead of every other value. This makes it extremely attractive, since all comers have an equal chance at selection regardless of race, color, creed, or financial status. On the other hand, it offends our notions of efficiency and fairness since it makes no distinctions among such things as the strength of the desires of the candidates, their potential survival, and their quality of life. In this sense it is a mindless method of trying to solve society's dilemma which is caused by its unwillingness or inability to spend enough resources to make a lottery unnecessary. By making this macro spending decision evident to all, it also undermines society's view of the pricelessness of human life. A first-come, first-served system is a type of natural lottery since referral to a transplant program is generally random in time. Nonetheless, higher income groups have quicker access to referral networks and thus have an inherent advantage over the poor in a strict first-come, first-served system.[8,9]

The Customary Approach

Society has traditionally attempted to avoid explicitly recognizing that we are making a choice not to save individuals lives because it is too expensive to do so. As long as such decisions are not explicitly acknowledged, they can be tolerated by society. For example, until recently there was said to be a general understanding among general practitioners in Britain that individuals over age 55 suffering from end-stage kidney disease not be referred for dialysis or transplant. In 1984, however, this unwritten practice became highly publicized, with figures that showed a rate of new cases of end-stage kidney disease treated in Britain at 40 per million (versus the U.S. figure of 80 per million) resulting in 1500–3000 "unnecessary deaths" annually.[10] This has, predictably, led to movements to enlarge the National Health Service budget to expand dialysis services to meet this need, a more socially acceptable solution than permitting the now publicly recognized situation to continue.

In the U.S., the customary approach permits individual physicians to select their patients on the basis of medical criteria or clinical suitability. This, however, contains many hidden social worth criteria. For example, one criterion, common in the transplant literature, requires an individual to have sufficient family support for successful aftercare. This discriminates against individuals without families and those who have become alienated from their families. The criterion may be relevant, but it is hardly medical.

Similar observations can be made about medical criteria that include IQ, mental illness, criminal records, employment, indigence, alcoholism, drug addiction, or geographical location. Age is perhaps more difficult, since it may be impressionistically related to outcome. But it is not medically logical to assume that an individual who is 49 years old is necessarily a better medical candidate for a transplant than one who is 50 years old. Unless specific examination of the characteristics of older persons that make them less desirable candidates is undertaken, such a cut off is arbitrary, and thus devalues the lives of older citizens. The same can be said of blanket exclusions of alcoholics and drug addicts.

In short, the customary approach has one great advantage for society and one great disadvantage: it gives us the illusion that we do not have to make choices; but the cost is mass deception, and when this deception is uncovered, we must deal with it either by universal entitlement or by choosing another method of patient selection.

A Combination of Approaches

A socially acceptable approach must be fair, efficient, and reflective of important social values. The most important values at stake in organ transplantation are fairness itself, equity in the sense of equality, and the value of life. To promote efficiency, it is important that no one receive a transplant unless they want one and are likely to obtain significant benefit from it in the sense of years of life at a reasonable level of functioning.

Accordingly, it is appropriate for there to be an initial screening process that is based *exclusively* on medical criteria designed to measure the probability of a successful transplant, i.e., one in which the patient survives for at least a number of years and is rehabilitated. There is room in medical criteria for social worth judgments, but there is probably no way to avoid this completely. For example, it has been noted that "in many respects social and medical criteria are inextricably intertwined" and that therefore medical criteria might "exclude the poor and disadvantaged because health and socioeconomic status are highly interdependent."[11] Roger Evans gives an example. In the End Stage Renal Disease Program, "those of lower socioeconomic status are likely to have multiple comorbid health conditions such as diabetes, hepatitis, and hypertension" making them both less desirable candidates and more expensive to treat.[11]

To prevent the gulf between the haves and have nots from widening, we must make every reasonable attempt to develop medical criteria that are objective and independent of social worth categories. One minimal way to approach this is to require that medical screening be reviewed and approved by an ethics committee with significant public

representation, filed with a public agency, and made readily available to the public for comment. In the event that more than one hospital in a state or region is offering a particular transplant service, it would be most fair and efficient for the individual hospitals to perform the initial medical screening themselves (based on the uniform, objective criteria), but to have all subsequent nonmedical selection done by a method approved by a single selection committee composed of representatives of all hospitals engaged in a particular transplant procedure, as well as significant representation of the public at large.

As this implies, after the medical screening is performed, there may be more acceptable candidates in the "pool" than there are organs or surgical teams to go around. Selection among waiting candidates will then be necessary. This situation occurs now in kidney transplantation, but since the organ matching is much more sophisticated than in hearts and livers (permitting much more precise matching of organ and recipient), and since dialysis permits individuals to wait almost indefinitely for an organ without risking death, the situations are not close enough to permit use of the same matching criteria. On the other hand, to the extent that organs are specifically tissue- and size-matched and fairly distributed to the best matched candidate, the organ distribution system itself will resemble a natural lottery.

When a pool of acceptable candidates is developed, a decision about who gets the next available, suitable organ must be made. We must choose between using a conscious, value-laden, social worth selection criterion (including a committee to make the actual choice), or some type of random device. In view of the unacceptability and arbitrariness of social worth criteria being applied, implicitly or explicitly, by committee, this method is neither viable nor proper. On the other hand, strict adherence to a lottery might create a situation where an individual who has only a one-in-four chance of living five years with a transplant (but who could survive another six months without one) would get an organ before an individual who could survive as long or longer, but who will die within days or hours if he or she is not immediately transplanted. Accordingly, the reasonable approach seems to be to allocate organs on a first-come, first-served basis to members of the pool but permit individuals to "jump" the queue if the second level selection committee believes they are in immediate danger of death (but still have a reasonable prospect for long-term survival with a transplant) and the person who would otherwise get the organ can survive long enough to be reasonably assured that he or she will be able to get another organ.

The first-come, first-served method of basic selection (after a medical screen) seems the preferred method because it most closely approximates the randomness of a straight lottery without the obviousness of making equity the only promoted value. Some unfairness is introduced by the fact that the more wealthy and medically astute will likely get into the pool first, and thus be ahead in line, but this advantage should decrease sharply as public awareness of the system grows. The possibility of unfairness is also inherent in permitting individuals to jump the queue, but some flexibility needs to be retained in the system to permit it to respond to reasonable contingencies.

We will have to face the fact that should the resources devoted to organ transplantation be limited (as they are now and are likely to be in the future), at some point it is likely that significant numbers of individuals will die in the pool waiting for a transplant. Three things can be done to avoid this: (1) medical criteria can be made stricter, perhaps by adding a more rigorous notion of "quality" of life to longevity and prospects for rehabilitation; (2) resources devoted to transplantation and organ procurement can be increased; or (3) individuals can be persuaded not to attempt to join the pool.

Of these three options, only the third has the promise of both conserving resources and promoting autonomy. While most persons medically eligible for a transplant would probably want one, some would not—at least if they understood all that was involved, including the need for a lifetime commitment to daily immunosuppression medications, and periodic medical monitoring for rejection symptoms. Accordingly, it makes public policy sense to publicize the risks and side effects of transplantation, and to require careful explanations of the procedure be given to prospective patients before they undergo medical screening. It is likely that by the time patients come to the transplant center they have made up their minds and would do almost anything to get the transplant. Nonetheless, if there are patients who, when confronted with all the facts, would voluntarily elect not to proceed, we enhance both their own freedom and the efficiency and cost-effectiveness of the transplantation system by screening them out as early as possible.

Conclusion

Choices among patients that seem to condemn some to death and give others an opportunity to survive will always be tragic. Society has developed a number of mechanisms to make such decisions more acceptable by camouflaging them. In an era of scarce resources and conscious cost containment, such mechanisms will become public, and they will be usable only if they are fair and efficient. If they are not so perceived, we will shift from one mechanism to another in an effort to continue the illusion that tragic choices really don't have to be made, and that we can simultaneously move toward equity of access, quality of services, and cost containment without any challenges to our values. Along with the prostitute, the playboy, and the poet, we all need to be involved in the development of an access model to extreme and expensive medical technologies with which we can live.

Notes

1. Calabresi G, Bobbitt P: *Tragic Choices.* New York: Norton, 1978.
2. Fletcher J: Our shameful waste of human tissue. In: Cutler DR (ed): *The Religious Situation.* Boston: Beacon Press, 1969; 223–252.
3. Quoted in Fox R, Swazey J: *The Courage to Fail.* Chicago: Univ of Chicago Press, 1974; 232.
4. Sanders D, Dukeheimer J: Medical advance and legal lag: haemodialysis and kidney transplantation. *UCLA L Rev* 1968; 15: 357.
5. Rettig RA: The policy debate on patient care financing for victims of end stage renal disease. *Law & Contemporary Problems* 1976; 40: 196.
6. President's Commission for the Study of Ethical Problems in Medicine: Securing Access to Health Care. US Govt Printing Office, 1983; 25.
7. Annas GJ: Ethics committees on neonatal care: substantive protection or procedural diversion? *Am J Public Health* 1984; 74: 843–845.
8. Bayer R: Justice and health care in an era of cost containment: allocating scarce medical resources. *Soc Responsibility* 1984; 9: 37–52.
9. Annas GJ: Allocation of artificial hearts in the year 2002: *Minerva v National Health Agency. Am J Law Med* 1977; 3: 59–76.
10. Commentary: UK's poor record in treatment of renal failure. *Lancet,* July 7, 1984; 53.
11. Evans R: Health care technology and the inevitability of resource allocation and rationing decisions, Part 11. *JAMA* 1983; 249: 2208, 2217.

SCENARIO

The microsurgical team at Benton Public Hospital consisted of twenty-three people. Five were surgeons, three were anesthesiologists, three were internists, two were radiologists, and the remaining members were various sorts of nurses and technicians.

Early Tuesday afternoon on a date late in March, the members of the team that had to be sterile were scrubbing while the others were preparing to start operating on Mr. Hammond Cox. Mr. Cox was a fiftynine- year-old unmarried African American who worked as a janitor in a large apartment building. While performing his duties Mr. Cox had caught his hand in the mechanism of a commercial trash compactor. The bones of his wrist had been crushed and blood vessels severed.

The head of the team, Dr. Herbert Lagorio, believed it was possible to restore at least partial functioning to Mr. Cox's hand. Otherwise, the hand would have to be amputated.

Mr. Cox had been drunk when it happened. When the police ambulance brought him to the emergency room, he was still so drunk that a decision was made to delay surgery for almost an hour to give him a chance to burn up some of the alcohol he had consumed. As it was, administering anesthesia to Mr. Cox would incur a greater-than-average risk. Furthermore, blood tests had shown that Mr. Cox already suffered from some degree of liver damage. In both short- and long-range terms, Mr. Cox was not a terribly good surgical risk.

Dr. Lagorio was already scrubbed when Dr. Carol Levine, a resident in emergency medicine, had him paged.

"This had better be important," he told her. "I've got a patient prepped and waiting."

"I know," Dr. Levine said. "But they just brought in a thirty-five-year-old white female with a totally severed right hand. She's a biology professor at Columbia and was working late in her lab when some maniac looking for drugs came in and attacked her with a cleaver."

"What shape is the hand in?"

"Excellent. The campus cops were there within minutes, and there was ice in the lab. One of the cops had the good sense to put the hand in a plastic bag and bring it with her."

"Is she in good general health?"

"It seems excellent," Dr. Levine said.

"This is a real problem."

"You can't do two cases at once?"

"No way. We need everybody we've got to do one."

"How about sending her someplace else?"

"No place else is set up to do what has to be done."

"So what are you going to do?"

1. Does a first-come, first-served criterion require that Mr. Cox receive the surgery?

2. Can the chance of a successful outcome in each case be used as a criterion without violating the notion that all people are of equal inherent worth?

3. Should the fact that Mr. Cox's injury is the consequence of his own negligence be considered in determining to whom Dr. Lagorio ought to devote his attention?

4. In your view, who should have the potential benefits of the surgery? Give reasons to support your view.

V. EUGENICS

In Chapter 15, we examined the history of genetics and the eugenics movement. We found that the study of human genetics wasn't always dealt with ethically. How does this historical view of eugenics fit with today's view?

READING
Eugenics

Like other organisms, we are the products of millions of years of evolutionary development. This process has taken place through the operation of natural selection on randomly produced genetic mutations. Individual organisms are successful in an evolutionary sense when they contribute a number of genes to the gene pool of their species proportionately greater than the number contributed by others.

Most often, this means that the evolutionarily successful individuals are those with the largest number of offspring. These are the individuals favored by natural selection. That is, they possess the genes for certain properties that are favored by existing environmental factors. (This favoring of properties is natural selection.) The genes of "favored" individuals will thus occur with greater frequency than the genes of others in the next generation. If the same environmental factors continue to operate, these genes will spread through the entire population.

Thanks to Darwin and the biologists who have come after him, we now have a sound understanding of the evolutionary process and the mechanisms by which it operates. This understanding puts us in a position to intervene in

evolution. We no longer have to consider ourselves subject to the blind working of natural selection, and if we wish, we can modify the course of human evolution. As the evolutionary biologist Theodosius Dobzhansky expressed the point: "Evolution need no longer be a destiny imposed from without; it may conceivably be controlled by man, in accordance with his wisdom and values."

Those who advocate eugenics accept exactly this point of view. They favor social policies and practices that, over time, offer the possibility of increasing the number of genes in the human population responsible for producing or improving intelligence, beauty, musical ability, and other traits we value.

The aim of increasing the number of favorable genes in the human population is called *positive eugenics*. By contrast, *negative eugenics* aims at decreasing the number of undesirable or harmful genes. Those who advocate negative eugenics are most interested in eliminating or reducing from the population genes responsible for various kinds of genetic diseases.

Both positive and negative eugenics require instituting some sort of control over human reproduction. Several kinds of policies and procedures have been advocated, and we will discuss a few of the possibilities.

Negative and Positive Eugenics

The discussion of genetic screening, counseling, prenatal genetic diagnosis, and embryo selection makes it unnecessary to repeat here information about the powers we possess for predicting and diagnosing genetic diseases. It is enough to recall that, given information about the genetic makeup and background of potential parents, a large number of genetic diseases can be predicted with a certain degree of probability as likely to occur in a child of such parents. Or the presence of the genes can be determined by genetic analysis of the chromosomes. This is true of such diseases as PKU, sickle cell, hemophilia, Huntington's disease, Tay–Sachs, and muscular dystrophy.

When genetic information isn't adequate for a reliable prediction or direct determination, information about the developing fetus can often be obtained by employing one of several procedures of prenatal diagnosis. Even when information is adequate for a reliable prediction, whether the fetus has a certain disease can be determined by prenatal testing. Thus, in addition to the genetic disorders named previously, prenatal tests can be performed for such developmental defects as neural tube anomalies and Down syndrome. Also, other tests can be performed on ova, sperm, or embryos.

A proponent of negative eugenics might advocate that a screening process for all or some currently detectable genetic diseases or dispositions (or developmental impairments) be required by law. When the probability of the occurrence of a disease is high (whatever figure that might be taken to be), then the potential parents might be encouraged to have no children. Indeed, the law might require that such a couple either abstain from having children or rely on embryo selection and prescribe a penalty for going against the decision of the screening board.

If those carrying the genes for some genetic diseases could be prevented from having children, over time the incidence of the diseases would decrease. In cases when the disease is the result of a dominant gene (as it is in Huntington's disease), the disease would eventually disappear. (It would appear again with new mutations, however.)

When the disease is of the sort that can be detected only after a child is conceived, if the results of a prenatal diagnosis show the developing fetus has a heritable disease, an abortion might be encouraged. Or a couple identified as at risk might be encouraged to seek artificial insemination and embryo testing and transfer.

Short of a law requiring abortion, a variety of social policies might be adopted to make abortion or embryo selection an attractive option. (For example, the cost of an abortion might be paid for by government funds or women choosing abortion might be financially rewarded. Or the costs of embryo selection might be paid for under a federal program.) The aborting of a fetus found to have a transmissible genetic disease would not only prevent the birth of an impaired infant, it would also eliminate a potential carrier of the genes responsible for the disease.

Similarly, the sterilization of people identified as having genes responsible for certain kinds of physical or mental impairments would prevent them from passing on these defective genes. In this way, the number of such genes in the population would be proportionately reduced.

Currently, no state or federal laws make it a crime for couples who are genetically a bad risk to have children. Yet a tendency toward more genetic regulation may be developing. Screening newborns for certain genetic diseases that respond well to early treatment is an established practice. Also, genetic testing programs are frequently offered in communities to encourage people to seek information about particular diseases.

At present, genetic testing (for adults) and counseling are voluntary. They aim at providing information and then leave reproductive decisions up to the individuals concerned. Most often, they are directed toward the immediate goal of decreasing the number of children suffering from birth defects and genetic diseases. Yet genetic testing and counseling might also be viewed as a part of negative eugenics. To the extent they discourage the birth of children carrying deleterious genes, they also discourage the spread of those genes in the human population.

Obviously, genetic testing and genetic counseling programs might also be used to promote positive eugenics. Individuals possessing genes for traits society values might be encouraged to have large numbers of children. In this way, genes for those traits would increase in relative frequency in the population.

No programs of positive eugenics currently operate in the United States. It is easy to imagine however, how a variety of social and economic incentives (such as government bonuses) might be introduced as part of a plan to promote the spread of certain genes by rewarding favored groups of people for having children.

Use of Desirable Germ Cells

Developments in reproductive technology have opened up possibilities once considered so remote as to be the stuff of science fiction. Artificial insemination by the use of frozen sperm is already commonplace. So too is the use of donor eggs and embryos. While some of the embryos may be donated by couples who don't need or want them, some are produced in infertility clinics by combining sperm from commercial sperm banks with donor ova. The developing embryos can be divided into several genetically identical embryos, and before long it may be possible to clone a human being from a single body cell.

Those wishing to have a child now have the option of selecting donor eggs or sperm from individuals with traits considered desirable. Alternatively, they may select a frozen embryo on the basis of descriptions of the gamete contributors. They may also turn to physicians who may offer them embryos they've created from sperm and eggs obtained from what they judge to be outstanding traits.

We have available to us right now the means to practice both negative and positive eugenics at the level of both the individual and the society. If we wished, we could encourage groups of individuals to avoid having their own biological children and, instead, make use of the "superior" sperm, ova, and embryos currently offered at sperm banks and infertility centers. In this way, we could increase the number of genes for desirable traits in the population. (See Chapter 6.)

Ethical Difficulties with Eugenics

Critics have been quick to point out that the proposals mentioned suffer from serious drawbacks. First, negative eugenics isn't likely to make much of a change in the species as a whole. Most hereditary diseases are genetically recessive and so occur only when both parents possess the same defective gene. Even though a particular couple might be counseled (or required) not to have children, the gene will still be widespread in the population among people we would consider wholly normal. For a similar reason, sterilization and even embryo selection would have few long-range effects.

Also, the uncomfortable fact is that geneticists have estimated that, on the average, everyone carries recessive genes for five genetic defects or diseases. Genetic counseling and the use of the techniques of assisted reproduction may help individuals, but negative eugenics doesn't promise much for the population as a whole.

Positive eugenics can promise little more. It's difficult to imagine we would all agree on what traits we'd like to see increased in the human species. But even if we could, it's not clear we'd be able to increase them in any simple way.

For one thing, we have little understanding of the genetic basis of traits such as "intelligence," "honesty," "musical ability," "beauty" and so on. It's clear, however, there isn't just a single gene for them, and the chances are they are the result of a complicated interplay between genetic endowment and social and environmental factors. Consequently, the task of increasing their frequency is quite different from that of, say, increasing the frequency of short-horned cattle. Furthermore, desirable traits may be accompanied by less desirable ones, and we may not be able to increase the first without also increasing the second.

Quite apart from biological objections, eugenics also raises questions of a moral kind. Have we indeed become the "business manager of evolution," as Julian Huxley once claimed? If so, do we have a responsibility to future generations to improve the human race? Would this responsibility justify requiring genetic screening and testing? Would it justify establishing a program of positive eugenics? Affirmative answers to these questions may generate conflicts with notions of individual dignity and self-determination.

Of the ethical theories we have discussed, it seems likely that only utilitarianism might be construed as favoring a program of positive eugenics. The possibility of increasing the frequency of desirable traits in the human species might, in terms of the principle of utility, justify placing restrictions on reproduction. Yet the goal of an improved society or human race might be regarded as too distant and uncertain to warrant the imposition of restrictions that would increase current human unhappiness.

As far as negative eugenics is concerned, the principle of utility could be appealed to in order to justify social policies that would discourage or prohibit parents who are carriers of the genes for serious diseases from having children. The aim here need not be the remote one of improving the human population but the more immediate one of preventing the increase in sorrows and pain that would be caused by an impaired child.

Natural law doctrines of Roman Catholicism forbid abortion, sterilization, and embryo selection. Thus, these means of practicing negative eugenics are ruled out. Also, the natural law view that reproduction is a natural function of sexual intercourse seems, at least prima facie, to rule out negative eugenics as a deliberate policy altogether. It could be argued, however, that voluntary abstinence from sexual intercourse or some other acceptable form of birth control would be a legitimate means of practicing negative eugenics.

Ross's prima facie duty of causing no harm might be invoked to justify negative eugenics. If there is good reason to believe a child is going to suffer from a genetic disease, we may have a duty to prevent the child from being born. Similarly, Rawls's theory might permit a policy that would require the practice of some form of negative eugenics for the benefit of its immediate effects of preventing suffering and sparing all the cost of supporting those with genetic diseases.

It is difficult to determine what sort of answer to the question of negative eugenics might be offered in terms of Kant's ethical principles. Laws regulating conception or forced abortion or sterilization might be considered to violate the dignity and autonomy of individuals. Yet moral agents as rational decision makers require information on which to base their decisions. Thus, programs of genetic screening and counseling might be considered to be legitimate.

SCENARIO

In 1983 a group of Orthodox Jews in New York and Israel initiated a screening program with the aim of eliminating from their community diseases transmitted as recessive genes. The group called itself Dor Yeshorim, "the generation of the righteous."

Because Orthodox Jews do not approve of abortion in most instances, the program does not employ prenatal testing. Instead, high school students are given a blood test to determine if they carry the genes for Tay–Sachs, cystic fibrosis, or Gaucher's disease. Each student is given a six-digit identification number, and if two students consider dating, they are encouraged to call a hotline. They are told either that they are "compatible" or that they each carry a recessive gene for one of the three diseases. Couples who are carriers are offered genetic counseling.

During 1993, 8000 people were tested, and eighty-seven couples who were considering marriage decided against it, after they learned that they were both carriers of recessive genes. The test costs $25, and the program is supported in part by funds from the Department of Health and Human Services. Some view the Dor Yeshorim program as a model that might be followed by other groups or by society in general.

The tests were initially only for Tay–Sachs, but over time the other two diseases were added. Current plans are to continue to add tests for even more diseases. However, some critics regard it as a mistake to have moved from testing for almost invariably lethal, untreatable diseases like Tay–Sachs to testing for cystic fibrosis. Individuals may feel pressured into being tested, and those who are carriers of one or more disease-predisposing genes may become unmarriageable social outcasts. Considering that genes for most diseases manifest themselves in various degrees of severity, many individuals may suffer social rejection for inadequate reasons. For example, Gaucher's disease, which involves an enzyme defect producing anemia and an enlarged liver and spleen, manifests itself only after age forty-five in half the diagnosed cases. Further, although the disease may be fatal, it often is not, and the symptoms can be treated.

1. Is the Dor Yeshorim screening program a form of eugenics? If so, does this make it unacceptable?
2. Is the program a good model for a national screening program? If not, why not?
3. Is it reasonable to screen for nonlethal genetic diseases?
4. What are the dangers inherent in any screening program?

Glossary

ABO blood types Three alleles of the isoagglutinin gene on chromosome 9 encode antigens on the surface of cells that determine the A, B, AB, and O blood types.

Acute myeloblastic leukemia (AML) A cancer of the blood in which too many immature white blood cells accumulate in the blood and bone marrow.

Adenine One of two nitrogen-containing purine bases found in nucleic acids, along with guanine.

Adenosine deaminase An enzyme encoded by a gene on chromosome 20. Mutation in this gene results in a disorder of the immune system called severe combined immunodeficiency (SCID).

Adult stem cell Undifferentiated cells found throughout the body that have the capacity to form a limited number of cell types to repair and renew adult tissues and organs.

Albinism A genetic condition characterized by the lack of melanin in the skin, hair, and eyes.

Allele One of the possible alternative forms of a gene, usually distinguished from other alleles by its phenotypic effects.

Allele frequency The frequency with which alleles of a given gene are present in a population.

Allelic expansion Increase in gene size caused by an increase in the number of trinucleotide sequences.

Allergens Antigens that provoke an inappropriate immune response.

Alpha globin One of the oxygen-carrying proteins found in embryonic, fetal, and adult hemoglobin. Each hemoglobin molecule is composed of two molecules of alpha globin and two molecules of beta globin.

Amicus curiae A legal term referring to a legal brief written by a group or person not involved in the case that is presented to a court of law with information on that case.

Amino acid One of the twenty monomers that are linked together to form proteins. Each amino acid carries two chemical groups: an amino group and a carboxylic acid group.

Amniocentesis A method of sampling the fluid surrounding the developing fetus by inserting a hollow needle and withdrawing suspended fetal cells and fluid; used in diagnosing fetal genetic and developmental disorders; usually performed in the sixteenth week of pregnancy.

Amniotic fluid A fluid composed mostly of fetal urine that surrounds and protects the fetus during its development. The fluid and cells in the fluid shed from the fetus are used in prenatal genetic diagnosis.

Anaphase I A stage in meiosis during which members of a homologous chromosome pair begin to separate from each other, creating haploid daughter cells.

Anaphase II A stage in meiosis where centromeres split and the daughter chromosomes begin to separate.

Anaphylactic shock A severe allergic response in which histamine is released into the circulatory system, causing rapid distress to the circulatory and respiratory system. Can be fatal.

Androgen receptor gene A gene on the X chromosome that when mutant, causes XY individuals to develop into phenotypic females.

Androgens Male sex hormones and related chemicals with male sex hormone activity.

Anencephaly A developmental disorder in which parts of the brain (cerebrum and cerebellum) do not form. Most affected fetuses die before birth; those born with anencephaly survive for only a few days.

Aneuploidy A chromosomal number that is not an exact multiple of the haploid set.

Animal models Organisms with biology, physiology, or genetic similarity to humans. Used in experiments to study human disorders and to develop tests and treatments.

Antibody A class of proteins produced by B cells that bind to foreign molecules (antigens) and inactivate them.

Anticodon A group of three nucleotides in a tRNA molecule that pairs with a complementary sequence (known as a codon) in an mRNA molecule.

Antigens Molecules carried or produced by microorganisms that initiate antibody production.

Arrhythmias Abnormal heart beats that can affect either the atria or the ventricles.

Artificial uterus A device used to grow an embryo or fetus to term outside the body of a female.

Ashkenazi Jews A population of Jews originating in Germany in about the tenth century, which migrated eastward to Poland, Hungary, and other areas of Eastern Europe and Russia. Intermarriage within the community has formed a clearly defined gene pool making this ethnic group a unique genetic subgroup.

Aspermia Inability to discharge semen at orgasm. Not to be confused with azoospermia, the absence of sperm in the semen.

Assemblers Software used to assemble long strings of DNA sequence data from shorter fragments and analyze the sequence for protein-coding sequences, mutations, etc.

Assumed consent A legal term meaning that consent for medical treatment or removal of organs for transplant is assumed without the written or oral consent of the patient.

Autoimmune disorder A disorder in which someone's immune system makes antibodies that attack cells and tissues of the body.

Autosomal dominant In humans, a pattern of inheritance in which a gene located on chromosomes 1–22 confers a distinct phenotype in the heterozygous and homozygous state.

Autosomal recessive In humans, a pattern of inheritance in which a gene located on chromosomes 1–22 confers a distinct phenotype only in the homozygous state.

Autosomes The chromosomes other than the sex-determining chromosomes. In humans chromosomes 1–22 are autosomes, and the X and Y chromosomes are sex chromosomes.

B cell A white blood cell that is part of the immune system. B cells produce antibodies.

Barr body A densely staining mass in the somatic nuclei of mammalian females. An inactivated X chromosome.

Behavioral genetics A branch of genetics that studies the genetic and environmental factors and their interactions that control or influence various forms of behavior.

Bell curve Also called a bell-shaped curve, or Gaussian distribution. This is the curve that results from a normal distribution of a data set. The title of a controversial book that linked IQ levels with race.

Benign In medicine, something that is not dangerous to health. Benign tumors for example, grow, but do not spread to other parts of the body.

Beta globin One of the oxygen-carrying proteins found in embryonic, fetal, and adult hemoglobin. Each hemoglobin molecule is composed of two molecules of alpha globin and two molecules of beta globin.

Biohistorian A scholar who integrates human evolution, anthropology, archeology and related fields to study human history and individuals.

Bioinformatics The use of computers and software to acquire, store, analyze, and visualize the information from genomics.

Biotechnology The use of recombinant DNA technology to produce commercial goods and services.

Blastocyst A stage in embryonic development consisting of an outer layer of cells, a hollow, fluid-filled interior and a cluster of cells called the inner cell mass, which form the embryo.

Blastomere A cell produced in the early stages of embryonic development; one of the cells of a blastocyst.

BRCA1 A gene located on chromosome 17. The normal allele of the gene controls cell growth. The mutant allele is associated with a hereditary predisposition to breast cancer.

BRCA2 A gene located on chromosome 13. The normal allele of the gene controls cell growth. The mutant allele is associated with a hereditary susceptibility to cancers of the breast, ovary, and prostate gland.

Bulbourethral glands In males, glands that secrete a mucus-like substance that provides lubrication for intercourse.

Bully whippet Dogs of this breed that are homozygous for a mutation in the myostatin gene have greatly increased muscle mass, are larger than normal, and run much faster than normal whippets.

Burkitt's lymphoma A fast-growing cancer of white blood cells (B cells) most often found in children or young adults.

Cancer A malignant growth caused by uncontrolled cell division. Cancer cells also have the ability to spread (metastasize) to other parts of the body.

Cancer cluster A much higher than expected number of cancer cases found in a population in a certain geographic area or over a certain time span.

Carcinogen A substance that induces cancer in humans or any other organism.

Carrier frequency The frequency of heterozygotes carrying a specific allele in a population.

CC-CKR5 A gene located on human chromosome 3 that encodes a receptor found on the surface of white blood cells. The receptor is used by HIV for entry into the cell. People homozygous for a mutation in this gene are resistant to infection with HIV.

Cell cycle The sequence of events that takes place between successive mitotic divisions.

Cell line A specific type of cell that can be grown in the laboratory. (HeLa cells, derived from a cervical cancer, is a cell line.)

Centromere A region of a chromosome to which microtubule fibers attach during cell division. The location of a centromere gives a chromosome its characteristic shape.

Cervix The lower neck of the uterus opening into the vagina.

CFTR (Cystic fibrosis transmembrane regulator). A gene on human chromosome 7 that encodes a plasma membrane protein that regulates cellular chloride transport. Mutations in this gene result in cystic fibrosis.

Chimeric immune system An immune system that originates from two different individuals. Usually created by combining cells from embryos of two different species. Used in experimental animals to prevent rejection of organ transplants across species (xenotransplantation).

Chorion A two-layered structure formed during embryonic development from cells called trophoblasts.

Chorionic villi Finger-like outgrowths of the chorion that anchor the developing embryo to the uterus.

Chorionic villus sampling (CVS) A method of prenatal diagnosis using fetal chorionic cells collected by inserting a catheter through the vagina or abdominal wall into the uterus. Used in diagnosing biochemical and cytogenetic defects in the embryo. Usually performed in the eighth or ninth week of pregnancy.

Chromosomes The threadlike structures in the nucleus that carry genetic information.

Chronic myelogenous leukemia A cancer of white blood cells caused by an exchange of parts between chromosomes 9 and 22.

Codominant Traits that have full phenotypic expression of both members of a gene pair in the heterozygous condition.

Codon Triplets of nucleotides in mRNA that encode the information for a specific amino acid in a protein.

Combined DNA Index System (CODIS) A panel of 13 DNA autosomal markers used to prepare DNA profiles of individuals. Using all 13 markers, the chance that any one individual has a particular combination is about 1 in 100 trillion.

Complete androgen insensitivity (AIS) A recessive, X-linked disorder caused by a mutation in a gene encoding a receptor for testosterone or other male hormone. Affected individuals are genetic males (XY sex chromosome), but have a female phenotype.

Concordance Presence of a trait in both members of a pair of twins.

Consanguinity Blood relatives by being descended from a common ancestor. Matings between first cousins is considered a consanguineous relationship and is indicated on a pedigree by a double horizontal line connecting the two individuals.

Conversion A legal term meaning removal of a person's private property.

Cystic fibrosis A fatal recessive genetic disorder associated with abnormal secretions of the exocrine glands.

Cytokinesis The process of cytoplasmic division that accompanies cell division.

Cytoplasm The material in a cell located between the nuclear and plasma membranes. It contains fluid (cytosol) and organelles.

Cytosine One of three nitrogen-containing pyrimidine bases found in nucleic acids along with thymine and uracil.

Deoxyribonucleic acid (DNA) A molecule consisting of anti-parallel strands of polynucleotides that is the primary carrier of genetic information.

Deoxyribose One of two pentose sugars found in nucleic acids. Deoxyribose is found in DNA, ribose in RNA.

Dermatoglyphics Study of the patterns of ridges on the fingers, palms, toes, and feet. These ridges form distinctive patterns of genetically determined loops, whorls, and arches.

Diploid (2*n*) number The condition in which each chromosome is represented twice and is a member of a homologous pair.

Discovery rule A legal term that refers to the judicial ruling requiring the prosecution in a legal trial to turn over all their evidence to the defense.

DNA fingerprinting Use of variations in the length of mini-satellites to identify individuals.

DNA polymerase An enzyme that catalyzes the synthesis of DNA using a template DNA strand and nucleotides.

DNA profile The pattern of short tandem repeat (STR) alleles used to identify individuals.

Dominant trait The trait expressed in the F1 (or heterozygous) condition.

Drosophila Commonly called the fruit fly. One species, *Drosophila melanogaster* is a favorite model organism of geneticists, developmental biologists and population biologists.

Dystrophin A protein important for the structural integrity of muscle cells. This protein is defective or absent in individuals with the X-linked forms of muscular dystrophy.

Egg The product of female meiosis; the female gamete.

Embryonic stem cells Pluripotent cells derived from the inner cell mass of an embryo that can differentiate into any of the 250 or so cell types found in the adult body.

Endometrium The inner lining of the uterus that is shed at menstruation if fertilization has not occurred.

Endoplasmic reticulum (ER) A system of cytoplasmic membranes arranged into sheets and channels that function in synthesizing and transporting gene products.

Endorphin A peptide hormone produced in the brain that binds to opiate receptors, reduces the perception of pain, and affects emotions.

Enfuvirtide A drug used to treat HIV infection. It disrupts fusion of HIV with the target white blood cell, preventing infection.

Epididymis Tubules in the testis where sperm are stored.

Erectile dysfunction The inability to have or maintain an erection. This is one of the causes of male infertility.

Essential amino acids Amino acids that cannot be synthesized in the body and must be supplied in the diet.

Euploidy Cells or organisms with a complete haploid or diploid genome. In humans, gametes containing 23 chromosomes have a euploid number representing the haploid state. Cells with 46 chromosomes have a euploid number representing the diploid state.

Expressed genes Those genes that are active in an organism's cells. They produce the phenotype.

Factor VIII A protein clotting factor that is missing or defective in the X-linked disorder, hemophilia A.

Fallopian tubes A structure that connects the mammalian ovary to the uterus. Eggs released from the ovary enter the fallopian tube, are fertilized there, and move down the tube to implant in the uterus where development takes place.

Fertilization The fusion of two gametes to form a zygote.

Folic acid A B-vitamin shown to prevent neural tube defects in developing embryos when present in the diet in sufficient amounts.

Follicles A developing egg surrounded by an outer layer of follicle cells, contained in the ovary.

Founder effect The establishment of a population by a small number of individuals whose genotypes are a subset of the alleles in the parental population. A form of genetic drift.

Fragile X An X chromosome that carries a non-staining gap, or break, at band q27; associated with mental retardation in males.

Fructose A simple sugar found in fruits and vegetables. It is one of the main blood sugars in humans.

G1 The first phase of the cell cycle following mitosis. It is a period of growth during which cell size doubles.

G1/S transition The border between G1 and the next phase of the cell cycle (S phase), during which DNA synthesis occurs. This represents an important control point in the cell cycle; failure to regulate the cell cycle at this point often results in cancer.

G2 The phase of the cell cycle which follows the S phase. It is a period during which DNA repair occurs and in which the cell prepares for division.

G2/M transition The border between G2 and mitosis (M). This is a control point in the cell cycle.

Gene The fundamental units of heredity.

Gene regulation The process that determines which genes will be expressed and how much gene product will be produced.

Gene therapy Procedure in which normal genes are transplanted into humans carrying defective copies as a means for treating genetic diseases.

Genetic screening The testing of members of a population to identify individuals at risk of a genetic disorder, or at risk for transmitting a genetic disorder to their offspring.

Genetic testing The use of biochemical or molecular techniques to identify whether an individual has a genetic disorder or carries a mutant allele for a disorder.

Genome The set of genetic information carried by an individual.

Genomic library A set of clones representing the entire genome of an organism.

Genomics The study of the organization, function, and evolution of genomes.

Genotype The specific genetic constitution of an organism.

Germ-line therapy Gene transfer to gametes or the cells that produce them. Transfers a gene to all cells in the next generation, including germ cells.

Golgi apparatus Membranous organelles composed of a series of flattened sacs. They sort, modify, and package proteins synthesized in the ER.

Guanine One of two nitrogen-containing purine bases found in nucleic acids, along with adenine.

Hairy cell leukemia A rare form of blood cell cancer in which abnormal B cells with hair-like extensions are found in the bone marrow, spleen, and blood.

Haplogroups A group of similar haplotypes that share a single nucleotide polymorphism (SNP) and are derived from a common ancestor. In humans, Y chromosomes and mitochondrial DNA is most often studied to identify haplogroups.

Haplotype A cluster of closely linked genes or markers that are inherited together. In the immune system, the HLA alleles on chromosome 6 are a haplotype.

Hardy-Weinberg Law The statement that allele and genotype frequencies remain constant from generation to generation when the population meets certain assumptions.

HeLa cells A cell line derived from Henrietta Lack's cervical cancer. One of the first permanent human cell lines, widely used in research.

Hemizygous A gene present on the X chromosome that is expressed in males in both the recessive and dominant condition.

Hemoglobin The oxygen-carrying protein of the blood. Adult hemoglobin is made up of two alpha globins and two beta globins. Each globin molecule is complexed with heme, an organic molecule which contains iron, and to which the oxygen binds.

Hemolytic disease of the newborn (HDN) A condition of immunological incompatibility between mother and fetus that occurs when the mother is Rh⁻ and the fetus is Rh⁺.

Hemophilia Inherited diseases characterized by blood clotting defects. The most common forms are hemophilia A and hemophilia B, encoded by genes on the X chromosome.

Herbicide A chemical that kills or inhibits the growth of plants.

Heterozygous Carrying two different alleles for one or more genes.

Histones DNA-binding proteins that help compact and fold DNA into chromosomes.

HLA complex A cluster of genes on human chromosome 6 that encode genes of the immune system. These genes have many alleles, generating a large number of possible HLA haplotypes.

Homo neanderthalensis A species of humans that lived from about 300,000 years ago to about 30,000 years ago in the Near East, Europe, and North Africa.

Homo sapiens The only living and youngest species of humans. Arose in Africa about 200,000 years ago, and spread from there to all parts of the globe.

Homologous chromosomes Chromosomes that physically associate (pair) during meiosis. Homologous chromosomes have identical gene loci.

Homozygous Having identical alleles for one or more genes.

Human chorionic gonadotropin (hCG) A hormone produced by the placenta that maintains the production of progesterone by the corpus luteum. hCG is what is detected in home pregnancy tests.

Human Genome Project An international effort, started in 1990 to sequence, identify, map, and establish the functions of all the genes carried in the human genome.

Human papillomavirus (HPV) A group of viruses, some of which can cause cervical cancer.

Huntingtin The gene and its encoded protein that in mutant form is associated with Huntington disease.

Huntington disease (HD) An autosomal dominant disorder associated with progressive neural degeneration and dementia and early death.

Hyperacute rejection A rapid, massive antibody-mediated rejection of a transplanted organ by the immune system of the recipient. Can be caused by pre-existing antibodies or cross-species transplants.

Immunoglobulins (Ig) The five classes of proteins to which antibodies belong.

Immunosuppressive drugs Drugs that reduce or eliminate the body's ability to mount an immune response and reject transplanted organs.

Implantation The attachment of the early embryo to the lining of the uterus. Usually takes place about 6–7 days after fertilization.

In vitro fertilization (IVF) A procedure in which gametes are collected and fertilized in a dish in the laboratory, and the resulting zygote is implanted in the uterus for development.

Informed consent A legal term referring to the necessity to tell patients all the information they need to know about their treatments.

Inner cell mass A cluster of cells in the blastocyst that gives rise to the embryonic body and is the source of embryonic stem cells.

Intelligence quotient (IQ) A score derived from standardized tests that is calculated by dividing the individual's mental age (determined by the test) by his or her chronological age and multiplying the quotient by 100.

Interphase The period of time in the cell cycle between mitotic divisions.

Junk science A term used by some legal writers that refers to science brought into a courtroom that has not passed the Frye test (based on the questions of Frye v. U.S.).

Karyotype A complete set of chromosomes from a cell that has been photographed during cell division and arranged in a standard sequence.

Knockout mouse A genetically engineered mouse that has had one or more of its genes mutated to be inactive. Used in biomedical research to study the effects of genes on development and phenotypes.

Laparoscopy A procedure in which a small slit is made in the abdominal wall, and a fiber optic device is inserted to inspect the interior of the abdomen. Often used to view the uterus, ovaries, and fallopian tubes, and to collect ovulated eggs.

Law of independent assortment One of Mendel's discoveries in genetics; the random distribution into gametes of genes located on different chromosomes during meiosis.

Law of segregation One of Mendel's discoveries in genetics; the separation of members of a chromosome pair into different gametes during meiosis.

Linkage A condition in which two or more genes do not show independent assortment. Rather, they tend to be inherited together. Such genes are located on the same chromosome. By measuring the degree of recombination between linked genes, the distance between them can be determined.

Lutenizing hormone (LH) A hormone produced by the pituitary gland that in females stimulates the maturation of an ovarian follicle.

Lymphokine A group of proteins produced by white blood cells that play a role in the immune response.

Lysosomes Membrane-enclosed organelles that contain digestive enzymes.

Mad-cow disease A prion disease of cattle, also known as bovine spongiform encephalopathy, or BSE.

Malignant A property of tumors that invade and destroy nearby tissue and have the ability to spread to other parts of the body; a cancerous growth.

Map In genetics, the order and distance between genes on a chromosome. A genetic map is based on the frequency of recombination between genes; a physical map is based on the location of genes to specific places on a chromosome.

Markers An allele, a DNA sequence, or chromosome structural feature that can be used to follow a chromosome or segment of a chromosome during genetic analysis. Positional cloning uses DNA markers to identify the chromosome carrying a gene of interest.

Meiosis The process of cell division during which one cycle of chromosomal replication is followed by two successive cell divisions to produce four haploid cells.

Memory cells A long-lived B cell produced after exposure to an antigen. Memory cells help defend the body against a second infection by the same antigen.

Meninges The membranes that surround and protect the brain and the spinal cord.

Messenger RNA (mRNA) A single-stranded complementary copy of the nucleotide sequence in a gene.

Metaphase I The stage of meiosis at which paired homologous chromosomes align at the equator of the cell. At anaphase I, members of each chromosome pair separate from each other and migrate to opposite poles of the cell.

Metaphase II The stage of meiosis at which unpaired chromosomes align at the equator of the cell. At anaphase II, the centromere splits, converting sister chromatids to chromosomes, which separate from each other and migrate to opposite poles of the cell.

Metastasis The process by which cancer cells move to new locations in the body and form new malignant tumors.

Minisatellites Nucleotide sequences 14 to 100 base pairs long organized into clusters of varying lengths; used in construction of DNA fingerprints.

Mitochondria (singular: mitochondrion) Membrane-bound organelles, present in the cytoplasm of all eukaryotic cells, that are the sites of energy production within the cells.

Mitochondrial DNA A circular DNA molecule present in multiple copies that contains the genome of the organelle.

Mitosis Form of cell division that produces two cells, each of which has the same complement of chromosomes as the parent cell.

Monoamine oxidase type A An enzyme encoded by a gene on the X chromosome that degrades neurotransmitters, specifically norepinephrine and serotonin. Mutations in this gene that reduce activity of the gene product are associated with aggressive behavior.

Monosomy A condition in which one member of a chromosome pair is missing ($2n-1$).

Monozygotic (MZ) twins Twins derived from a single fertilization involving one egg and one sperm; such twins are genetically identical.

Multifactorial traits Traits that result from the interaction of one or more environmental factors and two or more genes.

Multiple myeloma A cancer of the plasma cells of the immune system. Plasma cells are specialized B cells that produce antibodies.

Muscular dystrophy A group of genetic diseases associated with progressive degeneration of muscles. Two of these, Duchenne and Becker muscular dystrophy, are inherited as X-linked, allelic, recessive traits.

Mutagens Physical or chemical agents that cause mutations in DNA.

Mutation A heritable alteration in DNA.

Myometrium The muscular wall of the uterus, which surrounds the endometrium.

Nail-patella syndrome A rare genetic disorder that causes abnormalities of the joints, bones, and fingernails. Affected individuals are born with abnormal or missing kneecaps and thumbnails.

Neanderthals See *Homo neanderthalensis*.

Negative eugenics A part of eugenics that aims to lower or eliminate reproduction among individuals judged to be genetically inferior.

Neural tube defects A defect of the central nervous system affecting the spinal cord, brain, and often, the skull. It is caused by failure of the neural tube to close during early development.

Neurofibromatosis (NF) Two disorders, one caused by a mutant allele on chromosome 17 called neurofibromatosis 1 (NF-1), the other called neurofibromatosis 2 (NF-2), caused by a mutant allele on chromosome 22. NF-1 is inherited as an autosomal dominant condition that affects developmental events in the nervous system and skin. Affected individuals develop tumors, called neurofibromas along peripheral nerves.

Neurotransmitters Chemicals released into the synapse, a gap between nerve cells during the transmission of a nerve impulse from one neuron to another.

Nondisjunction In meiosis, the failure of homologous chromosomes to properly separate in meiosis I (primary nondisjunction) or the failure of sister chromatids to properly separate in meiosis II (secondary nondisjunction). The result is gametes with abnormal numbers of chromosomes.

Non-expressed sequences DNA sequences representing inactive genes or sequences that are not transcribed.

Nuclear transfer The process of transferring a nucleus from one cell to another. Methods include injection with a micropipette, cell fusion using a virus, chemicals, or electrical impulses.

Nucleolus (plural: nucleoli) A nuclear region that functions in the synthesis of ribosomes.

Nucleosomes A bead-like structure composed of histones wrapped with DNA. The basic unit of chromosome structure.

Nucleotide The basic building block of DNA and RNA. Each nucleotide consists of a base, a phosphate, and a sugar.

Nucleus The membrane-bounded organelle in eukaryotic cells that contains the chromosomes.

ob **gene** In humans, a gene on chromosome 7 that encodes the hormone leptin, which plays an important role in inhibiting food intake and stimulating energy expenditure. Defects in leptin production cause obesity.

Oncogenes Genes that induce or continue uncontrolled cell proliferation.

OncoMouse® A genetically engineered mouse developed and patented by Harvard University. Mice of this strain carry an activated oncogene which make them susceptible to cancer. This strain is used in cancer research.

Oocyte Female gamete.

Oogenesis The process of oocyte production.

Ovaries Female gonads that produce oocytes and female sex hormones.

Pedigree A diagram listing the members and ancestral relationships in a family; used in the study of human heredity.

Peptide bond A covalent chemical link between the carboxyl group of one amino acid and the amino group of another amino acid.

Pharmacogenetics A branch of genetics concerned with the inheritance of differences in the response to drugs.

Pharmacogenomics Analyzes genes and proteins to identify targets for therapeutic drugs.

Phenotype The observable properties of an organism.

Phenylalanine One of the essential amino acids in the human diet. Essential amino acids cannot be synthesized in the body, but must be included in the diet. Phenylalanine accumulates in the blood of those affected with phenylketonuria, an autosomal recessive genetic disorder, and causes mental retardation.

Phenylalanine hydroxylase (PAH) The enzyme that metabolizes phenylalanine and converts it to the amino acid tyrosine. This enzyme is defective in most cases of the autosomal recessive disorder phenylketonuria.

Phenylthiocarbamide (PTC) A compound that tastes bitter to about 70% of a population. Being a taster or non-taster of this and related compounds is genetically based.

Philadelphia chromosome An abnormal chromosome produced by translocation of parts of the long arms of chromosomes 9 and 22.

Phosphate group A compound containing phosphorus chemically bonded to four oxygen molecules.

Plaintiff A legal term referring to one participant in a trial. This might be the state in a criminal trial or one party in a civil trial who is suing another.

Plasma membrane The lipid and protein-containing membrane surrounding the cytoplasm of all cells.

Plasmid An extrachromosomal circular double stranded DNA molecule found in many species of bacteria. Plasmids can carry genetic information, such as antibiotic resistance. Genetically engineered plasmids are used as vectors in recombinant DNA research.

Pluripotent The ability of a stem cell to form all or most of the cell types in the body.

Point mutation A form of mutation involving a single nucleotide of DNA.

Polygenic traits Traits controlled by two or more genes.

Polymerase chain reaction (PCR) A method for amplifying DNA segments using cycles of denaturation, annealing to primers, and DNA-polymerase directed DNA synthesis.

Polyploidy A chromosomal number that is a multiple of the normal haploid chromosomal set.

Population frequency The frequency of a specific allele in a population.

Positional cloning A recombinant DNA-based method of mapping and cloning genes with no prior information about the gene product or its function.

Positive eugenics A part of eugenics that encourages reproduction among individuals judged to be genetically superior or genetically advantaged.

Preimplantation genetic diagnosis (PGD) A method of testing embryos created by *in vitro* fertilization (IVF) for genetic defects. In this procedure, a single cell (a blastomere) is removed from the embryo at the 6–8 cell stage and tested for genetic disorders.

Prenatal diagnosis Tests used in pregnancy to diagnose a genetic disorder or developmental abnormality in a fetus.

Presymptomatic testing Genetic testing to identify a disease-causing mutation before any symptoms appear.

Primers A short nucleotide sequence used to start DNA synthesis by DNA polymerase. Used in the PCR reaction.

Primary structure The amino acid sequence in a polypeptide chain.

Proband First affected family member who seeks medical attention for a genetic disorder.

Promoter A region of a DNA molecule to which RNA polymerase binds and initiates transcription.

Prophase I A stage in meiosis during which the chromosomes become visible, pair with their homologue, and split longitudinally except at the centromere.

Prophase II The stage of meiosis when unpaired haploid chromosomes consisting of sister chromatids joined at a common centromere first appear.

Prostaglandins Locally acting chemical messengers that stimulate contraction of the female reproductive system to assist in sperm movement.

Prostate gland A gland in males that secretes a milky, alkaline fluid that neutralizes acidic vaginal secretions and enhances sperm viability.

Protein A molecule consisting of one or more chains of amino acids.

Proteomics The study of the expressed proteins present in a cell at a given time under a given set of circumstances.

Psychiatrist A person with an M.D. degree and additional training in the treatment of mental illnesses.

Psychologist A person with advanced training (often a Ph.D) who studies the mind and behavior.

Purine A class of double-ringed organic bases found in nucleic acids.

Pyrimidine A class of single-ringed organic bases found in nucleic acids.

Quantitative trait loci (QTLs) Two or more genes that act on a single polygenic trait.

Recessive trait The trait unexpressed in the F1 but reexpressed in some members of the F2 generation.

Recombinant DNA molecule A molecule created by linking together DNA fragments from two or more different organisms.

Recombinant DNA technology A series of techniques in which DNA fragments are linked to self-replicating vectors to create recombinant DNA molecules, which are replicated in a host cell.

Rejection A part of the immune response in which transplanted organs or tissues are attacked and destroyed by the immune system.

Reproductive cloning A process that transfers a nucleus from an adult to an embryonic cell with the intent of creating an individual genetically identical to the donor of the nucleus. A form of cloning.

Restriction enzyme A bacterial enzyme that cuts DNA at specific sites.

Restriction fragment length polymorphism (RFLP) Variation in the length of DNA fragments produced after treatment with restriction enzymes. The variation is caused by nucleotide differences that create or destroy restriction enzyme recognition and cutting sites.

Rh blood group A blood group first identified in rhesus monkeys. Recessive individuals are Rh$^-$ and produce no antigen, Rh$^+$ individuals produce antigens. The basis of the condition called hemolytic disease of the newborn (HDN).

Ribonucleic acid (RNA) A nucleic acid molecule that contains the pyrimidine uracil and the sugar ribose. The several forms of RNA function in gene expression.

Ribose One of two pentose sugars found in nucleic acids. Deoxyribose is found in DNA, ribose in RNA.

RNA A nucleic acid that contains the sugar ribose and the pyrimidine uracil. There are several forms of RNA, many of which play important roles in gene expression.

RNA polymerase The enzyme that synthesizes RNA from a DNA template.

Ribosomes Cytoplasmic particles composed of two subunits that are the site of protein synthesis.

S phase The stage of the cell cycle during which DNA synthesis and chromosomal replication take place.

Schizophrenia A behavioral disorder characterized by disordered thought processes and withdrawal from reality. Genetic and environmental factors are involved in this disease.

Secondary amenorrhea Stopping menstruation in a woman who has previously menstruated.

Secondary sex characteristics Masculine (beard, deep voice) or feminine (breasts, body shape) features that appear at puberty under control of sex hormones.

Seminal vesicles Glands that secrete fructose and prostaglandins into the sperm.

Seminiferous tubules Small, tightly coiled tubes inside the testes where sperm are produced.

Sequencer A machine that automatically determines the nucleotide sequence of a DNA molecule. Used in genome projects.

Serotonin A brain chemical that plays a role in transmitting nerve impulses from one cell to another in the nervous system.

Severe combined immunodeficiency disease (SCID) A genetic disorder in which affected individuals have no immune response; both the cell-mediated and antibody-mediated responses are missing.

Sex chromosomes In humans, the X and Y chromosomes that are involved in sex determination.

Sex-limited genes Loci that produce a phenotype in only one sex.

Sex ratio The proportion of males to females, which changes throughout the life cycle. The ratio is close to 1:1 at fertilization but the female to male ratio increases as a population ages.

Sex selection The method or process of determining the sex chromosome constitution of a gamete or embryo with the goal of having a child of a certain sex.

Short tandem repeat (STR) Short nucleotide sequences 2 to 9 base pairs long organized into clusters of varying lengths; used in the construction of DNA profiles.

Sickle cell anemia A recessive genetic disorder associated with an abnormal type of hemoglobin, a blood transport protein.

Single nucleotide polymorphism (SNP) A genetic variation caused by a difference in a single nucleotide.

Sperm Male gamete.

Sperm sorting The process of sorting sperm into those carrying an X chromosome and those carrying a Y chromosome. A first step in sex selection.

Spermatids The cells formed by meiosis II that will differentiate to form sperm.

Spermatocytes Diploid cells that undergo meiosis to form haploid spermatids.

Spermatogenesis The process of sperm production.

Spina bifida A neural tube defect that leads to an abnormality in the spinal cord or the covering surrounding the spinal cord.

Start codon A codon present in mRNA that signals the location for translation to begin. The codon AUG functions as a start codon.

Stimulus (plural: stimuli) In biology, a detectable change in either the internal or the external environment.

Stop codon A codon present in mRNA that signals the end of a growing polypeptide chain. The codons UAG, UGA, and UAA function as stop codons.

Streptococcus pneumoniae A species of bacteria with encapsulated and non-encapsulated strains that was used to experimentally demonstrate that DNA carries genetic information.

Surrogacy A method of assisted reproduction in which a woman enters into a contract to carry a child for the other party to the contract.

Synapse The gap between two nerve cells that nerve impulses move across.

T-cells A type of lymphocyte that undergoes maturation in the thymus and mediates cellular immunity.

T4 helper cells A sub-type of one of the cells of the immune system. T4 cells assist in the initiation of an immune response, and are often called the "on switch" of the immune system.

Tamoxifen An anticancer drug that acts by blocking the action of estrogen on cell growth.

Tay-Sachs disease A recessively inherited metabolic genetic disorder that leads to early childhood death.

Telomere A specialized structure at the end of a chromosome.

Telophase I The last stage of meiosis I when the chromosomes uncoil and become less visible and the nuclear membrane re-forms. Cytokinesis follows telophase, forming two haploid cells.

Telophase II The last stage of meiosis II, during which the haploid chromosomes uncoil.

Telophase The last stage of mitosis, during which division of the cytoplasm occurs and the nuclear membrane re-forms.

Termination sequence The nucleotide sequence at the end of a gene that signals the end of transcription.

Testes Male gonads that produce spermatozoa and sex hormones.

Testosterone A steroid hormone produced by the testis; the male sex hormone.

Thymine One of three nitrogen-containing pyrimidine bases found in nucleic acids, along with uracil and cytosine.

Tort A legal term that refers to a wrongful act that may lead to a law suit.

Transcription The process of transferring genetic information from the nucleotide sequence of a DNA molecule into the complementary nucleotide sequence of an RNA molecule by the enzyme RNA polymerase.

Transfer RNA (tRNA) A small RNA molecule that contains a binding site for a specific type of amino acid and a three base segment known as an anticodon that recognizes a specific base sequence in messenger RNA.

Transformation The process of transferring genetic information between cells by DNA molecules.

Transgenic crops Food crops created by the transfer of genes between species by recombinant DNA technology to give the crop plant a new and useful trait such as herbicide resistance.

Transgenic organism A bacterium, plant, or animal that has been genetically modified to carry one or more genes from an organism of a different species.

Translation Conversion of information encoded in the nucleotide sequence of an mRNA molecule into the linear sequence of amino acids in a protein.

Trinucleotide repeats A form of mutation associated with the expansion in copy number of a nucleotide triplet in or near a gene.

Trisomy 21 Aneuploidy involving the presence of an extra copy of chromosome 21, resulting in Down syndrome.

Tumor suppressor genes Genes encoding proteins that suppress cell division.

Tyrosine One of the twenty amino acids used in protein synthesis. The amino acid that is produced from phenylalanine.

Ultrasonography A method used to visualize the fetus (or body parts) using a transducer that produces sound waves.

Universal donor In blood types, someone who can give blood to anyone else. In the ABO blood system, type O individuals are universal donors.

Universal recipient In blood types, someone who can receive blood from anyone else. In the ABO system, type AB individuals are universal recipients.

Uracil One of three nitrogen-containing pyrimidine bases found in nucleic acids, along with thymine and cytosine.

Urethra A tube that passes from the bladder and opens to the outside. It functions in urine transport and, in males, also carries sperm.

Uterus A hollow, pear-shaped muscular organ where a fertilized egg will develop.

Vaccine A preparation of weakened or dead pathogens that elicits an immune response when injected into the body.

Vagina The opening that receives the penis during intercourse and also serves as the birth canal.

Vas deferens A duct connected to the epididymis, which sperm travels through.

Vasectomy A contraceptive procedure for men in which each vas deferens is cut and sealed to prevent the transport of sperm.

Vectors Self-replicating DNA molecules that are used to transfer foreign DNA segments between host cells.

Whippet A dog breed that has individuals carrying a mutation in a gene (myostatin) that when homozygous leads to greatly enlarged muscle mass. See bully whippet.

Wrongful birth suit A legal term referring to a type of law suit. In this law suit a family sues a doctor claiming because they did not have a prenatal test, their child was born into a life of pain and suffering.

Wrongful life suit A legal term referring to a type of law suit. In this law suit a child sues a doctor claiming he or she should never have been born into a life of pain and suffering.

X chromosome One of the sex-determining chromosomes in humans. Males have an X and a Y chromosome, and females have two X chromosomes.

Xenotransplant Organ transplants between species.

X-linked The pattern of inheritance that results from genes carried on the X chromosome.

X-ray diffraction A method of analyzing molecular structure by passing X rays through a crystal. The resulting pattern of diffracted X rays can be used to determine the three dimensional structure of the molecular component of the crystal. Used in determining the structure of DNA.

Y chromosome One of the sex-determining chromosomes in humans. Males have an X and a Y chromosome, and females have two X chromosomes.

Y-linked The pattern of inheritance that results from genes located only on the Y chromosome.

Zygote The fertilized egg that develops into a new individual.

Index

Page numbers in *italics* indicate material presented in figures and tables.